Lijuan Li and Feng Liu

# Group Search Optimization for Applications in Structural Design

# Adaptation, Learning, and Optimization, Volume 9

**Series Editor-in-Chief**

Meng-Hiot Lim
Nanyang Technological University, Singapore
E-mail: emhlim@ntu.edu.sg

Yew-Soon Ong
Nanyang Technological University, Singapore
E-mail: asysong@ntu.edu.sg

---

Further volumes of this series can be found on our homepage: springer.com

Vol. 1. Jingqiao Zhang and Arthur C. Sanderson
Adaptive Differential Evolution, 2009
ISBN 978-3-642-01526-7

Vol. 2. Yoel Tenne and Chi-Keong Goh (Eds.)
Computational Intelligence in
Expensive Optimization Problems, 2010
ISBN 978-3-642-10700-9

Vol. 3. Ying-ping Chen (Ed.)
Exploitation of Linkage Learning in Evolutionary Algorithms, 2010
ISBN 978-3-642-12833-2

Vol. 4. Anyong Qing and Ching Kwang Lee
Differential Evolution in Electromagnetics, 2010
ISBN 978-3-642-12868-4

Vol. 5. Ruhul A. Sarker and Tapabrata Ray (Eds.)
Agent-Based Evolutionary Search, 2010
ISBN 978-3-642-13424-1

Vol. 6. John Seiffertt and Donald C. Wunsch
Unified Computational Intelligence for Complex Systems, 2010
ISBN 978-3-642-03179-3

Vol. 7. Yoel Tenne and Chi-Keong Goh (Eds.)
Computational Intelligence in Optimization, 2010
ISBN 978-3-642-12774-8

Vol. 8. Bijaya Ketan Panigrahi, Yuhui Shi, and Meng-Hiot Lim (Eds.)
Handbook of Swarm Intelligence, 2011
ISBN 978-3-642-17389-9

Vol. 9. Lijuan Li and Feng Liu
Group Search Optimization for Applications in Structural Design, 2011
ISBN 978-3-642-20535-4

Lijuan Li and Feng Liu

# Group Search Optimization for Applications in Structural Design

Springer

Prof. Lijuan Li
Guangdong University of Technology
Faculty of Civil and Transportation Engineering
No. 100 Waihuan Xi Road
Guangzhou
China
E-mail: lilj@gdut.edu.cn

Prof. Feng Liu
Guangdong University of Technology
Faculty of Civil and Transportation Engineering
No. 100 Waihuan Xi Road
Guangzhou
China
E-mail: fliu@gdut.edu.cn

ISBN 978-3-642-26847-2 ISBN 978-3-642-20536-1 (eBook)

DOI 10.1007/978-3-642-20536-1

Adaptation, Learning, and Optimization ISSN 1867-4534


Softcover reprint of the hardcover 1st edition 2011



*Typeset & Cover Design:* Scientific Publishing Services Pvt. Ltd., Chennai, India.

Printed on acid-free paper

9 8 7 6 5 4 3 2 1

springer.com

# Preface

Civil engineering structures such as buildings, bridges, stadiums, and offshore structures play an import role in our daily life. However, constructing these structures requires lots of budget. Thus, how to cost-efficiently design them satisfying all required design constraints is an important factor to structural engineers. Traditionally, mathematical gradient-based optimal techniques have been applied to the design of optimal structures. While, many practical engineering optimal problems are very complex and hard to solve by traditional method. In the past few decades, nature-inspired computation, such as evolutionary algorithm has attracted more and more attention. Nature serves as a fertile source of concepts, principles, and mechanisms for designing artificial computation system to tackle complex problems.

Swarm intelligence algorithm, which was inspired by the social behaviour of animals such as fish schooling and bird flocking, was developed in recent years because they do not require conventional mathematical assumptions and thus possess better global search abilities than the traditional optimization algorithms. The most efficient algorithms are particle swarm optimizer and group search optimizer. These intelligent based algorithms are very suitable for continuous and discrete design variable problems such as ready-made structural members and have been vigorously applied to various structural design problems and obtained good results. This book gathers our latest research work related with particle swarm optimizer algorithm and group search optimizer algorithm as well as their application to structural optimal design. The aim of the book is to provide a reference for researchers and engineers for sharing of latest developments in the research and application of structural optimal design with swarm intelligent algorithms. The readers can understand the full spectrum of the algorithms and apply the algorithms to their own research problems.

We would like to express our sincere appreciations to National Natural Science Foundation of China (project number: 10772052) and the Natural Science Foundation of Guangdong Province (project numbers: 8151009001000042, 9151009001000059) for founding the research work. Thanks are also given to Springer Berlin / Heidelberg for publishing this book.

February 2011

Lijuan Li
Feng Liu

# Contents

# Chapter 1
# Introduction of Swarm Intelligent Algorithms

It is fairly accepted fact that one of the most important human activities is decision making. It does not matter what field of activity one belongs to. Whether it is political, military, economic or technological, decisions have a far reaching influence on our lives. Optimization techniques play an important role in structural design, the very purpose of which is to find the best ways so that a designer or a decision maker can derive a maximum benefit from the available resources. The methods of optimization can be divided into two category such as traditional optimization algorithms and modern optimization algorithms. The traditional optimization algorithms turn into an independent subject began in 1947 when Dantzig [1, 2] proposed the simplex method for solving general linear optimization problems. From then on, study on the optimization method is booming. Many methods of optimization are proposed [3] sequentially as follow: unconstrained optimization methods, large-scale unconstrained optimization methods, nonlinear least squares methods, linear constrained optimization methods, nonlinear constrained optimization methods and so on. These traditional mathematical gradient-based optimal techniques have been applied to the design of optimal structures [4]. While, many practical engineering optimal problems are very complex and hard to solve by traditional method [5].

With the development of modern technology especially computer technology, also due to the traditional optimization algorithms have some limitations difficult to overcome. People began to seek and propose some efficient optimization methods which were called modern intelligent optimization algorithms. In the past few decades, nature-inspired computation has attracted more and more attention [6]. Nature serves as a fertile source of concepts, principles, and mechanisms for designing artificial computation system to tackle complex problems. Among them, the most successful are evolutionary algorithms [7], which draw inspiration from evolution by natural selection. The several different types of evolutionary algorithms include genetic algorithms, artificial neural network algorithm, genetic programming, evolutionary programming and evolutionary strategies.

Genetic algorithm is a widely applied and efficient method of random search and optimization which develops based on the theory of evolution. Its main features are the groups search strategy and the information exchange between individuals, moreover, the search does not depend on gradient information. It is developed by professor Holland [8] from University of Michigan USA in the early 70s. Professor Holland published the first monograph which discussed the genetic

L. Li & F. Liu: Group Search Optimization for Applications in Structural Design, ALO 9, pp. 1–6.
springerlink.com 

algorithm systematically named "Adaptation in Natural and Artificial Systems". Holland elaborated the basic theory and method of genetic algorithm systematically in his book, and proposed the template theory which is extremely important for the development of the theory of genetic algorithm. Artificial neural network is an engineering system based on the understanding of the human brain structure and operational mechanism, and then simulates the structure and intelligent behavior. Some scholars had proposed the first mathematical model of artificial neural network in the early 40s of last century [9], and from then on people pioneered the study of neuroscience theory. Subsequently, more scholars had proposed perceptual model, making the artificial neural network technology to a higher level.

Artificial intelligence went through the prosperity in the 80s of last century [10]. People experiencing hard problems because of the method had not overcome the limitations of classical computing thinking. Prospects for the study of artificial intelligence once again lost its luster. At the same time, as people kept understanding of the nature of life, life science is developed at an unprecedented rapid speed. The study of artificial intelligence began out of the shackles of classical logic computing, and people from the bold exploration of new channels of non-classical computation. As a pioneer in artificial intelligence, Minsky [11] thinks "we should be inspiration from biology rather than physics……", the research of heuristic calculation of biology became a new direction of artificial intelligence. In this context, social animals (such as ant colonies, bees, birds, etc.) from the organizational behavior had aroused widespread attention; many scholars established the mathematical modeling of this behavior and its simulation for computer. This creates a so-called "swarm intelligence" [12]. The beauty lies in social animals is that the behavior of individuals is very simple, but when they work together they will able to emergent a very complex (intelligent) behavior characteristics. For example: the ability of individual ants is extremely limited, but the composition of these simple ant colony will able to complete such as nesting, foraging, migration, nest cleaning and other complex behaviors; a group of blind behavior of bees can build seems fine cellular; birds without centralized control are able to synchronize flight and so on [13].

People founded the bionics in the mid-20th century [14], and proposed many new methods for solving complex optimization problems which based on the mechanism of biological evolution, such as genetic algorithms, evolutionary programming and evolutionary strategies. Swarm intelligence algorithm, as a new evolutionary computing technology had focused more and more researchers. It has a very special connection with artificial life, in particular, evolutionary strategy and genetic algorithm. The group in swarm intelligence means "a group agent that can communicate directly with each other or indirect communication (by changing the local environment), this group could solve the main problems with distributed collaboration", while swarm intelligence refers to “No intelligence agents have shown the main characteristics of intelligent behavior by cooperation with each other". Swarm intelligence provides the basis that people finds solutions for solving complex distributed problems in the premise of absence of centralized control and does not provide a global model [15].

Modern swarm intelligent optimization algorithms are very active and developed rapidly in recent years. In swarm intelligence optimization family, the most typical ones include ant colony optimization (ACO) [16], particle swarm optimization (PSO) [17] and group search optimization (GSO) [18]. The ant colony optimization algorithm is the simulation of the process of ants' foraging behavior. The particle swarm optimization derived from the swarm behavior of bird flocking or fish schooling. Broadly speaking, the two swarm intelligent algorithm mentioned were inspired by some aspects of animal behavior, which is a scientific study of everything animals do. The group search optimization was inspired by animal searching behavior which may be described as an active movement by which an animal finds or attempts to find resources such as food, mates or nesting sites.

Particle swarm optimization was firstly invented together by Kennedy, who is American social psychologists, and Eberhart, who is an electric engineer, in 1995 [19]. It is a new random optimization algorithm based on swarm intelligence algorithm. This algorithm is derived from the simulation of foraging birds' behaviors which is optimum tools based on iterating. It initialized to a set of random solutions by system and search for the optimal value through the iterating. The birds' searching bound is corresponding to the range of design variables. The food that birds find is corresponding to the optimal solution of the objective function. Every bird, that is, each particle's position in the searching space is corresponding to a feasible solution in the design space. PSO dues to its algorithm is simple and easy to implement, without gradient information, fewer parameters and other characteristics in the continuous optimization problems and discrete optimization problems have shown good results, especially owing to its natural real coding feature is suited to dealing with real optimization problem. In recent years, it becomes the hot research area in the field of international intelligence optimization research [20]. As a novel optimization search algorithm, during its emergence in over ten years, researchers have mainly focused on the research about the structure of algorithms and improvement of performance, including parameter settings, particle diversity, population structure and algorithm integration [21].

PSO has a very broad application in engineering design and optimization. Particle swarm algorithm is applied to evolution of neural networks, extraction of fuzzy neural network rule, circuit design, digital filter design, semiconductor device synthesis, layout optimization, the controller parameters optimization, system identification and state estimation [22-24]. In the field of electric power system, particle swarm optimization algorithm is used to achieve energy optimization, control voltage, improve reliability of power and optimize allocation of capacitor problems. In the robot control, PSO is applied in vibration inhibition path planning of the robot and mobile robot path planning. In the field of transportation, PSO is applied in route planning, the layout of base station and optimization problems. In the field of computer, PSO is applied in task allocation, pattern recognition, image processing and data excavation and other issues. In the field of industrial production, PSO is applied in raw materials hybrid optimization and the computer-controlled grinding optimization. In the field of biology, PSO is applied in biomedical image registration, the geometric arrangement of image data, the gene classification and other issues. In the field of electromagnetism, PSO is applied in solving the nonlinear magnetic medium magnetic fields and optimization of multi-layer plane shield in the electromagnetic field.

Group search optimization was first proposed by He et al. [25], the method is inspired from the behavior of biological communities in search of resources and developing. Group-living is a widespread phenomenon in the animal kingdom; group members benefit from sharing information or cooperating with each other by living together. One consequence of sharing information is that group searching allows group members to increase patch finding rates as well as to reduce the variance of search success. This has usually led to the adoption of two foraging strategies [14] within groups: (1) producing, e.g., searching for food; and (2) joining (scrounging), e.g., joining resources uncovered by others. Joining is a ubiquitous trait found in most social animals such as birds, fish, spiders and lions. In order to analyze the optimal policy for joining, two models have been proposed: Information-Sharing (IS) and Producer-Scrounger (PS). The IS model assumes foragers search concurrently for their own resource, whilst searching for opportunities to join. On the other hand, foragers in the PS model are assumed to use producing or joining strategies exclusively. The GSO is a population-based optimization algorithm and employs producer-scrounger model and animal scanning mechanism. Producer-scrounger model for designing optimum searching strategies was inspired by animal searching behavior and group living theory. In order not to entrap in local minima, GSO also employs "rangers" foraging strategies. The population of the GSO algorithm is called a group and each individual in the population is called a member just as PSO does. There are three kinds of members in the group: (1) producer, performs producing strategies, searching for food; (2) scrounger, performs scrounging strategies, joining resources uncovered by others; (3) ranger, employs random walks searching strategies for randomly distributed resources. At each iteration, the member who located the most promising resource is producer, a number of members except producer in the group are selected as scroungers, and the remaining members are rangers.

Compared with the particle swarm optimization algorithm, group search optimization is a relatively new swarm intelligent optimization algorithm, so the application research of group search optimization algorithm is still in the initial stage in the field of international research, mainly used in nuclear technology, cancer treatment research and the field of structural optimization [26]. As swarm intelligence optimization algorithm, it is self-evident for the applied prospect in every field, that needs researchers make further study for it.

## References


1. Dantzig, G.B.: On the need for a systems optimization laboratory, in optimization methods for resource allocation. In: Cottle, R.W., Krarup, J. (eds.) Proceedings of a 1971 NATO Conference. English Universities Press, London (1974)
2. Albers, D.J., Reid, C.: An interview with George B. Dantzig: the father of linear programming. The College Mathematics Journal 17(4), 292–314 (1986)
3. Gleick, J.: Breakthrough in problem solving, New York Times, November 18 (1984)

4. Rao, V.R., Iyengar, N.G.R., Rao, S.S.: Optimization of wing structures to satisfy strength and frequency requirement. Computers and Structures 10(4), 669–674 (1979)
5. Nanakorn, P., Konlakarn, M.: An adaptive penalty function in genetic algorithms for structural design optimization. Computers and Structures 79(29-30), 2527–2539 (2001)
6. Fogel, D.B.: The advantage of evolutionary computation. In: Proceeding of Biocomputing Emergent Computation, pp. 1–11. World Scientific Press, Singapore (1997)
7. Coello Coello, C.A.: Theoretical and numerical constraint-handling techniques used with evolutionary algorithms: A survey of the state of the art. Computer Methods in Applied Mechanics and Engineering 191(11-12), 1245–1287 (2002)
8. Holland, J.H.: Adaptation in natural and artificial systems. The University of Michigan Press (1975)
9. McCulloch, W., Pitts, W.: A logical calculus of the ideas immanent in nervous activity. Bulletin of Mathematical Biophysics 5, 115–133 (1943)
10. Hopfield, J.J.: Neural networks and physical systems with emergent collective computational abilities. Proceeding of National Academy of Sciences of the USA 79(8), 2554–2558 (1982)
11. Minsky, M., Papert, S.: Perceptrons: An introduction to computational geometry. MIT Press, Cambridge (1969)
12. Dorigo, M.: Ottimizzazione, apprendimento automatico, ed algoritmi basati su metafora naturale (Optimization, learning and natural algorithms). Ph.D. Thesis, Politecnico di Milano, Italy (1992) (in Italian)
13. Fogel, L.J., Owens, A.J., Walsh, M.J.: Artificial intelligence through simulated evolution. John Wiley, New York (1996)
14. Barnard, C.J., Sibly, R.M.: Producers and scroungers: a general model and its application to captive flocks of house sparrows. Animal Behavior 29(2), 543–550 (1981)
15. Bonabeau, E., Dorigo, M., Theraulza, G.: Inspiration for optimization from social insect behavior. Nature 406(4), 39–42 (2000)
16. Dorigo, M., Di Caro, G., Gambardella, L.: Ant algorithms for discrete optimization. Artificial Life 5(3), 137–172 (1999)
17. Kennedy, J., Eberhart, R.C.: Swarm intelligence. Morgan Kaufmann Publishers, San Francisco (2001)
18. He, S., Wu, Q.H., Saunders, J.R.: Group search optimizer: an optimization algorithm inspired by animal searching behavior. IEEE Transactions on Evolutionary Computation 13(5), 973–990 (2009)
19. Kennedy, J., Eberhart, R.C.: Particle swarm optimization. In: IEEE International Conference on Neural Networks, vol. 4, pp. 1942–1948. IEEE Press, Piscataway (1995)
20. Plevris, V., Papadrakakis, M.: A hybrid particle swarm-gradient algorithm for global structural optimization. Computer-Aided Civil and Infrastructure Engineering 26(1), 48–68 (2011)
21. Kaveh, A., Talatahari, S.: Particle swarm optimizer, ant colony strategy and harmony search scheme hybridized for optimization of truss structures. Computers & Structures 87(5-6), 267–283 (2009)
22. Wu, Q.H., Cao, Y.J., Wen, J.Y.: Optimal reactive power dispatch using an adaptive genetic algorithm. Electrical Power and Energy Systems 20(8), 563–569 (1988)
23. Van den Bergh, F., Engelbrecht, A.P.: Training product unit networks using cooperative particle swarm optimizers. In: Proc. of the Third Genetic and Evolutionary Computation Conference, San Francisco, USA (2001)

24. Perez, R.E., Behdinan, K.: Particle swarm approach for structural optimization. Computers and Structures 85(19-20), 1579–1588 (2007)
25. He, S., Wu, Q.H., Saunders, J.R.: A group search optimizer for neural network training. Springer, Heidelberg (2006)
26. Shen, H., Zhou, Y.H., Niu, B., Wu, Q.H.: An improved group search optimizer for mechanical design optimization problems. Progress in Natural Science 19(1), 91–97 (2009)

# Chapter 2
# Application of Particle Swarm Optimization Algorithm to Engineering Structures

**Abstract.** In this chapter, we present an approach that integrates the finite element method (FEM) with a particle swarm optimization (PSO) algorithm to deal with structural optimization problems. The proposed methodology is concerned with two main aspects. First, the problem definition must be established, expressing an explicit relationship between design variables and objective functions as well as constraints. The second aspect is to resolve the minimization problem using the PSO technique, including the use of finite element method. In this chapter, particle swarm optimizer is extended to solve structural design optimization problems involving problem-specific constraints and mixed variables such as integer, binary, discrete and continuous variables. The standard PSO algorithm is very efficient to solve global optimization problems with continuous variables, especially the PSO is combined with the FEM to deal with the constraints related with the boundary conditions of structures controlled by stresses or displacements. The proposed algorithm has been successfully used to solve structure design problems. The calculation results show that the proposed algorithm is able to achieve better convergence performance and higher accuracy in comparison with other conventional optimization methods used in civil engineering.

## 2.1 Introduction

In the past decades, many optimization algorithms have been applied to solve structural design optimization problems. Among them, evolutionary algorithms (EAs) such as genetic algorithms (GAs), evolutionary programming (EP) and evolution strategies (ES) are attractive because they do not apply mathematical assumptions to the optimization problems and have better global search abilities over conventional optimization algorithms [1].

Most structural optimal design problems are hard to solve for both conventional optimization algorithms and EAs, because they involve problem-specific constraints. To handle these constrains, many different approaches have been proposed. Normally, constrained problems are solved as unconstrained. The most common approach of them is penalty functions [2]. However the major drawback of using penalty functions is that they require additional tuning parameters. In

L. Li & F. Liu: Group Search Optimization for Applications in Structural Design, ALO 9, pp. 7–20.
springerlink.com 

particular, the penalty coefficients have to be fine tuned in order to balance the objective and penalty functions. Inappropriate penalty coefficients will make the optimization problem intractable [3]. Another difficulty for solving structural optimization problems is that structural optimal design problems may contain integer, discrete, and continuous variables, which are referred to as mixed-variable nonlinear optimization problems. Recently a new EA called particle swarm optimization (PSO) have been proposed. Originally PSO was proposed to handle continuous optimization problems. Now the standard PSO has been extended to handle mixed-variable nonlinear optimization problems more effectively [4].

This paper's objective is to offer a practical methodology for optimization of truss structures. It takes advantage of FEM software and PSO minimization strategy. It is desired that this methodology can be practically implemented in structural optimization.

## 2.2 Problem Statement

The aim of structure optimization is to determine the values for some design variables:

$$X = \{x_1, x_2, \cdots, x_n\}$$

where $x_i$ denote properties such as cross-section of bars, inertia, elastic modulus of materials, co-ordinates of nodes or bar joints, thickness of plates and other magnitudes, including economical or aesthetics aspects, if the latter can ever be quantified; and $n$ is the total number of design variables.

Minimizing an objective function:

$$f(X) = f(x_1, x_2, \cdots, x_n)$$

Where $f$ is an objective function. It could be the weight, the cost or any other relevant objective with respect to the designer's/design criteria. The minimization is, at the same, subject to m design constraints:

$$g(X) = g_i(x_1, x_2, \cdots x_n) = 0, \quad i = 1 \cdots m$$

Almost all constraints refer to the maximum allowed stresses or displacements, according to normative and material capabilities.

A structural design optimization problem can be formulated as a nonlinear programming problem. In contrast to generic nonlinear problems which only contain continuous or integer variables, a structural design optimization usually involves continuous, binary, discrete and integer variables. The binary variables are usually involved in the formulation of the design problem to select alternative options. The discrete variables are used to represent standardization constraints such as the diameters of standard sized bars. The integer variables are referred as to the numbers of objects which are design variables, such as the number of bars. To solve the mixed-variable nonlinear optimization problem, Sandgren [5] and Shih [6] have proposed nonlinear branch and bound algorithms based on the integer programming. Cao and Wu [7] developed mixed-variable evolutionary programming with different mutation operators associated with different types of variables. Considering the mixed variables, the structural optimization formulation can be expressed as follows:

$$\min f(X) \tag{2.1}$$

subject to:

$$h_i(X) = 0 \qquad i = 1, 2, \cdots m$$

$$g_i(X) \geq 0 \qquad i = m+1, \cdots p$$

where $f(X)$ is the scalar objective function, $h_i(X)$ and $g_i(X)$ are the equality and inequality constraints, respectively.

The variables vector $X \in R^N$ represents a set of design variables which can be written as:

$$X = \begin{pmatrix} X^C \\ X^B \\ X^I \\ X^D \end{pmatrix} = \left[x_1^C, \cdots, x_{n_C}^C, x_1^B, \cdots, x_{n_B}^B, x_1^I, \cdots, x_{n_I}^I, x_1^D, \cdots, x_{n_D}^D\right]^T$$

where

$$\begin{aligned} & x_i^{Cl} \leq x_i^C \leq x_i^{Cu}, \quad i = 1, 2, \cdots n_C \\ & x_i^B \in \left\{x_i^{Bl}, x_i^{Bu}\right\}, \quad i = 1, 2, \cdots n_B \\ & x_i^{Il} \leq x_i^I \leq x_i^{Iu}, \quad i = 1, 2, \cdots n_I \\ & x_i^{Dl} \leq x_i^D \leq x_i^{Du}, \quad i = 1, 2, \cdots n_D \end{aligned} \tag{2.2}$$

where $X^C \in R^{n_C}$, $X^B \in R^{n_B}$, $X^I \in R^{n_I}$ and $X^D \in R^{n_D}$ denote feasible subsets of comprising continuous, binary, integer and discrete variables, respectively. $x_i^{Cl}$, $x_i^{Bl}$, $x_i^{Il}$ and $x_i^{Dl}$ are the lower bounds of the $i$ th variables of $X^C$, $X^B$, $X^I$ and $X^D$, respectively. $x_i^{Cu}$, $x_i^{Bu}$, $x_i^{Iu}$ and $x_i^{Du}$ are the upper bounds of the $i$th variables of $X^C$, $X^B$, $X^I$ and $X^D$, respectively. $n_C$, $n_B$, $n_I$ and $n_D$ are the numbers of continuous, binary, integer and discrete variables, respectively. The total number of variable is $N = n_C + n_B + n_I + n_D$.

## 2.3 Particle Swarm Optimizer

The PSO is a population-based optimization algorithm which was inspired by the social behavior of animals such as fish schooling and birds flocking [8, 9]. Similar to other evolutionary algorithms, it can solve a variety of hard optimization problems but with a faster convergence rate, specially, it can handle constrains with mixed variables. Another advantage is that it requires only a few parameters to be tuned which making it attractive from an implementation viewpoint.

In PSO, its population is called a swarm and each individual is called a particle. Each particle flies through the problem space to search for optima. Each particle represents a potential solution of solution space, all particles form a swarm. The best position passed through by a flying particle is the optimal solution of this particle and is called *pbest*, and the best position passed through by a swarm is considered as optimal solution of the global and is called *gbest*. Each particle updates itself by *pbest* and *gbest*. A new generation is produced by this updating. The quality of a particle is evaluated by value the adaptability of an optimal function.

In PSO, each particle can be regard as a point of solution space. Assume the number of particles in a group is $M$ , and the dimension of variable of a particle is $N$. The $i$ th particle at iteration $k$ has the following two attributes:

(1) A current position in an $N$-dimensional search space which represents a potential solution: $X_i^k = (x_{i,1}^k, \cdots x_{i,n}^k \cdots x_{i,N}^k)$ , where $x_{i,n}^k \in [l_n, u_n]$ is the $n$th dimensional variable, $1 \le n \le N$ , $l_n$ and $u_n$ are the lower and upper bounds for the $n$th dimension, respectively.

(2) A current velocity, $V_i^k = (v_{i,1}^k, \cdots v_{i,n}^k, \cdots v_{i,N}^k)$, which controls its fly speed and direction. $V_i^k$ is restricted to a maximum velocity $V_{\max}^k = (v_{\max,1}^k, \cdots v_{\max,n}^k, \cdots v_{\max,N}^k)$.

At each iteration, the swarm is updated by the following equations:

$$V_i^{k+1} = \omega V_i^k + c_1 r_1 (P_i^k - X_i^k) + c_2 r_2 (P_g^k - X_i^k) \tag{2.3}$$

$$X_i^{k+1} = X_i^k + V_i^{k+1} \tag{2.4}$$

where $P_i$ is the best previous position of the $i$th particle (also known as *pbest*) and $P_g$ is the global best position among all the particles in the swarm (also known as *gbest*). They are given by the following equations:

$$P_i = \begin{cases} P_i : f(X_i) \ge P_i \\ X_i : f(X_i) < P_i \end{cases} \tag{2.5}$$

$$P_g \in \{P_0, P_1, \cdots, P_M\} \mid f(P_g) = \min(f(P_0), f(P_1) \cdots f(P_M)) \tag{2.6}$$

where $f$ is the objective function, $M$ is the total number of particles. $r_1$ and $r_2$ are the elements generated from two uniform random sequences on the interval [0,1]: $r_1 \sim U(0,1)$; $r_2 \sim U(0,1)$ and $\omega$ is an inertia weight [10] which is typically chosen in the range of [0, 1]. A larger inertia weight facilitates global exploration and a smaller inertia weight tends to facilitate local exploration to fine-tune the current search area. Therefore the inertia weight $\omega$ is critical for the PSO's convergence behavior. A suitable value for the inertia weight $\omega$ usually provides balance between global and local exploration abilities and consequently results in a better optimum solution. Initially the inertia weight was kept constant. However some literatures indicated that it is better to initially set the inertia to a large value, in order to promote global exploration of the search space, and gradually decrease it to get more refined solutions. $c_1$ and $c_2$ are acceleration constants which also control how far a particle will move in a single iteration. The maximum velocity $V_{max}$ is set to be half of the length of the search space in one dimension.

The right first part of equation (2.3) represents the current position of particles, and has the function to balance the global and local search. The second part makes the particles have powerful global search abilities. The third part means particles share information themselves.

## 2.4 Mixed-Variable Handling Methods

In its basic form, the PSO can only handle continuous variables. To handle integer variables, simply truncating the real values to integers to calculate fitness value will not affect the search performance significantly [11]. The truncation is only performed in evaluating the fitness function. That is the swarm will 'fly' in a continuous search space regardless of the variable type. Binary variables, since they can be regarded as integer variables within the range of [0, 1], are not considered separately.

For discrete variables of the $i$th particle $X_i$, the most straightforward way is to use the indices of the set of discrete variables with $n_D$ elements.

$$X_i^D = \left[x_{i,1}^D, \cdots, x_{i,n_D}^D\right]$$

For particle $i$, the index value $j$ of the discrete variable $x_{i,j}^D$ is then optimized instead of the discrete value of the variable directly. In the population, the indices of the discrete variables of the $i$th particle should be the flout point variables before truncation. That is $j \in [1, n_D + 1)$, $n_D$ is the number of discrete variables. Hence, the fitness function of the $i$th particle $X_i$ can be expressed as follows:

$$f\left(X_i\right) \quad i = 1, 2, \cdots, M \tag{2.7}$$

where

$$X_i = \begin{cases} x_{i,j} : x_{i,j} \in X_i^C & j = 1, \cdots, n_C \\ INT\left(x_{i,j}\right) : x_{i,j} \in X_i^I \bigcup X_i^B & j = 1, \cdots, n_I + n_B \\ x_{i,INT(j)} : x_{i,INT(j)} \in X_i^D & j \in [1, n_D + 1) \end{cases} \tag{2.8}$$

where $X^C \in R^{n_C}$, $X^B \in R^{n_B}$, $X^I \in R^{n_I}$ and $X^D \in R^{n_D}$ denote the feasible subsets of continuous, binary, integer and discrete variables of particle $X_i$, respectively. $INT\left(x\right)$ denotes the greatest integer less than the real value $x$.

## 2.5 The Constraint Handling Method

EAs are heuristic optimization techniques which have been successfully applied to various optimization problems. However, they are not able to handle constrained optimization problems directly. In the past few years, much work has been done to improved EAs performance to deal with constrained optimization problems. By maintaining a feasible population, PSO algorithms have been applied to constrained optimization problems. The technique starts from a feasible initial population. A closed set of operators is used to maintain the feasibility of the solutions. Therefore, the subsequent solutions generated at each iteration are also feasible. Algorithms based on this technique are much more reliable than those based on a penalty approach. For structural design problems, reliability is crucial since almost all of the constraints need to be satisfied.

For the PSO algorithm, the intuitive idea to maintain a feasible population is for a particle to fly back to its previous position when it is outside the feasible region. This is the so called 'fly-back mechanism'. Since the population is initialized in the feasible region, flying back to a previous position will guarantee the solution to be feasible. From literatures, the global minima of optimal design problems are usually close to the boundaries of the feasible space, as shown in Figure 2.1. Flying back to its previous position when a particle violates the constraints will allow a new search closer to the boundaries. Figures 2.2 and 2.3 illustrate the search process of the 'fly-back mechanism'. In Figure 2.2, the *i*th particle would fly into the infeasible search space at the *k*th iteration. At the next iteration as shown in Figure 2.3, this particle is set back to its previous position $X_i^{k-1}$ and starts a new search. Assuming that the global best particle $P_g$ stays in the same position, the direction of the new velocity $V_i^{k+1}$ will still point to the boundary but closer to global best particle $P_g$. Since $P_g$ is inside the feasible space and $\omega V_i^k$ is smaller than $V_i^k$, the chance of particle $X_i$ flying outside the boundaries at the next iteration will be decreased. This property makes the particles more likely to explore the feasible search space near the boundaries. Therefore, such a 'fly-back mechanism' is suitable for structural design problems.

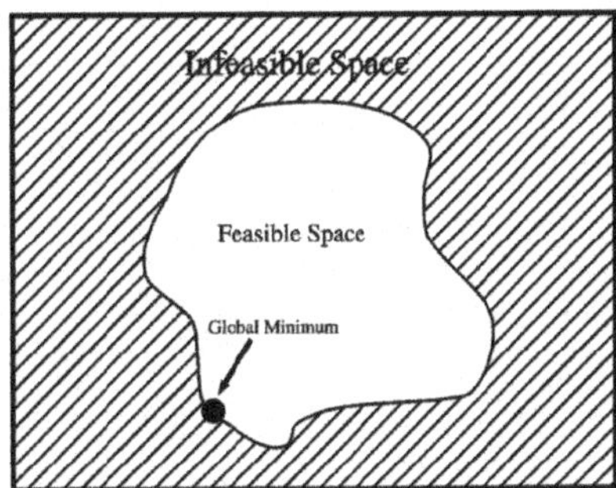


**Fig. 2.1** Global minimum in the feasible space

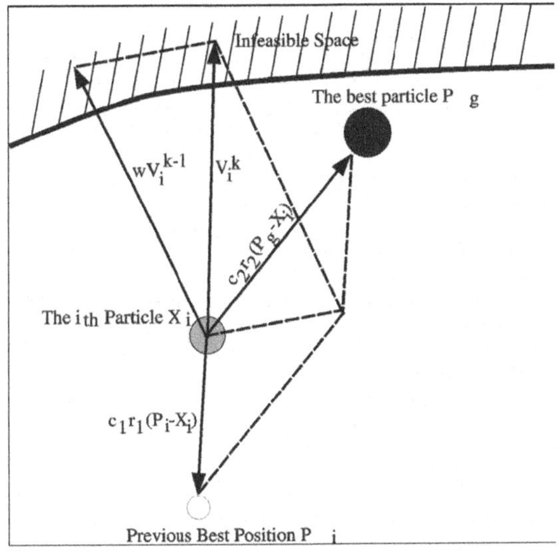


**Fig. 2.2** $X_i$ at iteration $k$ would fly outside the feasible search space

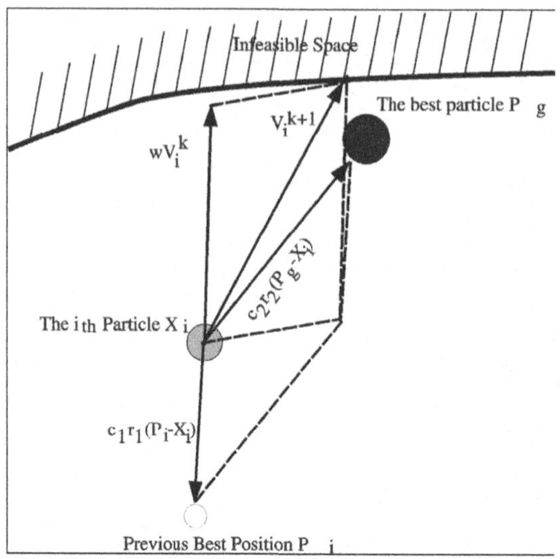


**Fig. 2.3** $X_i$ flies back to its previous position and starts a new search

## 2.6 Application of the PSO in Structural Optimization Problems

The proposed constraint handling technique requires a feasible initial population to guarantee that the solutions of successive generations are feasible. For structure optimization design problems, a feasible initial population usually is easily obtained since their feasible search spaces are usually large and feasible particles can be easily generated. Small size populations are preferred to minimize the time to find a feasible initial population.

For a practical structure optimization design problem, since the number of variables $N$ is big, it is necessary to use finite element method (FEM) [12, 13] to obtain a feasible initial population. The updated particles are also used by FEM to adjust particle position so that the stress and deformation constraints are not violated.

In standard PSO, parameters $\omega$, $c_1$ and $c_2$ are constant for all particles. While we try using matrix instead of constants so that each particle has different calculation parameter value, which augments search space and accelerates the convergence velocity.

The constraints handling is also improved to fit the structural optimization problem solving. Denote $X_i^k$ the $i$th particle in $k$th iteration, randomly take a particle $X_j^{k-1}$ from previous generation. If any variables in $X_i^k$ violate the constraints, then put $X_i^k = X_j^{k-1}$, and go into next iteration. Repeat equations (2.1)~(2.4) until global best position is founded. when $X_i^k$ violates constraints, do not put $X_i^k = X_i^{k-1}$ but $X_i^k = X_j^{k-1}$ will accelerate the convergence speed and prevent premature. Compared with penalty function method, this handling method does not need extra parameters, but requires that initial particles must meet the constraints and belong to feasible solution space. This demand can be easily met by using FEM.

Table 2.1 lists the calculation flow of the PSO algorithm for structural design problems:

**Table 2.1** The PSO algorithm calculation flow

```
Set k = 1;
Initialize positions and velocities of all particles;
  FOR (each particle i in the initial population)
  WHILE (the constraints are violated)
       Re-initialize current particle X_i
  END WHILE
END FOR
WHILE (the termination conditions are not met)
  FOR (each particle i in the swarm)
       Check feasibility: Check the feasibility of the current particle. If X_i^k
                    is outside the feasible region, then reset X_i^k to the
                    randomly chosen previous generation particle X_j^{k-1};
       Calculate fitness: Calculate the fitness value f(X_i^k) of current parti-
                    cle using Eq. (2.8);
       Update pbest: Compare the fitness value of pbest with f(X_i^k). If
                    f(X_i^k) is better than the fitness value of pbest, then
                    set pbest to the current position X_i^k;
       Update gbest: Find the global best position of the swarm. If the
                    f(X_i^k) is better than the fitness value of gbest, then
                    gbest is set to the position of the current particle
                    X_i^k;
       Update velocities: Calculate velocities V_i^k using Eq. (2.3);
       Update positions: Calculate positions X_i^k using Eq. (2.4);
  END FOR
  Set k = k +1
END WHILE
```

## 2.7 Examples

The two examples are used to show that the application of PSO to structural optimization problems is feasible and effective. Both of them take cross-sections of a truss structure as an optimization objective.

(1) Optimization design for a three-bar truss structure

A three bar truss structure is shown in Figure 2.4. The optimization objective is the volume of bars of truss. Denote the cross-section of bar $x_i$ as variable. There are two load conditions: 1) $P_1$=2000 kN, $P_2$=0 kN. 2) $P_1$=0 kN, $P_2$=2000 kN

Objective function is as follows:

$$\min V = \sum_{i=1}^{n} A_i \cdot l_i \qquad n = 3$$

$$s.t. \qquad [\sigma_-] \le \sigma_i \le [\sigma_+]$$

$$A_{\min} \le A_i \le A_{\max}$$

Material parameters are $A_{i\min} = 1.\times 10^{-5} m^2$, $A_{i\max} = 1.\times 10^{-3} m^2$, $[\sigma_+] = 2.\times 10^7$ kPa, $[\sigma_-] = 1.5\times 10^7$ kPa, $E = 2.\times 10^8$ kPa. Calculation parameters are $N = 3$, $M = 50$, $iter_{\max} = 300$. Where $V$ is total volume of truss, $l_i$ and $A_i$ is bar's length and area respectively, $\sigma_i$ is Bar's stress.

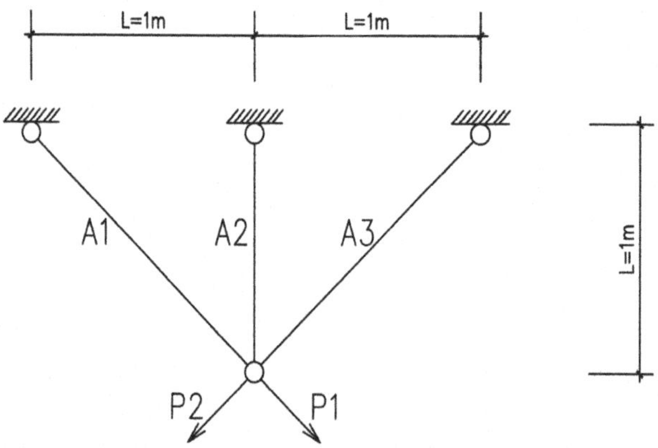


**Fig. 2.4** A three-bar truss structure subject to loads

Except for the PSO method, the Matlab optimal toolbox, the Stress ratio method and the analytical method were used to compare the solutions, which were expressed in Table 2.2.

Matlab toolbox is a traditional optimal method with faster convergence speed, but it requires definite objective function and constraint condition, which made it limited to deal with complex problems. Stress ratio method is a conventional optimal method used for truss structure. Table 2.1 shows that the PSO method has the best precision. By 30 iterations, optimal solution reatched 100% analytical solution. The Matlab toobox method and the Stress ratio method has 1.43% and 6.63% error respectively.

**Table 2.2** A three-bar truss structure optimization results

| Variables ($m^2$) | Optimal results | | | |
|---|---|---|---|---|
| | PSO | Matlab toolbox | Stress ratio method | Analytical solution |
| $x(1)$ | 0.79051e-4 | 0.8698e-4 | 0.9898e-4 | 0.789e-4 |
| $x(2)$ | 0.40308e-4 | 0.2166e-4 | 0.0144e-4 | 0.408e-4 |
| $x(3)$ | 0.79051e-4 | 0.8698e-4 | 0.9898e-4 | 0.789e-4 |
| $V$ ($m^3$) | 2.639e-4 | 2.6767e-4 | 2.8141e-4 | 2.639e-4 |
| Error (%) | 0 | 1.43 | 6.63 | |

(2) Optimization design for a 15-bar truss structure

A 15-bar truss structure is shown in Figure 2.5. The optimization objective is the volume of bars of truss. Denote the cross-section of bar $x_i$ as variables. The constraints are lower and upper bounds of cross-sections and stresses of bars.

The objective function is :

$$\min V = \sum_{i=1}^{n} A_i \cdot l_i \qquad n = 15$$

$$s.t. \qquad \sigma_i \le [\sigma_+] \qquad \sigma_i \ge 0$$

$$\sigma_i \ge [\sigma_-] \qquad \sigma_i \le 0$$

$$A_{\min} \le A_i \le A_{\max}$$

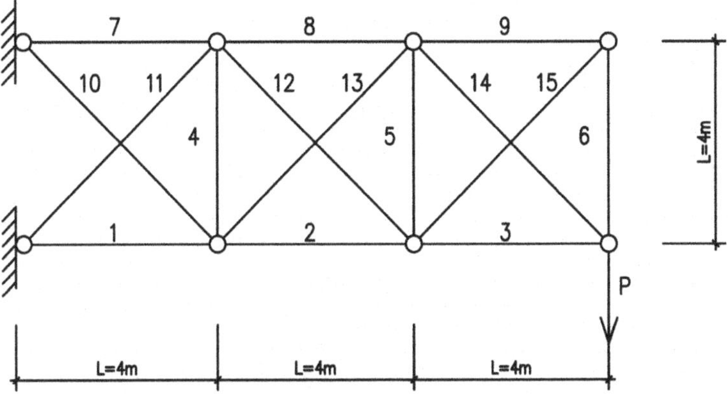


**Fig. 2.5** A 15-bar truss structure subjected to loads

Material parameters are $A_{i\min} = 8.\times10^{-5} m^2$ , $A_{i\max} = 1.\times10^{-3} m^2$ , $[\sigma_+] = 7.\times10^4 kPa$ , $[\sigma_-] = -3.5\times10^4 kPa$ , $E = 2.\times10^8$ kPa , $v$ : total volume of bars of truss, $l_i, A_i$ : Bar's length and area respectively, $\sigma_i$ : bar's stress.

The calculation parameters: $N = 15$ , $M = 100$ , $iter_{\max} = 300$ .

Except for the PSO method, Matlab optimal toolbox and Ansys program are used to compare the solutions, which were expressed in Table 2.3.

**Table 2.3** A fifteen-bar truss structure optimization results

| Variables ($m^2$) | Optimization results | | |
|---|---|---|---|
| | PSO | Matlab toolbox | Ansys |
| $x(1)$ | 0.00068933 | 0.00067985 | 0.00069575 |
| $x(2)$ | 0.00039448 | 0.00039634 | 0.00040504 |
| $x(3)$ | 0.00016872 | 0.00016884 | 0.00018412 |
| $x(4)$ | 8e-005 | 8e-005 | 8e-005 |
| $x(5)$ | 8e-005 | 8e-005 | 8.0419e-005 |
| $x(6)$ | 8e-005 | 8e-005 | 8e-005 |
| $x(7)$ | 0.00036965 | 0.00037436 | 0.00038024 |
| $x(8)$ | 0.00023148 | 0.0002304 | 0.00024087 |
| $x(9)$ | 8e-005 | 8e-005 | 8e-005 |
| $x(10)$ | 8.7203e-005 | 8e-005 | 8.9916e-005 |
| $x(11)$ | 0.00023734 | 0.00025073 | 0.00024610 |
| $x(12)$ | 8e-005 | 8e-005 | 8.0905e-005 |
| $x(13)$ | 0.00025026 | 0.00024761 | 0.00026109 |
| $x(14)$ | 0.0001193 | 0.00011939 | 0.00012775 |
| $x(15)$ | 0.00016581 | 0.00016529 | 0.00016720 |
| $V$ ($m^3$) | 0.014011 | 0.014014 | 0.014734 |

The Ansys program is a commercial software package [14]. It consumes long time and the optimal course is complicated. It is not easy convergent for complex problem. Moreover it can only deal with limited variables.

From Table 2.3 it can be seen that the PSO method is simple and has more optimal results compared with other two methods. Ansys analysis results is less precision and need more time to converge.

## 2.8 Conclusions

This chapter has presented a mixed-variable PSO algorithm to deal with structural optimization problems. The structural design optimization problem involves problem-specific constraints and mixed variables such as integer, binary, discrete and continuous variables. The PSO is combined with the FEM to deal with the constraints related with the boundary conditions of structures controlled by stresses or displacements.

The work presented in this chapter consists of three aspects. The first is that the FEM was used to generate a initial feasible particle swarm and deal with boundary conditions. The second is that the randomly 'fly back' scheme was used to guarantee the variety of particles to fit the need of civil engineering. The third is that the problem solving dimensions were not limited to the number of objective functions, variables and constraints, the proposed algorithm can be extended to larger problems with further modification required. The proposed algorithm has been successfully applied to solve two structural design problems through comprehensive simulation studies. The simulation results show that application of PSO to the structural optimization problems is feasible and effective, and the proposed algorithm is able to achieve better convergence performance and higher accuracy in comparison with other conventional optimization methods used in civil engineering.

## References


1. Coello Coello, C.A.: Theoretical and numerical constraint-handling techniques used with evolutionary algorithms: A survey of the state of the art. Computer Methods in Applied Mechanics and Engineering 191(11-12), 1245–1287 (2002)
2. Abdo, T., Rackwitz, R.: A new beta point algorithm for large time invariant reliability problems. in: reliability and optimization of structures. In: Proceedings of the Third WG 7.5 IFIP Conference, pp. 1–11 (1990)
3. Le Riche, R.G., Knopf-Lenoir, C., Haftka, R.T.: A segregated genetic algorithm for constrained structural optimization. In: Sixth International Conference on Genetic Algorithms, pp. 558–565. University of Pittsburgh, Morgan Kaufmann (1995)
4. He, S., Prempain, E., Wu, Q.H.: An improved particle swarm optimizer for mechanical design optimization problems. Engineering Optimization 36(5), 585–605 (2004)
5. Sandgren, E.: Nonlinear integer and discrete programming in mechanical design Optimization. Journal of Mechanical Design 112, 223–229 (1990)
6. Hajela, P., Shih, C.: Multiobjective optimum design in mixed-integer and discrete design variable problems. AIAA Journal 28(4), 670–675 (1989)
7. Cao, Y.J., Wu, Q.H.: A mixed variable evolutionary programming for optimization of mechanical design. International Journal of Engineering Intelligent Systems for Electrical Engineering and Communications 7(2), 77–82 (1999)
8. Eberhart, R.C., Kennedy, J.: A new optimizer using particles swarm theory. In: Sixth International Symposium on Micro Machine and Human Science (1995)
9. Kennedy, J., Eberhart, R.C.: Particle swarm optimization. In: IEEE International Conference on Neural Networks, vol. 4, pp. 1942–1948. IEEE Press, Los Alamitos (1995)

10. Shi, Y., Eberhart, R.C.: Fuzzy adaptive particle swarm optimization. In: Proceedings of the IEEE International Conference on Evolutionary Computation, pp. 101–106 (2001)
11. Parsopoulos, K.E., Vrahatis, M.N.: Recent approaches to global optimization problems through particle swarm optimization. Natural Computing 1, 235–306 (2002)
12. Wan, Y.H., Li, L.J., Li, Y.P.: Finite element method and programs. South China University Publication (January 2001) (in Chinese)
13. Kattan, P.I.: Matlab guide to finite elements, pp. 53–63. Springer, Heidelberg (2003)
14. Yi, R.: Suructure analysis using Ansys software. Beijing University Publication (2002) (in Chinese)

# Chapter 3
# Optimum Design of Structures with Heuristic Particle Swarm Optimization Algorithm

**Abstract.** This chapter introduces the application of an improved particle swarm algorithm to optimal structure design. The algorithm is named heuristic particle swarm optimization (HPSO). It is based on heuristic search schemes and the standard particle swarm algorithm. The efficiency of HPSO for pin connected structures with continuous variables and for pin connected structures and plates with discrete variables is compared with that of other intelligent algorithms, and the implementation of HPSO is presented in detail. An optimal result of a complex practical double-layer grid shell structure is presented to value the effectiveness of the HPSO.

## 3.1 Introduction

In the last 30 years, a great attention has been paid to structural optimization, since material consumption is one of the most important factors influencing building construction. Designers prefer to reduce the volume or weight of structures through optimization. Many traditional mathematical optimization algorithms have been used in structural optimization problems. The traditional optimal algorithms provide a useful strategy to obtain the global optimal solution in a simple model.

However, many practical engineering optimal problems are very complex and hard to solve by the traditional optimal algorithms. Recently, evolutionary algorithms (EAs), such as genetic algorithms (GAs), evolutionary programming (EP) and evolution strategies (ES) have become more attractive because they do not require conventional mathematical assumptions and thus possess better global search abilities than the conventional optimization algorithms [1]. For example, GAs have been applied for structural optimization problems [2, 3, 4].

A new evolutionary algorithm called particle swarm optimizer (PSO) was developed by Kennedy and Eberhart [5], which was inspired by the social behaviour of animals such as fish schooling and bird flocking. It is a population-based algorithm, which is based on the premise that social sharing of information among members of a species offers an evolutionary advantage. With respect to other algorithms such as evolutionary algorithms, a number of advantages make PSO

L. Li & F. Liu: Group Search Optimization for Applications in Structural Design, ALO 9, pp. 21–67.
springerlink.com 

an ideal candidate to be used in optimization tasks. The algorithm can handle continuous, discrete and integer variable types with ease. In addition, its easiness of implementation makes it more attractive for the applications of real-engineering optimization problems. Furthermore, it is a population-based algorithm, so it can be efficiently parallelized to reduce the total computational effort. The PSO has fewer parameters and is easier to implement than the GAs [6]. The PSO also shows a faster convergence rate than the other EAs for solving some optimization problems.

The foundation of PSO is based on the hypothesis that social sharing of information among conspecifics offers an evolutionary advantage. It involves a number of particles, which are initialized randomly in the search space of an objective function. These particles are referred to as swarm. Each particle of the swarm represents a potential solution of the optimization problem. The particles fly through the search space and their positions are updated based on the best positions of individual particles in each iteration. The objective function is evaluated for each particle and the fitness values of particles are obtained to determine which position in the search space is the best.

In each iteration, the swarm is updated using the following equations:

$$V_i^{k+1} = \omega V_i^k + c_1 r_1 \left(P_i^k - X_i^k\right) + c_2 r_2 \left(P_g^k - X_i^k\right) \tag{3.1}$$

$$X_i^{k+1} = X_i^k + V_i^{k+1} \tag{3.2}$$

where $X_i$ and $V_i$ represent the current position and the velocity of the ith particle respectively; $P_i$ is the best previous position of the ith particle (called pbest) and $P_g$ is the best global position among all the particles in the swarm (called gbest); $r_1$ and $r_2$ are two uniform random sequences generated from U(0, 1); and ω is the inertia weight used to discount the previous velocity of the particle persevered.

The PSO model is based on the following two factors:

(1) The autobiographical memory, which remembers the best previous position of each individual ($P_i$) in the swarm; and

(2) The publicized knowledge, which is the best solution ($P_g$) found currently by the population.

Angeline [7] pointed out that although PSO may outperform other evolutionary algorithms in the early iterations, its performance may not be competitive as the number of generations is increased. Recently, many investigations have been undertaken to improve the performance of the standard PSO (SPSO). He [8] et al. found that adding the passive congregation model to the SPSO may increase its performance. Therefore, they improved the SPSO with passive congregation (PSOPC), which can improve the convergence rate and accuracy of the SPSO efficiently.

## 3.2 Constraint Handling Method: Fly-Back Mechanism

Most structural optimization problems include the problem-specific constraints, which are difficult to solve using the traditional mathematical optimization algorithms [9]. Penalty functions have been commonly used to deal with constraints. However, the major disadvantage of using the penalty functions is that some tuning parameters are added in the algorithm and the penalty coefficients have to be tuned in order to balance the objective and penalty functions. If appropriate penalty coefficients cannot be provided, difficulties will be encountered in the solution of the optimization problems [10, 11]. To avoid such difficulties, a new method, called 'fly-back mechanism', was developed.

For most of the optimization problems containing constraints, the global minimum locates on or close to the boundary of a feasible design space. The particles are initialized in the feasible region. When the optimization process starts, the particles fly in the feasible space to search the solution. If any one of the particles flies into the infeasible region, it will be forced to fly back to the previous position to guarantee a feasible solution. The particle which flies back to the previous position may be closer to the boundary at the next iteration. This makes the particles to fly to the global minimum in a great probability. Therefore, such a 'fly-back mechanism' technique is suitable for handling the optimization problem containing the constraints. Compared with the other constraint handling techniques, this method is relatively simple and easy to implement. Some experimental results have shown that it can find a better solution with a fewer iterations than the other techniques.

## 3.3 A Heuristic Particle Swarm Optimization (HPSO)

The heuristic particle swarm optimizer (HPSO) [12] is based on the PSOPC and a harmony search (HS) scheme, and uses a 'fly-back mechanism' method to handle the constraints. The pseudo-code for the HPSO algorithm is listed in Table 3.1.

When a particle flies in the searching space, it may fly into infeasible regions. In this case, there are two possibilities. It may violate either the problem-specific constraints or the limits of the variables, as illustrated in Fig. 3.1. Because the 'fly-back mechanism' technique is used to handle the problem-specific constraints, the particle will be forced to fly back to its previous position no matter whether it violates the problem-specific constraints or the variable boundaries. If it flies out of the variable boundaries, the solution cannot be used even if the problem-specific constraints are satisfied. In our experiments, particles violate the variables' boundary frequently for some simple structural optimization problems. If the structure becomes complicated, the number of occurrences of violating tends to rise. In other words, a large amount of particles' flying behaviours are wasted, due to searching outside the variables' boundary. Although minimizing the maximum of the velocity can make fewer particles violate the variable boundaries, it may also

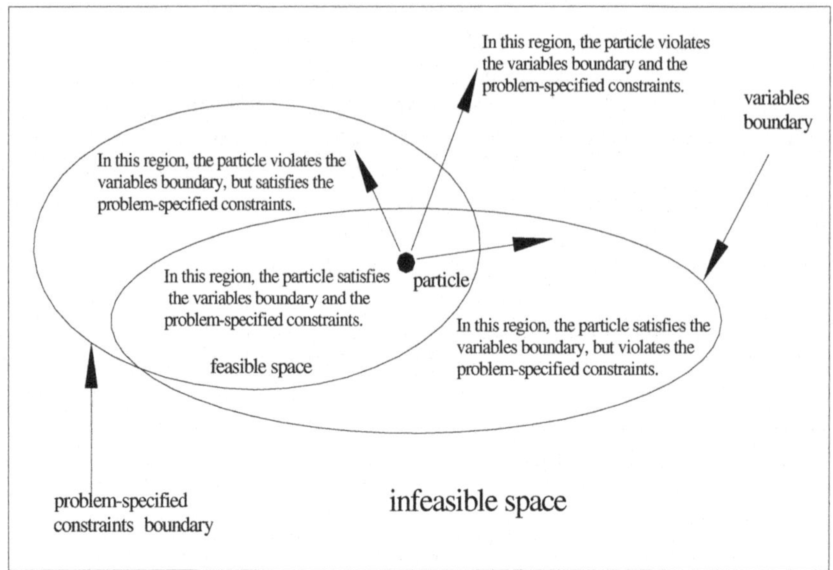


**Fig. 3.1** The particle may violate the problem-specific constraints or the variables' boundary

prevent the particles to cross the problem-specific constraints. Therefore, we hope that all of the particles fly inside the variable boundaries and then to check whether they violate the problem-specific constraints and get better solutions or not. The particles, which fly outside the variables' boundary, have to be regenerated in an alternative way. Here, we introduce a new method to handle these particles. It is derived from one of the ideas in a new meta-harmony algorithm called harmony search algorithm [13, 14].

Harmony search algorithm is based on natural musical performance processes that occur when a musician searches for a better state of harmony, such as during jazz improvisation [15]. The engineers seek for a global solution as determined by an objective function, just like the musicians seek to find musically pleasing harmony as determined by an aesthetic [16]. The harmony search algorithm includes a number of optimization operators, such as the harmony memory (HM), the harmony memory size (HMS), the harmony memory considering rate (HMCR), and the pitch adjusting rate (PAR). In this paper, the harmony memory (HM) concept has been used in the PSO algorithm to avoid searching trapped in local solutions. The other operators have not been employed. How the HS algorithm generates a new vector from its harmony memory and how it is used to improve the PSO algorithm will be discussed as follows.

**Table 3.1** The pseudo-code for the HPSO

```
Set k=1;
Randomly initialize positions and velocities of all particles;
    FOR (each particle i in the initial population)
        WHILE (the constraints are violated)
            Randomly re-generate the current particle X_i
        END WHILE
    END FOR
WHILE (the termination conditions are not met)
    FOR (each particle i in the swarm)
        Generate the velocity and update the position of the current particle (vector) X_i
        Check feasibility stage I: Check whether each component of the current vector
      violates its corresponding boundary or not. If it does, select the corresponding
      component of the vector from pbest swarm randomly.
        Check feasibility stage II: Check whether the current particle violates the problem
      specified constraints or not. If it does, reset it to the previous position X_ik-1.
        Calculate the fitness value f(X_ik) of the current particle.
        Update pbest: Compare the fitness value of pbest with f(X_ik). If the f(X_ik) is better
      than the fitness value of pbest, set pbest to the current position X_ik.
        Update gbest: Find the global best position in the swarm. If the f(X_ik) is better than
      the fitness value of gbest, gbest is set to the position of the current particle X_ik.
    END FOR
    Set k=k+1
END WHILE
```

In the HS algorithm, the harmony memory stores the feasible vectors, which are all in the feasible space. The harmony memory size determines how many vectors it stores. A new vector is generated by selecting the components of different vectors randomly in the harmony memory. Undoubtedly, the new vector does not violate the variables boundaries, but it is not certain if it violates the problem-specific constraints. When it is generated, the harmony memory will be updated by accepting this new vector if it gets a better solution and deleting the worst vector.

Similarly, the PSO stores the feasible and "good" vectors (particles) in the pbest swarm, as does the harmony memory in the HS algorithm. Hence, the vector (particle) violating the variables' boundaries can be generated randomly again by such a technique-selecting for the components of different vectors in the pbest swarm. There are two different ways to apply this technique to the PSO when any one of the components of the vector violates its corresponding variables' boundary. Firstly, all the components of this vector should be generated. Secondly, only this component of the vector should be generated again by such a technique. In our experiments, the results show that the former makes the particles moving to the local solution easily, and the latter can reach the global solution in relatively less number of iterations.

Therefore, applying such a technique to the PSOPC can improve its performance, although it already has a better convergence rate and accuracy than the PSO.

## 3.4 The Application of the HPSO on Truss Structures with Continuous Variables

In this section, five pin-connected structures commonly used in literature are selected as benchmark problems to test the HPSO. The proposed algorithm is coded in Fortran language and executed on a Pentium 4, 2.93GHz machine.

The examples given in the simulation studies include

- a 10-bar planar truss structure subjected to four concentrated loads as shown in Fig. 3.2;
- a 17-bar planar truss structure subjected to a single concentrated load at its free end as shown in Fig. 3.5;
- a 22-bar spatial truss structure subjected to three load cases;
- a 25-bar spatial truss structure subjected to two load cases;
- a 72-bar spatial truss structure subjected to two load cases.

All these truss structures are analyzed by the finite element method (FEM).

The PSO, PSOPC and HPSO schemes are applied respectively to all these examples and the results are compared in order to evaluate the performance of the new algorithm. For all these algorithms, a population of 50 individuals is used; the inertia weight $\omega$ decrease linearly from 0.9 to 0.4; and the value of acceleration constants $c_1$ and $c_2$ are set to be the same and equal to 0.8. The passive congregation coefficient $c_3$ is given as 0.6 for the PSOPC [8] and the HPSO algorithms. The maximum number of iterations is limited to 3000. The maximum velocity is set as the difference between the upper bound and the lower bound of variables, which ensures that the particles are able to fly into the problem-specific constraints' region.

### *3.4.1 Numerical Examples*

#### *(1) The 10-bar planar truss structure*

The 10-bar truss structure, shown in Fig. 3.2, has previously been analyzed by many researchers, such as Lee [16], Schmit [17], Rizzi [18], and Li [19]. The material density is 0.1 lb/in$^3$ and the modulus of elasticity is 10,000 ksi. The members are subjected to the stress limits of ±25 ksi. All nodes in both vertical and horizontal directions are subjected to the displacement limits of ±2.0 in. There are 10 design variables in this example and the minimum permitted cross-sectional area of each member is 0.1 in$^2$. Two cases are considered: Case 1, $P_1$=100 kips and $P_2$=0; Case 2, $P_1$=150 kips and $P_2$=50 kips.

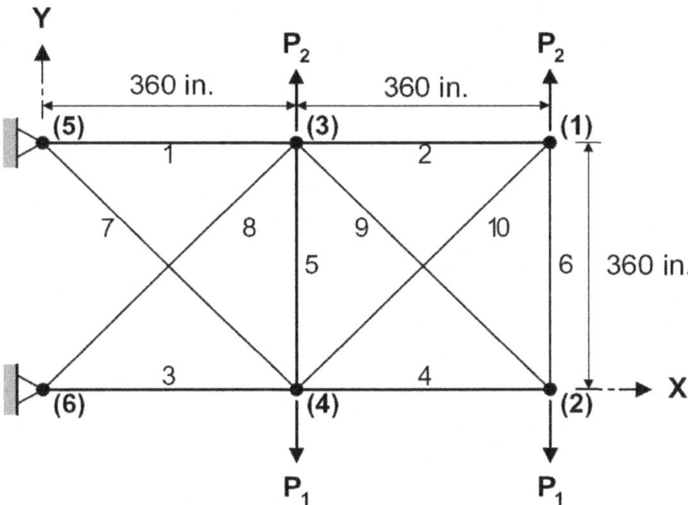


**Fig. 3.2** A 10-bar planar truss structure

For both load cases, the PSOPC and the HPSO algorithms achieve the best solutions after 3,000 iterations. However, the latter is closer to the best solution than the former after about 500 iterations. The HPSO algorithm displays a faster convergence rate than the PSOPC algorithm in this example. The performance of the PSO algorithm is the worst among the three. Tables 3.2 and 3.3 show the solutions. Figs. 3.3 and 3.4 provide a comparison of the convergence rates of the three algorithms.

**Table 3.2** Design results for the 10-bar planar truss structure (Case 1)

| Variables | | Optimal cross-sectional areas (in.$^2$) | | | | | |
|---|---|---|---|---|---|---|---|
| | | Schmit [17] | Rizzi [18] | Lee [16] | Li [19] PSO | Li [19] PSOPC | Li [19] HPSO |
| 1 | $A_1$ | 33.43 | 30.73 | 30.15 | 33.469 | 30.569 | 30.704 |
| 2 | $A_2$ | 0.100 | 0.100 | 0.102 | 0.110 | 0.100 | 0.100 |
| 3 | $A_3$ | 24.26 | 23.93 | 22.71 | 23.177 | 22.974 | 23.167 |
| 4 | $A_4$ | 14.26 | 14.73 | 15.27 | 15.475 | 15.148 | 15.183 |
| 5 | $A_5$ | 0.100 | 0.100 | 0.102 | 3.649 | 0.100 | 0.100 |
| 6 | $A_6$ | 0.100 | 0.100 | 0.544 | 0.116 | 0.547 | 0.551 |
| 7 | $A_7$ | 8.388 | 8.542 | 7.541 | 8.328 | 7.493 | 7.460 |
| 8 | $A_8$ | 20.74 | 20.95 | 21.56 | 23.340 | 21.159 | 20.978 |
| 9 | $A_9$ | 19.69 | 21.84 | 21.45 | 23.014 | 21.556 | 21.508 |
| 10 | $A_{10}$ | 0.100 | 0.100 | 0.100 | 0.190 | 0.100 | 0.100 |
| Weight (lb) | | 5089.0 | 5076.66 | 5057.88 | 5529.50 | 5061.00 | 5060.92 |

**Table 3.3** Design results for the 10-bar planar truss structure (Case 2)

| Variables | | Optimal cross-sectional areas (in.$^2$) | | | | | |
|---|---|---|---|---|---|---|---|
| | | Schmit [17] | Rizzi [18] | Lee [16] | Li [19] PSO | Li [19] PSOPC | Li [19] HPSO |
| 1 | $A_1$ | 24.29 | 23.53 | 23.25 | 22.935 | 23.743 | 23.353 |
| 2 | $A_2$ | 0.100 | 0.100 | 0.102 | 0.113 | 0.101 | 0.100 |
| 3 | $A_3$ | 23.35 | 25.29 | 25.73 | 25.355 | 25.287 | 25.502 |
| 4 | $A_4$ | 13.66 | 14.37 | 14.51 | 14.373 | 14.413 | 14.250 |
| 5 | $A_5$ | 0.100 | 0.100 | 0.100 | 0.100 | 0.100 | 0.100 |
| 6 | $A_6$ | 1.969 | 1.970 | 1.977 | 1.990 | 1.969 | 1.972 |
| 7 | $A_7$ | 12.67 | 12.39 | 12.21 | 12.346 | 12.362 | 12.363 |
| 8 | $A_8$ | 12.54 | 12.83 | 12.61 | 12.923 | 12.694 | 12.894 |
| 9 | $A_9$ | 21.97 | 20.33 | 20.36 | 20.678 | 20.323 | 20.356 |
| 10 | $A_{10}$ | 0.100 | 0.100 | 0.100 | 0.100 | 0.103 | 0.101 |
| Weight (lb) | | 4691.84 | 4676.92 | 4668.81 | 4679.47 | 4677.70 | 4677.29 |

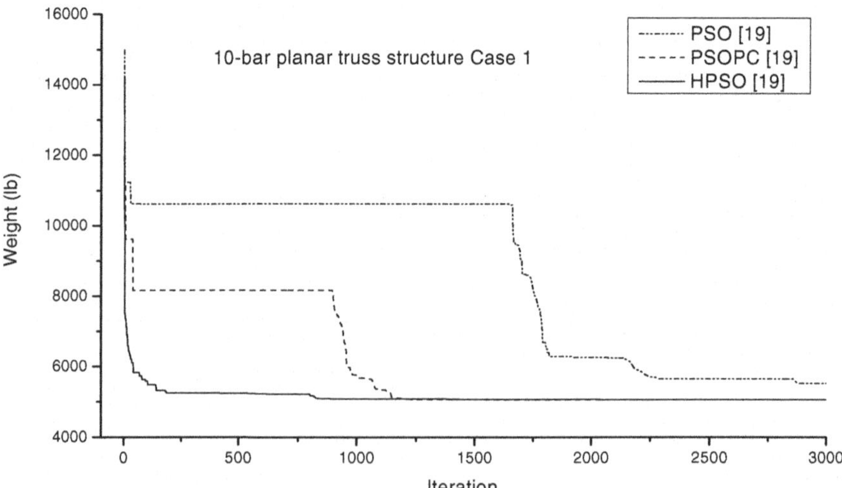


**Fig. 3.3** Convergence rates of the three algorithms for the 10-bar planar truss structure (Case 1)

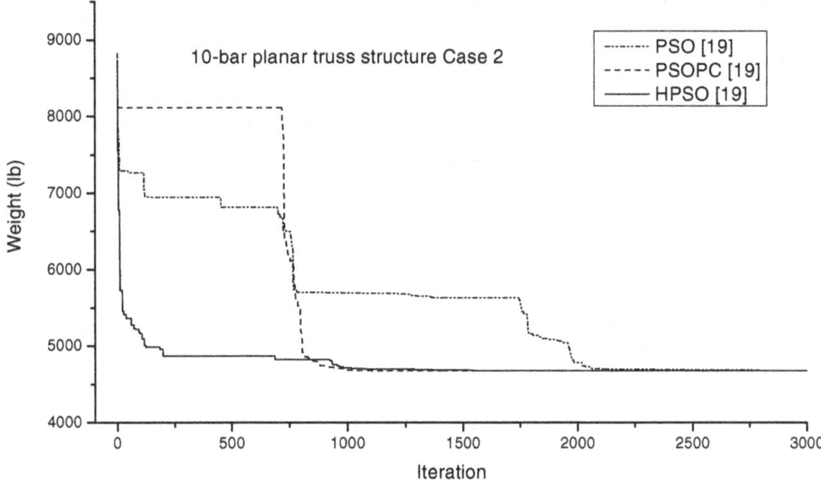


**Fig. 3.4** Convergence rates of the three algorithms for the 10-bar planar truss structure (Case 2)

The curves of stability of 10-bar truss structure with three different algorithms (PSO, PSOPC and HPSO) is shown in Fig. 3.5. The stability curves come from the best results of 100 independent calculation times. The standard deviation of PSO, PSOPC and HPSO is 664.07891, 12.84174 and 3.8402 respectively. It can be seen from Fig. 3.5 that the HPSO has the best stability.

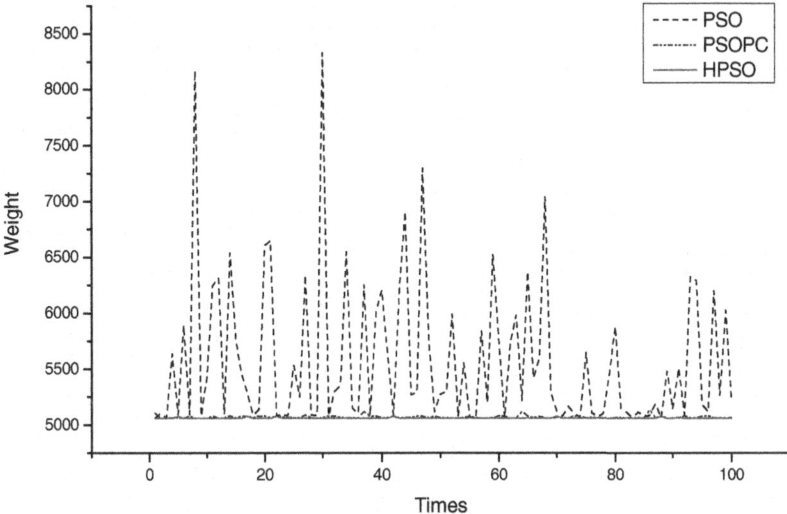


**Fig. 3.5** The stability of the 10-bar truss structure with three algorithms

### *(2) The 17-bar planar truss structure*

The 17-bar truss structure, shown in Fig. 3.6, had been analyzed by Khot [20], Adeli [21], Lee [16] and Li [19]. The material density is 0.268 lb/in.$^3$ and the modulus of elasticity is 30,000 ksi. The members are subjected to the stress limits of ±50 ksi. All nodes in both directions are subjected to the displacement limits of ±2.0 in.

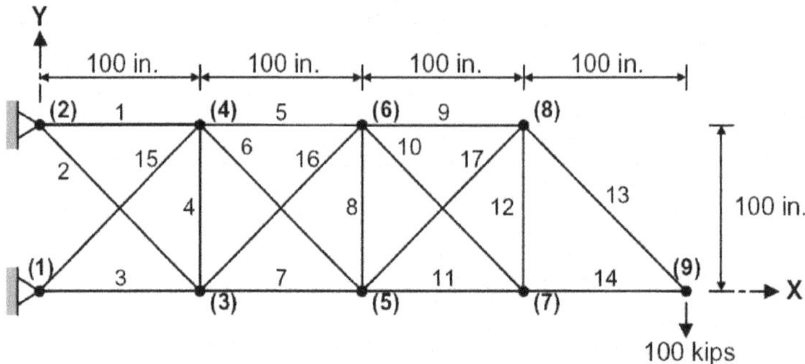


**Fig. 3.6** A 17-bar planar truss structure

**Table 3.4** Comparison of the designs for the 17-bar planar truss

| Variables | | Optimal cross-sectional areas (in.$^2$) | | | | | |
|---|---|---|---|---|---|---|---|
| | | Khot [20] | Adeli [21] | Lee [16] | Li [19] PSO | Li [19] PSOPC | Li [19] HPSO |
| 1 | $A_1$ | 15.930 | 16.029 | 15.821 | 15.766 | 15.981 | 15.896 |
| 2 | $A_2$ | 0.100 | 0.107 | 0.108 | 2.263 | 0.100 | 0.103 |
| 3 | $A_3$ | 12.070 | 12.183 | 11.996 | 13.854 | 12.142 | 12.092 |
| 4 | $A_4$ | 0.100 | 0.110 | 0.100 | 0.106 | 0.100 | 0.100 |
| 5 | $A_5$ | 8.067 | 8.417 | 8.150 | 11.356 | 8.098 | 8.063 |
| 6 | $A_6$ | 5.562 | 5.715 | 5.507 | 3.915 | 5.566 | 5.591 |
| 7 | $A_7$ | 11.933 | 11.331 | 11.829 | 8.071 | 11.732 | 11.915 |
| 8 | $A_8$ | 0.100 | 0.105 | 0.100 | 0.100 | 0.100 | 0.100 |
| 9 | $A_9$ | 7.945 | 7.301 | 7.934 | 5.850 | 7.982 | 7.965 |
| 10 | $A_{10}$ | 0.100 | 0.115 | 0.100 | 2.294 | 0.113 | 0.100 |
| 11 | $A_{11}$ | 4.055 | 4.046 | 4.093 | 6.313 | 4.074 | 4.076 |
| 12 | $A_{12}$ | 0.100 | 0.101 | 0.100 | 3.375 | 0.132 | 0.100 |
| 13 | $A_{13}$ | 5.657 | 5.611 | 5.660 | 5.434 | 5.667 | 5.670 |
| 14 | $A_{14}$ | 4.000 | 4.046 | 4.061 | 3.918 | 3.991 | 3.998 |
| 15 | $A_{15}$ | 5.558 | 5.152 | 5.656 | 3.534 | 5.555 | 5.548 |
| 16 | $A_{16}$ | 0.100 | 0.107 | 0.100 | 2.314 | 0.101 | 0.103 |
| 17 | $A_{17}$ | 5.579 | 5.286 | 5.582 | 3.542 | 5.555 | 5.537 |
| Weight (lb) | | 2581.89 | 2594.42 | 2580.81 | 2724.37 | 2582.85 | 2581.94 |

There are 17 design variables in this example and the minimum permitted cross-sectional area of each member is 0.1in.$^2$. A single vertical downward load of 100 kips at node 9 is considered. Table 3.4 shows the solutions and Fig. 3.7 compares the convergence rates of the three algorithms.

Both the PSOPC and HPSO algorithms achieve a good solution after 3,000 iterations and the latter shows a better convergence rate than the former, especially at the early stage of iterations. In this case, the PSO algorithm is not fully converged when the maximum number of iterations is reached.

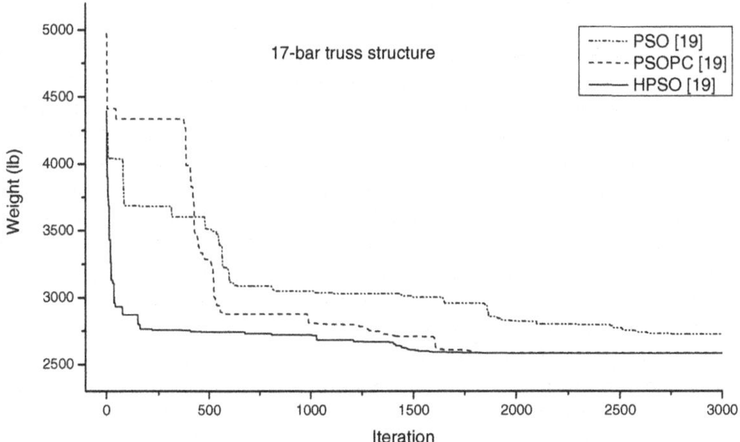


**Fig. 3.7** Convergence rates of the three algorithms for the 17-bar planar truss structure

## *(3) The 22-bar spatial truss structure*

The 22-bar spatial truss structure, shown in Fig. 3.8, had been studied by Lee [16] and Li [19]. The material density is 0.1 lb/in.$^3$ and the modulus of elasticity is 10,000 ksi. The stress limits of the members are listed in Table 3.5.

**Table 3.5** Member stress limits for the 22-bar spatial truss structure

| Variables | | Compressive stress limitations (ksi) | Tensile stress Limitation (ksi) |
|---|---|---|---|
| 1 | $A_1$ | 24.0 | 36.0 |
| 2 | $A_2$ | 30.0 | 36.0 |
| 3 | $A_3$ | 28.0 | 36.0 |
| 4 | $A_4$ | 26.0 | 36.0 |
| 5 | $A_5$ | 22.0 | 36.0 |
| 6 | $A_6$ | 20.0 | 36.0 |
| 7 | $A_7$ | 18.0 | 36.0 |

All nodes in all three directions are subjected to the displacement limits of ±2.0 in. Three load cases are listed in Table 3.6. There are 22 members, which fall into 7 groups, as follows: (1) $A_1$~$A_4$, (2) $A_5$~$A_6$, (3) $A_7$~$A_8$, (4) $A_9$~$A_{10}$, (5) $A_{11}$~$A_{14}$, (6) $A_{15}$~$A_{18}$, and (7) $A_{19}$~$A_{22}$. The minimum permitted cross-sectional area of each member is 0.1 in.$^2$.

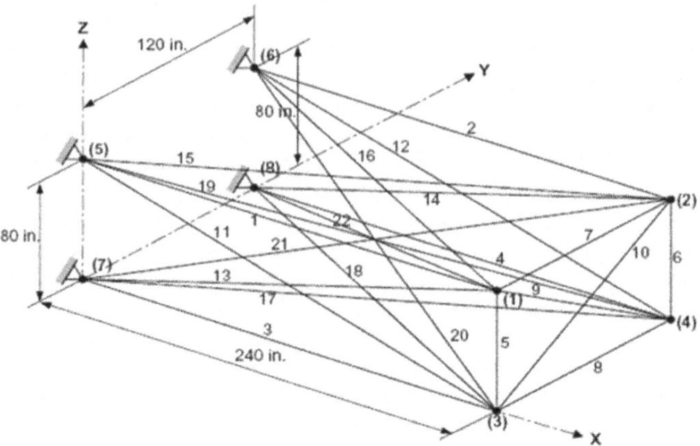


**Fig. 3.8** A 22-bar spatial truss structure

**Table 3.6** Load cases for the 22-bar spatial truss structure

| Node | Case 1 (kips) | | | Case 2 (kips) | | | Case 3 (kips) | | |
|---|---|---|---|---|---|---|---|---|---|
| | $P_X$ | $P_Y$ | $P_Z$ | $P_X$ | $P_Y$ | $P_Z$ | $P_X$ | $P_Y$ | $P_Z$ |
| 1 | -20.0 | 0.0 | -5.0 | -20.0 | -5.0 | 0.0 | -20.0 | 0.0 | 35.0 |
| 2 | -20.0 | 0.0 | -5.0 | -20.0 | -50.0 | 0.0 | -20.0 | 0.0 | 0.0 |
| 3 | -20.0 | 0.0 | -30.0 | -20.0 | -5.0 | 0.0 | -20.0 | 0.0 | 0.0 |
| 4 | -20.0 | 0.0 | -30.0 | -20.0 | -50.0 | 0.0 | -20.0 | 0.0 | -35.0 |

In this example, the HPSO algorithm has converged after 50 iterations, while the PSOPC and PSO algorithms need more than 500 and 1000 iterations respectively. The optimum results obtained by using the HPSO algorithm are significantly better than that obtained by the HS and the PSO algorithms. Table 3.7 shows the optimal solutions of the four algorithms and Fig. 3.9 provides the convergence rates of three of the four algorithms.

**Table 3.7** Comparison of the designs for the 22-bar spatial truss structure

| Variables | | Optimal cross-sectional areas (in.$^2$) | | | |
|---|---|---|---|---|---|
| | | Lee [16] | Li [19] PSO | Li [19] PSOPC | Li [19] HPSO |
| 1 | $A_1$ | 2.588 | 1.657 | 3.041 | 3.157 |
| 2 | $A_2$ | 1.083 | 0.716 | 1.191 | 1.269 |
| 3 | $A_3$ | 0.363 | 0.919 | 0.985 | 0.980 |
| 4 | $A_4$ | 0.422 | 0.175 | 0.105 | 0.100 |
| 5 | $A_5$ | 2.827 | 4.576 | 3.430 | 3.280 |
| 6 | $A_6$ | 2.055 | 3.224 | 1.543 | 1.402 |
| 7 | $A_7$ | 2.044 | 0.450 | 1.138 | 1.301 |
| Weight (lb) | | 1022.23 | 1057.14 | 977.80 | 977.81 |

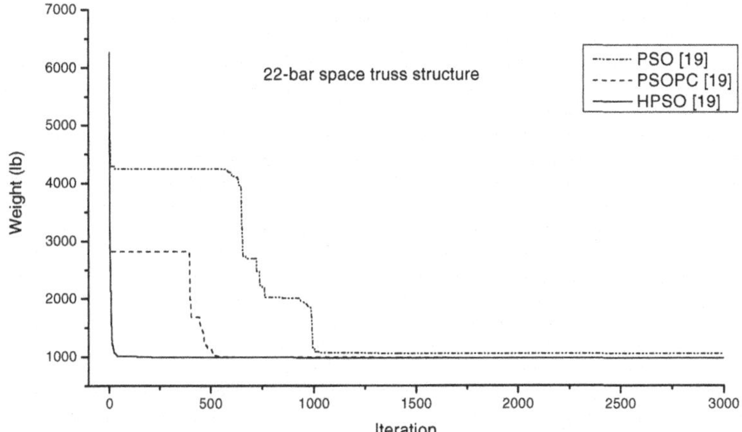


**Fig. 3.9** Convergence ratesof the three algorithms for the 22-bar spatial truss structure

### *(4) The 25-bar spatial truss structure*

The 25-bar spatial truss structure shown in Fig. 3.10 had been studied by several researchers, such as Schmit [17], Rizzi [18], Lee [16] and Li [19]. The material density is 0.1 lb/in.$^3$ and the modulus of elasticity is 10,000 ksi. The stress limits of the members are listed in Table 3.8. All nodes in all directions are subjected to the displacement limits of ±0.35 in. Two load cases listed in Table 3.9 are considered. There are 25 members, which are divided into 8 groups, as follows: (1) $A_1$, (2) $A_2$~$A_5$, (3) $A_6$~$A_9$, (4) $A_{10}$~$A_{11}$, (5) $A_{12}$~$A_{13}$, (6) $A_{14}$~$A_{17}$, (7) $A_{18}$~$A_{21}$ and (8) $A_{22}$~$A_{25}$. The minimum permitted cross-sectional area of each member is 0.01 in$^2$.

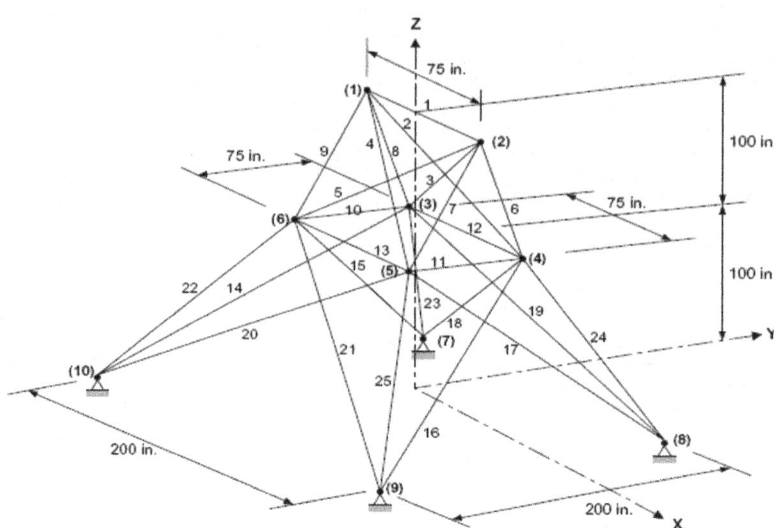


Fig. 3.10 A 25-bar spatial truss structure

Table 3.8 Member stress limits for the 25-bar spatial truss structure

| Variables | | Compressive stress limitations (ksi) | Tensile stress limitation (ksi) |
|---|---|---|---|
| 1 | $A_1$ | 35.092 | 40.0 |
| 2 | $A_2$ | 11.590 | 40.0 |
| 3 | $A_3$ | 17.305 | 40.0 |
| 4 | $A_4$ | 35.092 | 40.0 |
| 5 | $A_5$ | 35.902 | 40.0 |
| 6 | $A_6$ | 6.759 | 40.0 |
| 7 | $A_7$ | 6.959 | 40.0 |
| 8 | $A_8$ | 11.802 | 40.0 |

Table 3.9 Load cases for the 25-bar spatial truss structure

| Node | Case 1 | | | Case 2 | | |
|---|---|---|---|---|---|---|
| | $P_X$ (kips) | $P_Y$ (kips) | $P_Z$ (kips) | $P_X$ (kips) | $P_Y$ (kips) | $P_Z$ (kips) |
| 1 | 0.0 | 20.0 | -5.0 | 1.0 | 10.0 | -5.0 |
| 2 | 0.0 | -20.0 | -5.0 | 0.0 | 10.0 | -5.0 |
| 3 | 0.0 | 0.0 | 0.0 | 0.5 | 0.0 | 0.0 |
| 6 | 0.0 | 0.0 | 0.0 | 0.5 | 0.0 | 0.0 |

For this spatial truss structure, it takes about 1000 and 3000 iterations, respectively, for the PSOPC and the PSO algorithms to converge. However the HPSO algorithm takes only 50 iterations to converge. Indeed, in this example, the PSO algorithm did not fully converge when the maximum number of iterations is reached. Table 3.10 shows the solutions and Fig. 3.11 compares the convergence rate of the three algorithms.

**Table 3.10** Comparison of the designs for the 25-bar spatial truss structure

| Variables | | Optimal cross-sectional areas (in.$^2$) | | | | | |
|---|---|---|---|---|---|---|---|
| | | Schmit [17] | Rizzi [18] | Lee [16] | Li [19] PSO | Li [19] PSOPC | Li [19] HPSO |
| 1 | $A_1$ | 0.010 | 0.010 | 0.047 | 9.863 | 0.010 | 0.010 |
| 2 | $A_2$~$A_5$ | 1.964 | 1.988 | 2.022 | 1.798 | 1.979 | 1.970 |
| 3 | $A_6$~$A_9$ | 3.033 | 2.991 | 2.950 | 3.654 | 3.011 | 3.016 |
| 4 | $A_{10}$~$A_{11}$ | 0.010 | 0.010 | 0.010 | 0.100 | 0.100 | 0.010 |
| 5 | $A_{12}$~$A_{13}$ | 0.010 | 0.010 | 0.014 | 0.100 | 0.100 | 0.010 |
| 6 | $A_{14}$~$A_{17}$ | 0.670 | 0.684 | 0.688 | 0.596 | 0.657 | 0.694 |
| 7 | $A_{18}$~$A_{21}$ | 1.680 | 1.677 | 1.657 | 1.659 | 1.678 | 1.681 |
| 8 | $A_{22}$~$A_{25}$ | 2.670 | 2.663 | 2.663 | 2.612 | 2.693 | 2.643 |
| Weight (lb) | | 545.22 | 545.36 | 544.38 | 627.08 | 545.27 | 545.19 |

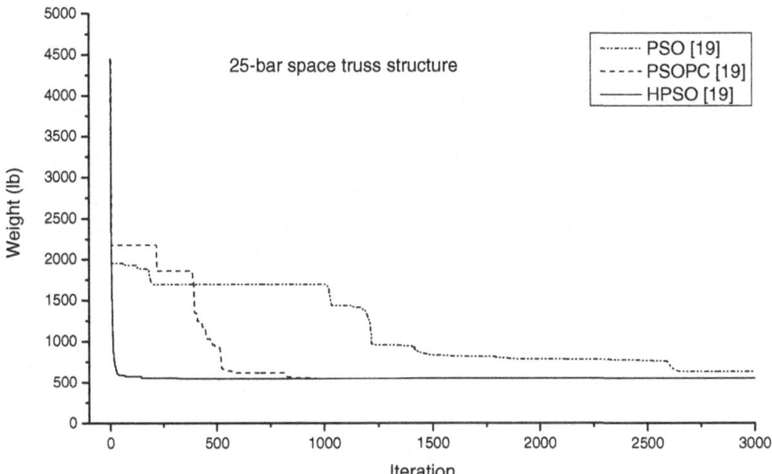


**Fig. 3.11** Convergence rates of the three algorithms for the 25-bar spatial truss structure

The curves of stability of 10-bar truss structure with three different algorithms (PSO, PSOPC and HPSO) is shown in Fig. 3.12. The stability curves come from the best results of 100 independent calculation times. The standard deviation of PSO, PSOPC and HPSO is 256.7491, 1.04208 and 0.02664 respectively. It can be seen from Fig. 3.12 that the HPSO has the best stability in this example.

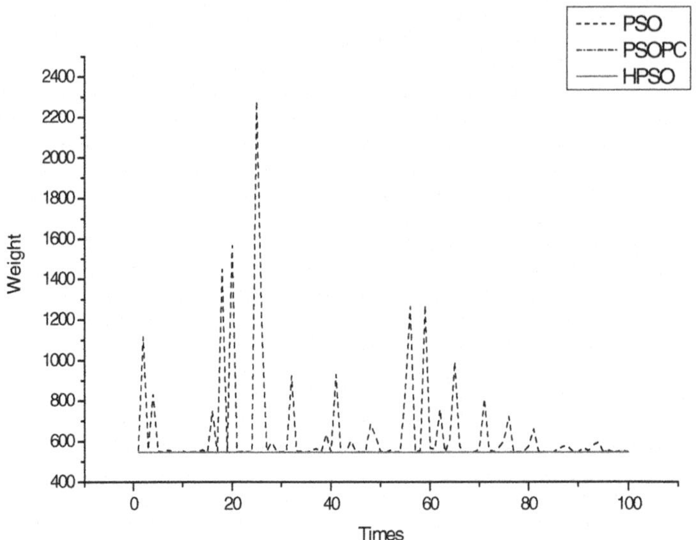


**Fig. 3.12** The stability of the 25-bar truss structure with three algorithms

## *(5) The 72-Bar Spatial Truss Structure*

The 72-bar spatial truss structure shown in Fig. 3.13 had also been studied by many researchers, such as Schmit [17], Khot [20], Adeli [21], Lee [16], Sarma [22] and Li [19]. The material density is 0.1 lb/in.$^3$ and the modulus of elasticity is 10,000 ksi. The members are subjected to the stress limits of ±25 ksi. The uppermost nodes are subjected to the displacement limits of ±0.25 in. in both the x and y directions. Two load cases are listed in Table 3.11. There are 72 members classified into 16 groups: (1) $A_1$~$A_4$, (2) $A_5$~$A_{12}$, (3) $A_{13}$~$A_{16}$, (4) $A_{17}$~$A_{18}$, (5) $A_{19}$~$A_{22}$, (6) $A_{23}$~$A_{30}$ (7) $A_{31}$~$A_{34}$, (8) $A_{35}$~$A_{36}$, (9) $A_{37}$~$A_{40}$, (10) $A_{41}$~$A_{48}$, (11) $A_{49}$~$A_{52}$, (12) $A_{53}$~$A_{54}$, (13) $A_{55}$~$A_{58}$, (14) $A_{59}$~$A_{66}$ (15) $A_{67}$~$A_{70}$, (16) $A_{71}$~$A_{72}$. For case 1, the minimum permitted cross-sectional area of each member is 0.1 in $^2$. For case 2, the minimum permitted cross-sectional area of each member is 0.01 in.$^2$.

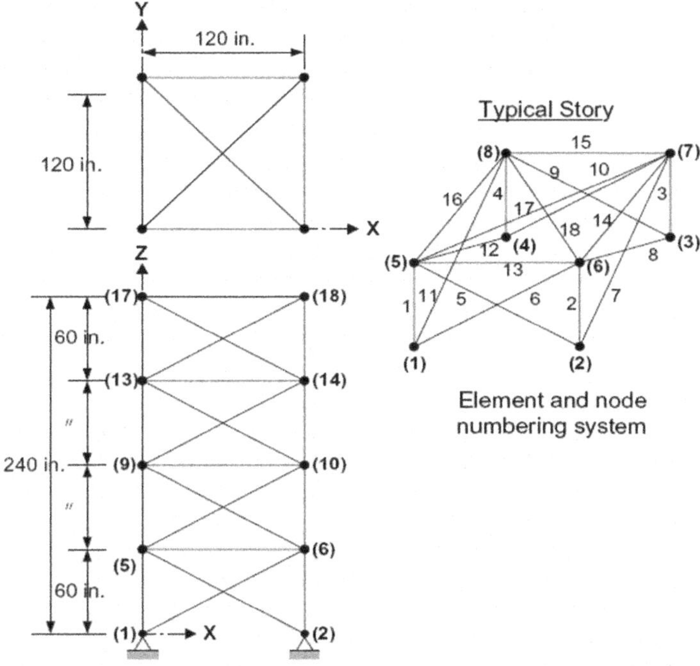


**Fig. 3.13** A 72-bar spatial truss structure

**Table 3.11** Load cases for the 72-bar spatial truss structure

| Node | Case 1 | | | Case 2 | | |
|---|---|---|---|---|---|---|
| | $P_X$ (kips) | $P_Y$ (kips) | $P_Z$ (kips) | $P_X$ (kips) | $P_Y$ (kips) | $P_Z$ (kips) |
| 17 | 5.0 | 5.0 | -5.0 | 0.0 | 0.0 | -5.0 |
| 18 | 0.0 | 0.0 | 0.0 | 0.0 | 0.0 | -5.0 |
| 19 | 0.0 | 0.0 | 0.0 | 0.0 | 0.0 | -5.0 |
| 20 | 0.0 | 0.0 | 0.0 | 0.0 | 0.0 | -5.0 |

For both the loading cases, the PSOPC and the HPSO algorithms can achieve the optimal solution after 2500 iterations. However, the latter shows a faster convergence rate than the former, especially at the early stage of iterations. The PSO algorithm cannot reach the optimal solution after the maximum number of iterations. The solutions of the two loading cases are given in Tables 3.12 and 3.13 respectively. Figs. 3.14 and 3.15 compare the convergence rate of the three algorithms for the two loading cases.

**Table 3.12** Comparison of the designs for the 72-bar spatial truss structure (Case 1)

| Variables | | Optimal cross-sectional areas (in.$^2$) | | | | | | |
|---|---|---|---|---|---|---|---|---|
| | | Schmit [17] | Adeli [21] | Khot [20] | Lee [16] | Li [19] PSO | Li [19] PSOPC | Li [19] HPSO |
| 1 | $A_1$~$A_4$ | 2.078 | 2.026 | 1.893 | 1.7901 | 41.794 | 1.855 | 1.857 |
| 2 | $A_5$~$A_{12}$ | 0.503 | 0.533 | 0.517 | 0.521 | 0.195 | 0.504 | 0.505 |
| 3 | $A_{13}$~$A_{16}$ | 0.100 | 0.100 | 0.100 | 0.100 | 10.797 | 0.100 | 0.100 |
| 4 | $A_{17}$~$A_{18}$ | 0.100 | 0.100 | 0.100 | 0.100 | 6.861 | 0.100 | 0.100 |
| 5 | $A_{19}$~$A_{22}$ | 1.107 | 1.157 | 1.279 | 1.229 | 0.438 | 1.253 | 1.255 |
| 6 | $A_{23}$~$A_{30}$ | 0.579 | 0.569 | 0.515 | 0.522 | 0.286 | 0.505 | 0.503 |
| 7 | $A_{31}$~$A_{34}$ | 0.100 | 0.100 | 0.100 | 0.100 | 18.309 | 0.100 | 0.100 |
| 8 | $A_{35}$~$A_{36}$ | 0.100 | 0.100 | 0.100 | 0.100 | 1.220 | 0.100 | 0.100 |
| 9 | $A_{37}$~$A_{40}$ | 0.264 | 0.514 | 0.508 | 0.517 | 5.933 | 0.497 | 0.496 |
| 10 | $A_{41}$~$A_{48}$ | 0.548 | 0.479 | 0.520 | 0.504 | 19.545 | 0.508 | 0.506 |
| 11 | $A_{49}$~$A_{52}$ | 0.100 | 0.100 | 0.100 | 0.100 | 0.159 | 0.100 | 0.100 |
| 12 | $A_{53}$~$A_{54}$ | 0.151 | 0.100 | 0.100 | 0.101 | 0.151 | 0.100 | 0.100 |
| 13 | $A_{55}$~$A_{58}$ | 0.158 | 0.158 | 0.157 | 0.156 | 10.127 | 0.100 | 0.100 |
| 14 | $A_{59}$~$A_{66}$ | 0.594 | 0.550 | 0.539 | 0.547 | 7.320 | 0.525 | 0.524 |
| 15 | $A_{67}$~$A_{70}$ | 0.341 | 0.345 | 0.416 | 0.442 | 3.812 | 0.394 | 0.400 |
| 16 | $A_{71}$~$A_{72}$ | 0.608 | 0.498 | 0.551 | 0.590 | 18.196 | 0.535 | 0.534 |
| Weight (lb) | | 388.63 | 379.31 | 379.67 | 379.27 | 6818.67 | 369.65 | 369.65 |

**Table 3.13** Comparison of the designs for the 72-bar spatial truss structure (Case 2)

| Variables | | Optimal cross-sectional areas (in.$^2$) | | | | | | |
|---|---|---|---|---|---|---|---|---|
| | | Adeli [21] | Sarma [22] Simple GA | Sarma [22] Fuzzy GA | Lee [16] | Li [19] PSO | Li [19] PSOPC | Li [19] HPSO |
| 1 | $A_1$~$A_4$ | 2.755 | 2.141 | 1.732 | 1.963 | 40.053 | 1.652 | 1.907 |
| 2 | $A_5$~$A_{12}$ | 0.510 | 0.510 | 0.522 | 0.481 | 0.237 | 0.547 | 0.524 |
| 3 | $A_{13}$~$A_{16}$ | 0.010 | 0.054 | 0.010 | 0.010 | 21.692 | 0.100 | 0.010 |
| 4 | $A_{17}$~$A_{18}$ | 0.010 | 0.010 | 0.013 | 0.011 | 0.657 | 0.101 | 0.010 |
| 5 | $A_{19}$~$A_{22}$ | 1.370 | 1.489 | 1.345 | 1.233 | 22.144 | 1.102 | 1.288 |
| 6 | $A_{23}$~$A_{30}$ | 0.507 | 0.551 | 0.551 | 0.506 | 0.266 | 0.589 | 0.523 |
| 7 | $A_{31}$~$A_{34}$ | 0.010 | 0.057 | 0.010 | 0.011 | 1.654 | 0.011 | 0.010 |
| 8 | $A_{35}$~$A_{36}$ | 0.010 | 0.013 | 0.013 | 0.012 | 10.284 | 0.010 | 0.010 |
| 9 | $A_{37}$~$A_{40}$ | 0.481 | 0.565 | 0.492 | 0.538 | 0.559 | 0.581 | 0.544 |
| 10 | $A_{41}$~$A_{48}$ | 0.508 | 0.527 | 0.545 | 0.533 | 12.883 | 0.458 | 0.528 |
| 11 | $A_{49}$~$A_{52}$ | 0.010 | 0.010 | 0.066 | 0.010 | 0.138 | 0.010 | 0.019 |
| 12 | $A_{53}$~$A_{54}$ | 0.643 | 0.066 | 0.013 | 0.167 | 0.188 | 0.152 | 0.020 |
| 13 | $A_{55}$~$A_{58}$ | 0.215 | 0.174 | 0.178 | 0.161 | 29.048 | 0.161 | 0.176 |
| 14 | $A_{59}$~$A_{66}$ | 0.518 | 0.425 | 0.524 | 0.542 | 0.632 | 0.555 | 0.535 |
| 15 | $A_{67}$~$A_{70}$ | 0.419 | 0.437 | 0.396 | 0.478 | 3.045 | 0.514 | 0.426 |
| 16 | $A_{71}$~$A_{72}$ | 0.504 | 0.641 | 0.595 | 0.551 | 1.711 | 0.648 | 0.612 |
| Weight (lb) | | 376.50 | 372.40 | 364.40 | 364.33 | 5417.02 | 368.45 | 364.86 |

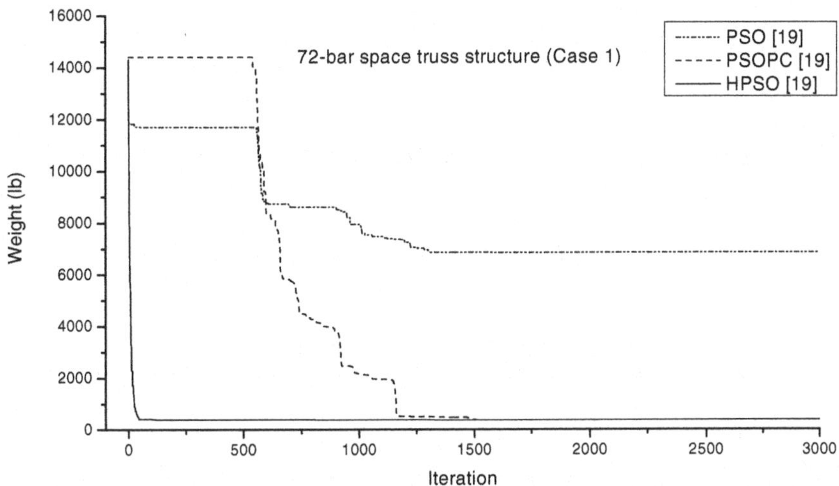


**Fig. 3.14** Convergence rates of the three algorithms for the 72-bar spatial truss structure (Case 1)

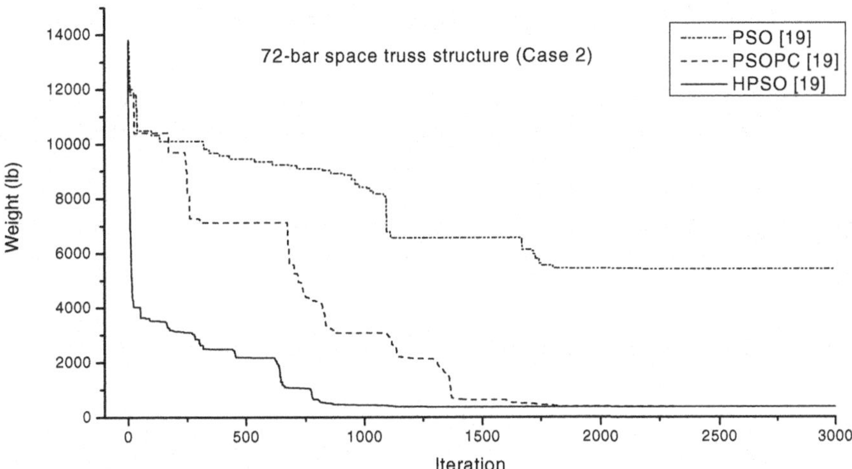


**Fig. 3.15** Convergence rates of the three algorithms for the 72-bar spatial truss structure (Case 2)

## 3.5 The Application of the HPSO on Truss Structures with Discrete Variables

In the past thirty years, many algorithms have been developed to solve structural engineering optimization problems. Most of these algorithms are based on the assumption that the design variables are continuously valued and the gradients of functions and the convexity of the design problem satisfied. However, in reality, the design variables of optimization problems such as the cross-section areas are discretely valued. They are often chosen from a list of discrete variables. Furthermore, the function of the problems is hard to express in an explicit form. Traditionally, the discrete optimization problems are solved mathematically by employing round-off techniques based on the continuous solutions. However, the solutions obtained by this method may be infeasible or far from the optimum solutions [23].

Most of the applications of the PSO algorithm to structural optimization problems are based on the assumption that the variables are continuous. Only in few papers PSO algorithm is used to solve the discrete structural optimization problems [24, 25].

In this section, the HPSO algorithm, which is based on the standard particle swarm optimize (SPSO) and the harmony search scheme, is applied to the discrete valued structural optimization problems.

### *3.5.1 Mathematical Model for Discrete Structural Optimization Problems*

A structural optimization design problem with discrete variables can be formulated as a nonlinear programming problem. In the size optimization for a truss structure, the cross-section areas of the truss members are selected as the design variables. Each of the design variables is chosen from a list of discrete cross-sections based on production standard. The objective function is the structure weight. The design cross-sections must also satisfy some inequality constraints equations, which restrict the discrete variables. The optimization design problem for discrete variables can be expressed as follows:

$$\min f\left(x^1, x^2, ..., x^d\right), \quad d = 1, 2, \cdots, D \tag{3.3}$$

subjected to: $g_q\left(x^1, x^2, ..., x^d\right) \le 0, \quad d = 1, 2, \cdots, D, \quad q = 1, 2, \cdots, M$

$x^d \in S_d = \{X_1, X_2, \cdots, X_p\}$

where $f\left(x^1, x^2, ..., x^d\right)$ is the truss's weight function, which is a scalar function. And $x^1, x^2, ..., x^d$ represent a set of design variables. The design variable $x^d$ belongs to a scalar $S_d$, which includes all permissive discrete variables $\{X_1, X_2, ... X_p\}$.

The inequality $g_q\left(x^1, x^2, ..., x^d\right) \le 0$ represents the constraint functions. The letter D and M are the number of the design variables and inequality functions respectively. The letter p is the number of available variables.

### 3.5.2 The Heuristic Particle Swarm Optimizer (HPSO) for Discrete Variables

The heuristic particle swarm optimizer (HPSO) algorithm, which is based on the PSOPC algorithm and the harmony search scheme, is introduced by Li [19] and is first used in continuous variable optimization problems. The HPSO algorithm then was used for discrete problems [26]. Similarly, The HPSO algorithm for the discrete valued variables can be expressed as follows:

$$V_i^{(k+1)} = \omega V_i^{(k)} + c_1 r_1\left(P_i^{(k)} - x_i^{(k)}\right) + c_2 r_2\left(P_g^{(k)} - x_i^{(k)}\right) + c_3 r_3\left(R_i^{(k)} - x_i^{(k)}\right) \tag{3.4}$$

$$x_i^{(k+1)} = INT\left(x_i^{(k)} + V_i^{(k+1)}\right) \quad 1 \le i \le n \tag{3.5}$$

where $\boldsymbol{x}_i$ is the vector of a particle's position, and $x_i^d$ is one component of this vector. After the (k+1)th iterations, if $x_i^d < x^d(LowerBound)$ or $x_i^d > x^d(UpperBound)$, the scalar $x_i^d$ is regenerated by selecting the corresponding component of the vector from pbest swarm randomly, which can be described as follows:

$$x_i^d = \left(P_b\right)_t^d, \; t = INT\left(rand\left(1, n\right)\right) \tag{3.6}$$

where $\left(P_b\right)_t^d$ denotes the $d$th dimension scalar of *pbest* swarm of the $t$th particle, and $t$ denotes a random integer number.

In this section, the HPSO algorithm is tested by five truss structures. The algorithm proposed is coded in FORTRAN language and executed on a Pentium 4, 2.93GHz machine.

The PSO, the PSOPC and the HPSO algorithms for discrete variables are applied to all these examples and the results are compared in order to evaluate the performance of the HPSO algorithm for discrete variables. For all these algorithms, a population of 50 individuals are used, the inertia weight ω, which starts at 0.9 and ends at 0.4, decreases linearly, and the value of acceleration constants $c_1$ and $c_2$ are set to 0.5 [27]. The passive congregation coefficient $c_3$ is set to 0.6 for the PSOPC and the HPSO algorithms. All these truss structures have been analyzed by the finite element method (FEM). The maximum velocity is set as the difference between the upper and the lower bounds, which ensures that the particles are able to fly across the problem-specific constraints' region. Different iteration numbers are used for different optimization structures, with smaller iteration number for smaller variable number structures and larger one for large variable number structures.

### 3.5.3 *Numerical Examples*

#### *(1) A 10-bar planar truss structure*

A 10-bar truss structure, shown in Fig. 3.16, has previously been analyzed by many researchers, such as Wu [24], Rajeev [28], Ringertz [29] and Li [26]. The material density is 0.1 lb/in$^3$ and the modulus of elasticity is 10,000 ksi. The members are subjected to stress limitations of ±25 ksi. All nodes in both directions are subjected to displacement limitations of ±2.0 in. $P_1$=105 lb, $P_2$=0. There are 10 design variables and two load cases in this example to be optimized. For case 1: the discrete variables are selected from the set D={1.62, 1.80, 1.99, 2.13, 2.38, 2.62, 2.63, 2.88, 2.93, 3.09, 3.13, 3.38, 3.47, 3.55, 3.63, 3.84, 3.87, 3.88, 4.18, 4.22, 4.49, 4.59, 4.80, 4.97, 5.12, 5.74, 7.22, 7.97, 11.50, 13.50, 13.90, 14.20, 15.50, 16.00, 16.90, 18.80, 19.90, 22.00, 22.90, 26.50, 30.00, 33.50} (in$^2$) ; For case 2: the discrete variables are selected from the set D={0.1, 0.5, 1.0, 1.5, 2.0, 2.5, 3.0, 3.5, 4.0, 4.5, 5.0, 5.5, 6.0, 6.5, 7.0, 7.5, 8.0, 8.5, 9.0, 9.5, 10.0, 10.5, 11.0, 11.5, 12.0, 12.5, 13.0, 13.5, 14.0, 14.5, 15.0, 15.5, 16.0, 16.5, 17.0, 17.5, 18.0, 18.5, 19.0, 19.5, 20.0, 20.5, 21.0, 21.5, 22.0, 22.5, 23.0, 23.5, 24.0, 24.5, 25.0, 25.5, 26.0, 26.5, 27.0, 27.5, 28.0, 28.5, 29.0, 29.5, 30.0, 30.5, 31.0, 31.5} (in$^2$). A maximum number of 1000 iterations is imposed.

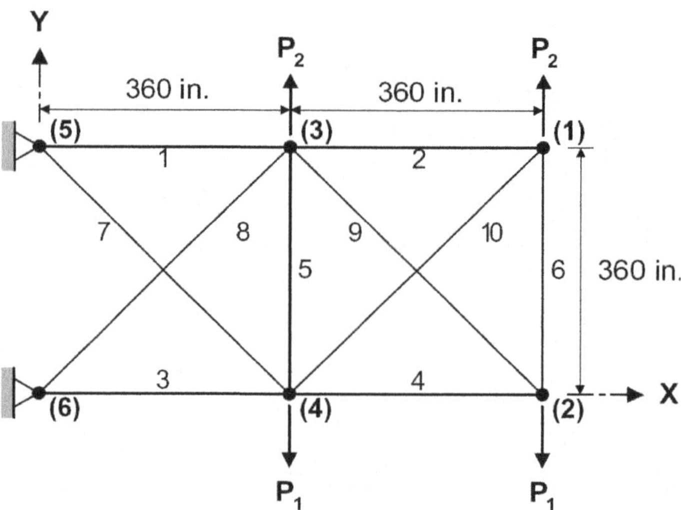


**Fig. 3.16** A 10-bar planar truss structure

Table 3.14 and Table 3.15 give the comparison of optimal design results for the 10-bar planar truss structure under two load cases respectively. Fig.3.17 and Fig.3.18 show the comparison of convergence rates for the 10-bar truss structure. From the Table 3.14 and Table 3.15, we find the results obtained by the HPSO algorithm are larger than those of Wu's [24]. However, it is found that Wu's results do not satisfy the constraints of this problem. It is believed that Wu's results need to be further valuated. For both cases of this structure, the PSO, PSOPC and HPSO algorithms have achieved the optimal solutions after 1,000 iterations. But the latter is much closer to the best solution than the former in the early iterations.

**Table 3.14** Design results for the 10-bar planar truss structure (case 1)

| Variables ($in^2$) | Wu [24] | Rajeev [28] | Li [26] PSO | Li [26] PSOPC | Li [26] HPSO |
|---|---|---|---|---|---|
| $A_1$ | 26.50 | 33.50 | 30.00 | 30.00 | 30.00 |
| $A_2$ | 1.62 | 1.62 | 1.62 | 1.80 | 1.62 |
| $A_3$ | 16.00 | 22.00 | 30.00 | 26.50 | 22.90 |
| $A_4$ | 14.20 | 15.50 | 13.50 | 15.50 | 13.50 |
| $A_5$ | 1.80 | 1.62 | 1.62 | 1.62 | 1.62 |
| $A_6$ | 1.62 | 1.62 | 1.80 | 1.62 | 1.62 |
| $A_7$ | 5.12 | 14.20 | 11.50 | 11.50 | 7.97 |
| $A_8$ | 16.00 | 19.90 | 18.80 | 18.80 | 26.50 |
| $A_9$ | 18.80 | 19.90 | 22.00 | 22.00 | 22.00 |
| $A_{10}$ | 2.38 | 2.62 | 1.80 | 3.09 | 1.80 |
| Weight (lb) | 4376.20 | 5613.84 | 5581.76 | 5593.44 | 5531.98 |

**Table 3.15** Design results for the 10-bar planar truss structure (case 2)

| Variables ($in^2$) | Wu [24] | Ringertz [29] | Li [26] PSO | Li [26] PSOPC | Li [26] HPSO |
|---|---|---|---|---|---|
| A1 | 30.50 | 30.50 | 24.50 | 25.50 | 31.50 |
| A2 | 0.50 | 0.10 | 0.10 | 0.10 | 0.10 |
| A3 | 16.50 | 23.00 | 22.50 | 23.50 | 24.50 |
| A4 | 15.00 | 15.50 | 15.50 | 18.50 | 15.50 |
| A5 | 0.10 | 0.10 | 0.10 | 0.10 | 0.10 |
| A6 | 0.10 | 0.50 | 1.50 | 0.50 | 0.50 |
| A7 | 0.50 | 7.50 | 8.50 | 7.50 | 7.50 |
| A8 | 18.00 | 21.0 | 21.50 | 21.50 | 20.50 |
| A9 | 19.50 | 21.5 | 27.50 | 23.50 | 20.50 |
| A10 | 0.50 | 0.10 | 0.10 | 0.10 | 0.10 |
| Weight (lb) | 4217.30 | 5059.9 | 5243.71 | 5133.16 | 5073.51 |

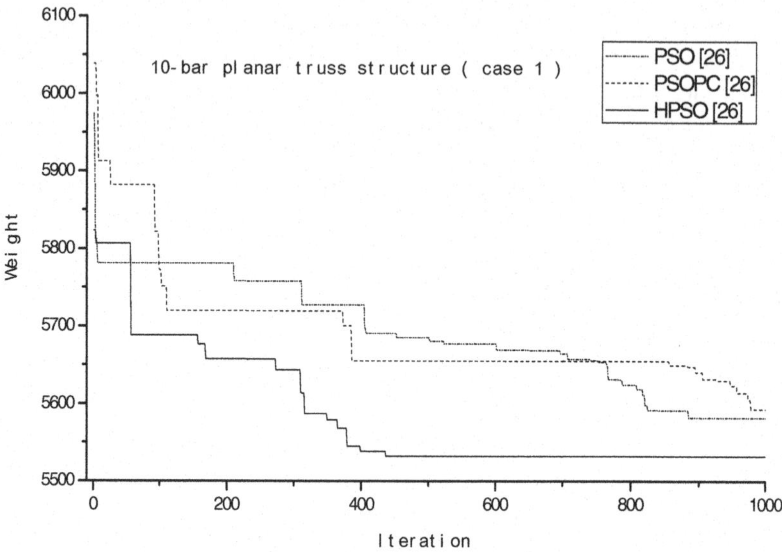


Fig. 3.17 Convergence rates for the 10-bar planar truss structure (Case 1)

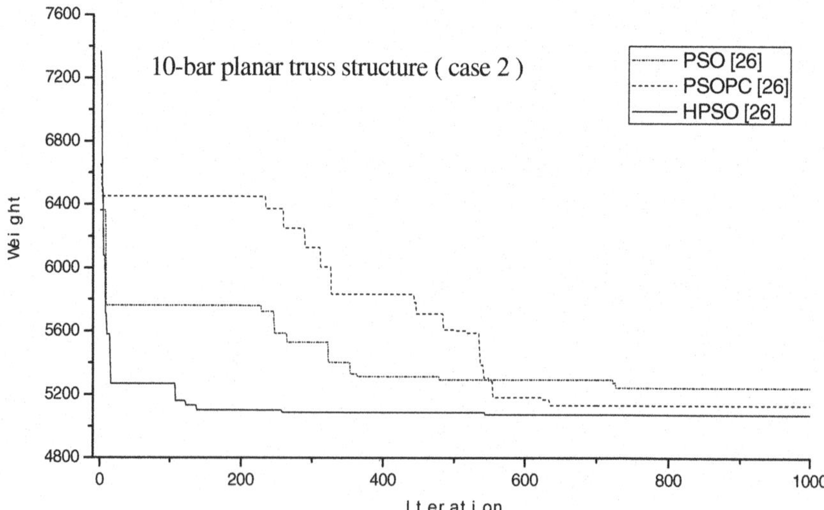


Fig. 3.18 Convergence rates for the 10-bar planar truss structure (Case 2)

## *(2) A 15-bar planar truss structure*

A 15-bar planar truss structure, shown in Fig. 3.19, has previously been analyzed by Zhang [30] and Li [26]. The material density is 7800 kg/m$^3$ and the modulus of elasticity is 200 GPa. The members are subjected to stress limitations of ±120 MPa. All nodes in both directions are subjected to displacement limitations of ±10mm. There are 15 design variables in this example. The discrete variables are selected from the set D= {113.2, 143.2, 145.9, 174.9, 185.9, 235.9, 265.9, 297.1, 308.6, 334.3, 338.2, 497.8, 507.6, 736.7, 791.2, 1063.7} (mm$^2$). Three load cases are considered: Case 1: $P_1$=35 kN, $P_2$=35 kN, $P_3$=35 kN; Case 2: $P_1$=35 kN, $P_2$=0 kN, $P_3$=35 kN; Case 3: $P_1$=35 kN, $P_2$=35 kN, $P_3$=0 kN. A maximum number of 500 iterations is imposed.

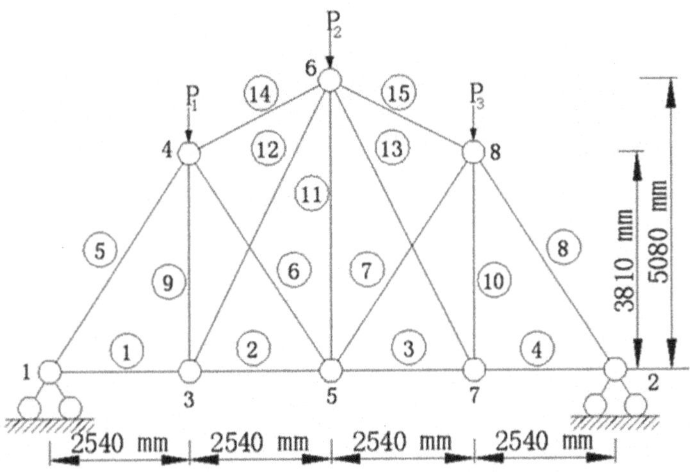


**Fig. 3.19** A 15-bar planar truss structure

Table 3.16 and Fig. 3.20 give the comparison of optimal design results and convergence rates of 15-bar planar truss structure respectively. It can be seen that, after 500 iterations, three algorithms have obtained good results, which are better than the Zhang's. The Fig. 3.20 shows that the HPSO algorithm has the fastest convergence rate, especially in the early iterations.

**Table 3.16** Comparison of optimal designs for the 15-bar planar truss structure

| Variables ($mm^2$) | Zhang [30] | Li [26] PSO | Li [26] PSOPC | Li [26] HPSO |
|---|---|---|---|---|
| $A_1$ | 308.6 | 185.9 | 113.2 | 113.2 |
| $A_2$ | 174.9 | 113.2 | 113.2 | 113.2 |
| $A_3$ | 338.2 | 143.2 | 113.2 | 113.2 |
| $A_4$ | 143.2 | 113.2 | 113.2 | 113.2 |
| $A_5$ | 736.7 | 736.7 | 736.7 | 736.7 |
| $A_6$ | 185.9 | 143.2 | 113.2 | 113.2 |
| $A_7$ | 265.9 | 113.2 | 113.2 | 113.2 |
| $A_8$ | 507.6 | 736.7 | 736.7 | 736.7 |
| $A_9$ | 143.2 | 113.2 | 113.2 | 113.2 |
| $A_{10}$ | 507.6 | 113.2 | 113.2 | 113.2 |
| $A_{11}$ | 279.1 | 113.2 | 113.2 | 113.2 |
| $A_{12}$ | 174.9 | 113.2 | 113.2 | 113.2 |
| $A_{13}$ | 297.1 | 113.2 | 185.9 | 113.2 |
| $A_{14}$ | 235.9 | 334.3 | 334.3 | 334.3 |
| $A_{15}$ | 265.9 | 334.3 | 334.3 | 334.3 |
| Weight (kg) | 142.117 | 108.84 | 108.96 | 105.735 |

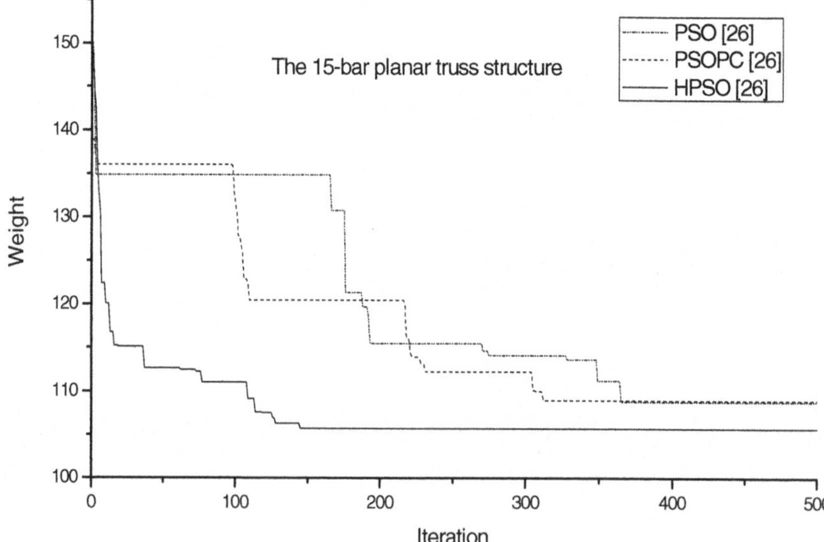


**Fig. 3.20** Comparison of convergence rates for the 15-bar planar truss structure

### (3) A 25-bar spatial truss structure

A 25-bar spatial truss structure, shown in Fig. 3.21, has been studied by Wu [24], Rajeev [28], Ringertz [29], Lee [13] and Li [26]. The material density is 0.1 lb/in.$^3$ and the modulus of elasticity is 10,000 ksi. The stress limitations of the members are ±40000 psi. All nodes in three directions are subjected to displacement limitations of ±0.35 in. The structure includes 25 members, which are divided into 8 groups, as follows: (1) $A_1$, (2) $A_2$~$A_5$, (3) $A_6$~$A_9$, (4) $A_{10}$~$A_{11}$, (5) $A_{12}$~$A_{13}$, (6) $A_{14}$~$A_{17}$, (7) $A_{18}$~$A_{21}$ and (8) $A_{22}$~$A_{25}$. There are three optimization cases to be implemented. Case 1: The discrete variables are selected from the set D= {0.1, 0.2, 0.3, 0.4, 0.5, 0.6, 0.7, 0.8, 0.9, 1.0, 1.1, 1.2, 1.3, 1.4, 1.5, 1.6, 1.7, 1.8, 1.9, 2.0, 2.1, 2.2, 2.3, 2.4, 2.6, 2.8, 3.0, 3.2, 3.4} (in$^2$). The loads are shown in Table 3.17; Case 2: The discrete variables are selected from the set D= {0.01, 0.4, 0.8, 1.2, 1.6, 2.0, 2.4, 2.8, 3.2, 3.6, 4.0, 4.4, 4.8, 5.2, 5.6, 6.0} (in$^2$). The loads are shown in Table 3.18. Case 3: The discrete variables are selected from the American Institute of Steel Construction (AISC) Code, which is shown in Table 3.19. The loads are shown in Table 3.18. A maximum number of 500 iterations is imposed for three cases.

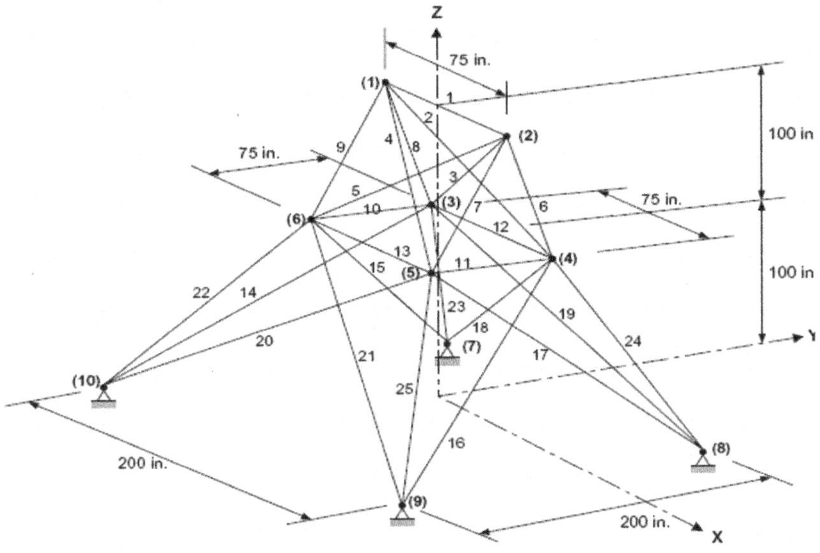


**Fig. 3.21** A 25-bar spatial truss structure

**Table 3.17** The load case 1 for the 25-bar spatial truss structure

| | Load Cases | Nodes | Loads | | |
|---|---|---|---|---|---|
| | | | $P_x$ (kips) | $P_y$ (kips) | $P_z$ (kips) |
| Case 1 | 1 | 1 | 1.0 | -10.0 | -10.0 |
| | | 2 | 0.0 | -10.0 | -10.0 |
| | | 3 | 0.5 | 0.0 | 0.0 |
| | | 6 | 0.6 | 0.0 | 0.0 |

**Table 3.18** The load case 2 and case 3 for the 25-bar spatial truss structure

| | Load Cases | Nodes | Loads | | |
|---|---|---|---|---|---|
| | | | $P_x$ (kips) | $P_y$ (kips) | $P_z$ (kips) |
| Case 2 & 3 | 2 | 1 | 0.0 | 20.0 | -5.0 |
| | | 2 | 0.0 | -20.0 | -5.0 |
| | 3 | 1 | 1.0 | 10.0 | -5.0 |
| | | 2 | 0.0 | 10.0 | -5.0 |
| | | 3 | 0.5 | 0.0 | 0.0 |
| | | 6 | 0.5 | 0.0 | 0.0 |

**Table 3.19** The available cross-section areas of the ASIC code

| No. | $in^2$ | $mm^2$ | No. | $in^2$ | $mm^2$ |
|---|---|---|---|---|---|
| 1 | 0.111 | 71.613 | 33 | 3.840 | 2477.414 |
| 2 | 0.141 | 90.968 | 34 | 3.870 | 2496.769 |
| 3 | 0.196 | 126.451 | 35 | 3.880 | 2503.221 |
| 4 | 0.250 | 161.290 | 36 | 4.180 | 2696.769 |
| 5 | 0.307 | 198.064 | 37 | 4.220 | 2722.575 |
| 6 | 0.391 | 252.258 | 38 | 4.490 | 2896.768 |
| 7 | 0.442 | 285.161 | 39 | 4.590 | 2961.284 |
| 8 | 0.563 | 363.225 | 40 | 4.800 | 3096.768 |
| 9 | 0.602 | 388.386 | 41 | 4.970 | 3206.445 |
| 10 | 0.766 | 494.193 | 42 | 5.120 | 3303.219 |
| 11 | 0.785 | 506.451 | 43 | 5.740 | 3703.218 |
| 12 | 0.994 | 641.289 | 44 | 7.220 | 4658.055 |
| 13 | 1.000 | 645.160 | 45 | 7.970 | 5141.925 |
| 14 | 1.228 | 792.256 | 46 | 8.530 | 5503.215 |
| 15 | 1.266 | 816.773 | 47 | 9.300 | 5999.988 |
| 16 | 1.457 | 939.998 | 48 | 10.850 | 6999.986 |
| 17 | 1.563 | 1008.385 | 49 | 11.500 | 7419.340 |
| 18 | 1.620 | 1045.159 | 50 | 13.500 | 8709.660 |
| 19 | 1.800 | 1161.288 | 51 | 13.900 | 8967.724 |
| 20 | 1.990 | 1283.868 | 52 | 14.200 | 9161.272 |
| 21 | 2.130 | 1374.191 | 53 | 15.500 | 9999.980 |
| 22 | 2.380 | 1535.481 | 54 | 16.000 | 10322.560 |
| 23 | 2.620 | 1690.319 | 55 | 16.900 | 10903.204 |
| 24 | 2.630 | 1696.771 | 56 | 18.800 | 12129.008 |
| 25 | 2.880 | 1858.061 | 57 | 19.900 | 12838.684 |
| 26 | 2.930 | 1890.319 | 58 | 22.000 | 14193.520 |
| 27 | 3.090 | 1993.544 | 59 | 22.900 | 14774.164 |
| 28 | 1.130 | 729.031 | 60 | 24.500 | 15806.420 |
| 29 | 3.380 | 2180.641 | 61 | 26.500 | 17096.740 |
| 30 | 3.470 | 2238.705 | 62 | 28.000 | 18064.480 |
| 31 | 3.550 | 2290.318 | 63 | 30.000 | 19354.800 |
| 32 | 3.630 | 2341.931 | 64 | 33.500 | 21612.860 |

Table 3.20, Table 3.21 and Table 3.22 show the comparison of optimal design results for the 25-bar spatial truss structure under three load cases. While Fig. 3.22, Fig. 3.23 and Fig. 3.24 show comparison of convergence rates for the 25-bar spatial truss structure under three load cases. For all load cases of this structure, three algorithms can achieve the optimal solution after 500 iterations. But Fig. 3.22, 3.23 and 3.24 show that the HPSO algorithm has the fastest convergence rate.

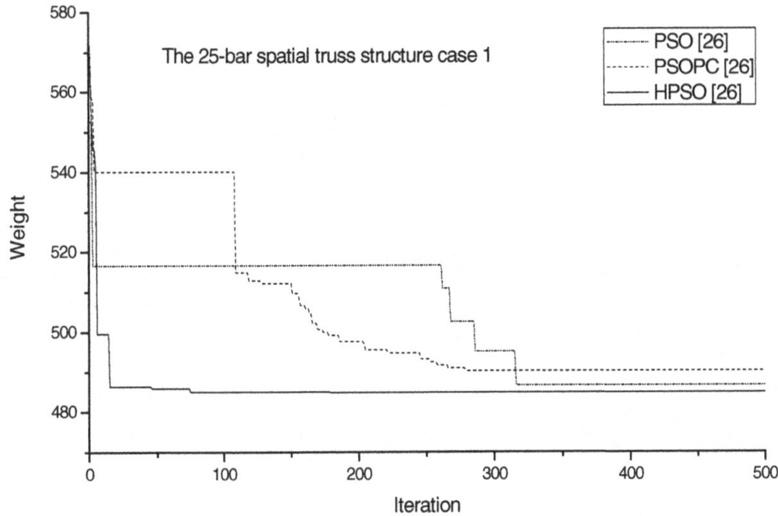


**Fig. 3.22** Comparison of convergence rates for the 25-bar spatial truss structure (case 1)

**Table 3.20** Comparison of optimal designs for the 25-bar spatial truss structure (case 1)

| Variables (in$^2$) | Case 1 | | | | | |
|---|---|---|---|---|---|---|
| | Wu [24] | Rajeev [28] | Lee [13] | Li [26] PSO | Li [26] PSOPC | Li [26] HPSO |
| $A_1$ | 0.1 | 0.1 | 0.1 | 0.4 | 0.1 | 0.1 |
| $A_2$~$A_5$ | 0.5 | 1.8 | 0.3 | 0.6 | 1.1 | 0.3 |
| $A_6$~$A_9$ | 3.4 | 2.3 | 3.4 | 3.5 | 3.1 | 3.4 |
| $A_{10}$~$A_{11}$ | 0.1 | 0.2 | 0.1 | 0.1 | 0.1 | 0.1 |
| $A_{12}$~$A_{13}$ | 1.5 | 0.1 | 2.1 | 1.7 | 2.1 | 2.1 |
| $A_{14}$~$A_{17}$ | 0.9 | 0.8 | 1.0 | 1.0 | 1.0 | 1.0 |
| $A_{18}$~$A_{21}$ | 0.6 | 1.8 | 0.5 | 0.3 | 0.1 | 0.5 |
| $A_{22}$~$A_{25}$ | 3.4 | 3.0 | 3.4 | 3.4 | 3.5 | 3.4 |
| Weight (lb) | 486.29 | 546.01 | 484.85 | 486.54 | 490.16 | 484.85 |

**Table 3.21** Comparison of optimal designs for the 25-bar spatial truss structure (case 2)

| Variables ($in^2$) | Case 2 | | | | | |
|---|---|---|---|---|---|---|
| | Wu [24] | Ringertz [29] | Lee [13] | Li [26] PSO | Li [26] PSOPC | Li [26] HPSO |
| $A_1$ | 0.4 | 0.01 | 0.01 | 0.01 | 0.01 | 0.01 |
| $A_2$~$A_5$ | 2.0 | 1.6 | 2.0 | 2.0 | 2.0 | 2.0 |
| $A_6$~$A_9$ | 3.6 | 3.6 | 3.6 | 3.6 | 3.6 | 3.6 |
| $A_{10}$~$A_{11}$ | 0.01 | 0.01 | 0.01 | 0.01 | 0.01 | 0.01 |
| $A_{12}$~$A_{13}$ | 0.01 | 0.01 | 0.01 | 0.4 | 0.01 | 0.01 |
| $A_{14}$~$A_{17}$ | 0.8 | 0.8 | 0.8 | 0.8 | 0.8 | 0.8 |
| $A_{18}$~$A_{21}$ | 2.0 | 2.0 | 1.6 | 1.6 | 1.6 | 1.6 |
| $A_{22}$~$A_{25}$ | 2.4 | 2.4 | 2.4 | 2.4 | 2.4 | 2.4 |
| Weight (lb) | 563.52 | 568.69 | 560.59 | 566.44 | 560.59 | 560.59 |

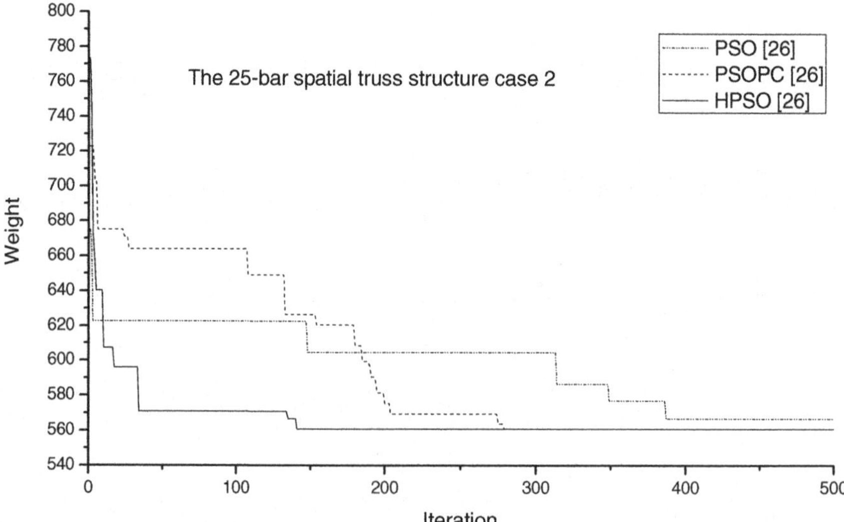


**Fig. 3.23** Comparison of convergence rates for the 25-bar spatial truss structure (case 2)

**Table 3.22** Comparison of optimal designs for the 25-bar spatial truss structure (case 3)

| Variables ($in^2$) | Case 3 | | | |
|---|---|---|---|---|
| | Wu [24] | Li [26] PSO | Li [26] PSOPC | Li [26] HPSO |
| $A_1$ | 0.307 | 1.0 | 0.111 | 0.111 |
| $A_2$~$A_5$ | 1.990 | 2.62 | 1.563 | 2.130 |
| $A_6$~$A_9$ | 3.130 | 2.62 | 3.380 | 2.880 |
| $A_{10}$~$A_{11}$ | 0.111 | 0.25 | 0.111 | 0.111 |
| $A_{12}$~$A_{13}$ | 0.141 | 0.307 | 0.111 | 0.111 |
| $A_{14}$~$A_{17}$ | 0.766 | 0.602 | 0.766 | 0.766 |
| $A_{18}$~$A_{21}$ | 1.620 | 1.457 | 1.990 | 1.620 |
| $A_{22}$~$A_{25}$ | 2.620 | 2.880 | 2.380 | 2.620 |
| Weight (lb) | 556.43 | 567.49 | 556.90 | 551.14 |

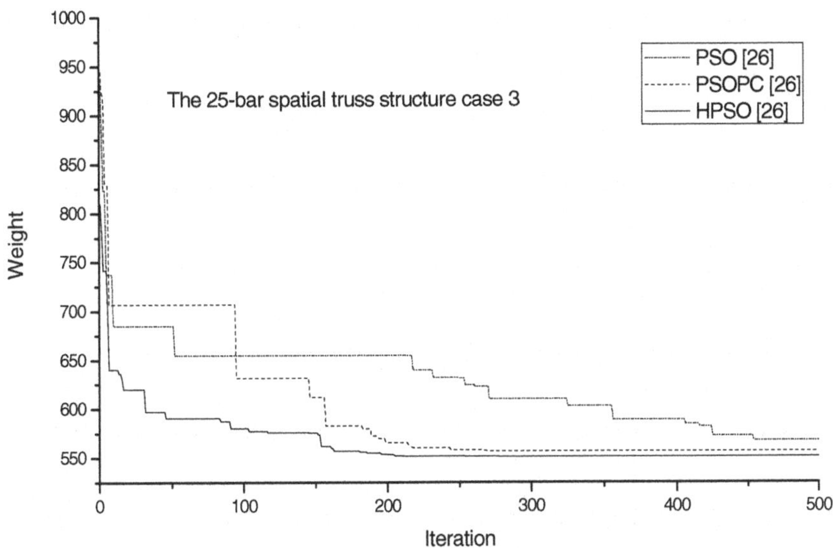


**Fig. 3.24** Comparison of convergence rates for the 25-bar spatial truss structure (case 3)

## *(4) A 52-bar planar truss structure*

A 52-bar planar truss structure, shown in Fig. 3.25, has been analysed by Wu [24] Lee [13] and Li [26]. The members of this structure are divided into 12 groups: (1) $A_1$~$A_4$, (2) $A_5$~$A_6$, (3) $A_7$~$A_8$, (4) $A_9$~$A_{10}$, (5) $A_{11}$~$A_{14}$, (6) $A_{15}$~$A_{18}$, and (7) $A_{19}$~$A_{22}$. The material density is 7860.0 kg/m$^3$ and the modulus of elasticity is $2.07\times10^5$ MPa. The members are subjected to stress limitations of ±180 MPa. Both of the loads, $P_x$ =100 kN, $P_y$ =200 kN are considered. The discrete variables are selected from the Table 3.19. A maximum number of 3,000 iterations is imposed.

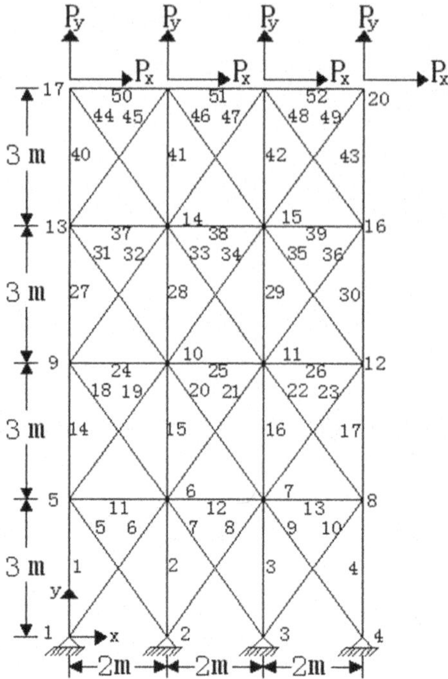

**Fig. 3.25** A 52-bar planar truss structure

**Table 3.23** Comparison of optimal designs for the 52-bar planar truss structure

| Variables ($mm^2$) | Wu [24] | Lee [13] | Li [26] PSO | Li [26] PSOPC | Li [26] HPSO |
|---|---|---|---|---|---|
| $A_1$~$A_4$ | 4658.055 | 4658.055 | 4658.055 | 5999.988 | 4658.055 |
| $A_5$~$A_{10}$ | 1161.288 | 1161.288 | 1374.190 | 1008.380 | 1161.288 |
| $A_{11}$~$A_{13}$ | 645.160 | 506.451 | 1858.060 | 2696.770 | 363.225 |
| $A_{14}$~$A_{17}$ | 3303.219 | 3303.219 | 3206.440 | 3206.440 | 3303.219 |
| $A_{18}$~$A_{23}$ | 1045.159 | 940.000 | 1283.870 | 1161.290 | 940.000 |
| $A_{24}$~$A_{26}$ | 494.193 | 494.193 | 252.260 | 729.030 | 494.193 |
| $A_{27}$~$A_{30}$ | 2477.414 | 2290.318 | 3303.220 | 2238.710 | 2238.705 |
| $A_{31}$~$A_{36}$ | 1045.159 | 1008.385 | 1045.160 | 1008.380 | 1008.385 |
| $A_{37}$~$A_{39}$ | 285.161 | 2290.318 | 126.450 | 494.190 | 388.386 |
| $A_{40}$~$A_{43}$ | 1696.771 | 1535.481 | 2341.93 | 1283.870 | 1283.868 |
| $A_{44}$~$A_{49}$ | 1045.159 | 1045.159 | 1008.38 | 1161.290 | 1161.288 |
| $A_{50}$~$A_{52}$ | 641.289 | 506.451 | 1045.16 | 494.190 | 792.256 |
| Weight (kg) | 1970.142 | 1906.76 | 2230.16 | 2146.63 | 1905.495 |

Table 3.23 and Fig. 3.24 give the comparison of optimal design results and convergence rates of 52-bar planar truss structure respectively. From Table 3.23 and Fig. 3.26, it can be observed that only the HPSO algorithm achieves the good optimal result. The PSO and PSOPC algorithms do not get optimal results when the maximum number of iterations is reached.

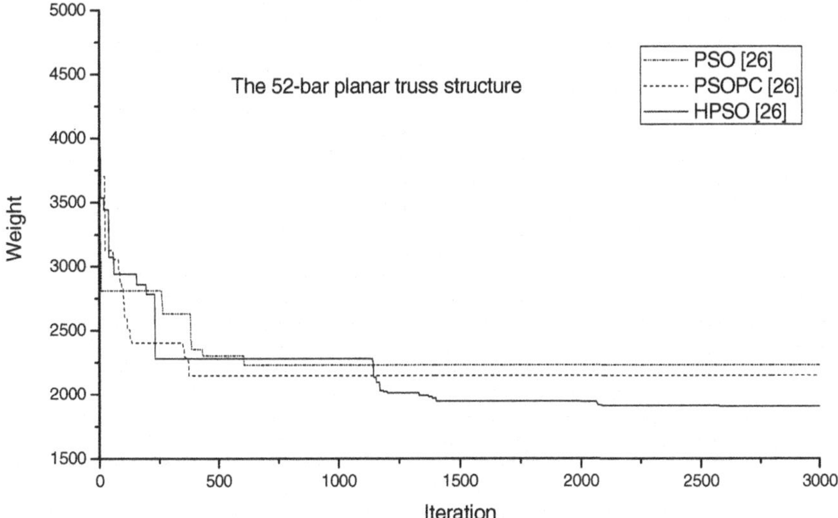


**Fig. 3.26** Comparison of convergence rates for the 52-bar planar truss structure

### *(5) A 72-bar spatial truss structure*

A 72-bar spatial truss structure, shown in Fig. 3.27, has been studied by Wu [24] Lee [13] and Li [26]. The material density is 0.1 lb/in.$^3$ and the modulus of elasticity is 10,000 ksi. The members are subjected to stress limitations of ±25 ksi. The uppermost nodes are subjected to displacement limitations of ±0.25 in. both in *x* and *y* directions. Two load cases are listed in Table 3.24. There are 72 members, which are divided into 16 groups, as follows: (1) $A_1$~$A_4$, (2) $A_5$~$A_{12}$, (3) $A_{13}$~$A_{16}$, (4) $A_{17}$~$A_{18}$, (5) $A_{19}$~$A_{22}$, (6) $A_{23}$~$A_{30}$ (7) $A_{31}$~$A_{34}$, (8) $A_{35}$~$A_{36}$, (9) $A_{37}$~$A_{40}$, (10) $A_{41}$~$A_{48}$, (11) $A_{49}$~$A_{52}$, (12) $A_{53}$~$A_{54}$, (13) $A_{55}$~$A_{58}$, (14) $A_{59}$~$A_{66}$ (15) $A_{67}$~$A_{70}$, (16) $A_{71}$~$A_{72}$. There are two optimization cases to be implemented. Case 1: The discrete variables are selected from the set D={0.1, 0.2, 0.3, 0.4, 0.5, 0.6, 0.7, 0.8, 0.9, 1.0, 1.1, 1.2, 1.3, 1.4, 1.5, 1.6, 1.7, 1.8, 1.9, 2.0, 2.1, 2.2, 2.3, 2.4, 2.5, 2.6, 2.7, 2.8, 2.9, 3.0, 3.1, 3.2} (in$^2$); Case 2: The discrete variables are selected from the Table 3.19. A maximum number of 1,000 iterations is imposed.

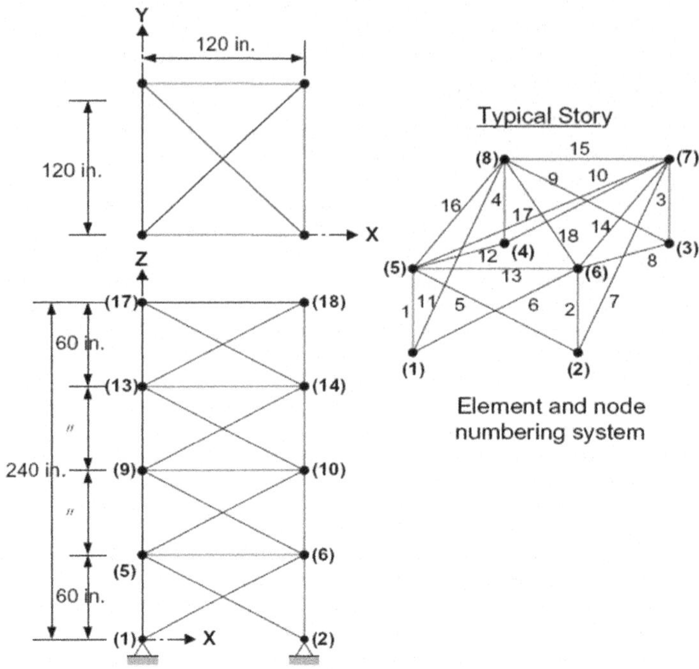


**Fig. 3.27** The 72-bar spatial truss structure

**Table 3.24** The load cases for the 72-bar spatial truss structure

| Nodes | Load Case 1 | | | Load Case 2 | | |
|---|---|---|---|---|---|---|
| | $P_X$ (kips) | $P_Y$ (kips) | $P_Z$ (kips) | $P_X$ (kips) | $P_Y$ (kips) | $P_Z$ (kips) |
| 17 | 5.0 | 5.0 | -5.0 | 0.0 | 0.0 | -5.0 |
| 18 | 0.0 | 0.0 | 0.0 | 0.0 | 0.0 | -5.0 |
| 19 | 0.0 | 0.0 | 0.0 | 0.0 | 0.0 | -5.0 |
| 20 | 0.0 | 0.0 | 0.0 | 0.0 | 0.0 | -5.0 |

Table 3.25 and Table 3.26 are the comparison of optimal design results.

**Table 3.25** Comparison of optimal designs for the 72-bar spatial truss structure (case 1)

| Variables (in$^2$) | Wu [24] | Lee [13] | Li [26] PSO | Li [26] PSOPC | Li [26] HPSO |
|---|---|---|---|---|---|
| $A_1$~$A_4$ | 1.5 | 1.9 | 2.6 | 3.0 | 2.1 |
| $A_5$~$A_{12}$ | 0.7 | 0.5 | 1.5 | 1.4 | 0.6 |
| $A_{13}$~$A_{16}$ | 0.1 | 0.1 | 0.3 | 0.2 | 0.1 |
| $A_{17}$~$A_{18}$ | 0.1 | 0.1 | 0.1 | 0.1 | 0.1 |
| $A_{19}$~$A_{22}$ | 1.3 | 1.4 | 2.1 | 2.7 | 1.4 |
| $A_{23}$~$A_{30}$ | 0.5 | 0.6 | 1.5 | 1.9 | 0.5 |
| $A_{31}$~$A_{34}$ | 0.2 | 0.1 | 0.6 | 0.7 | 0.1 |
| $A_{35}$~$A_{36}$ | 0.1 | 0.1 | 0.3 | 0.8 | 0.1 |
| $A_{37}$~$A_{40}$ | 0.5 | 0.6 | 2.2 | 1.4 | 0.5 |
| $A_{41}$~$A_{48}$ | 0.5 | 0.5 | 1.9 | 1.2 | 0.5 |
| $A_{49}$~$A_{52}$ | 0.1 | 0.1 | 0.2 | 0.8 | 0.1 |
| $A_{53}$~$A_{54}$ | 0.2 | 0.1 | 0.9 | 0.1 | 0.1 |
| $A_{55}$~$A_{58}$ | 0.2 | 0.2 | 0.4 | 0.4 | 0.2 |
| $A_{59}$~$A_{66}$ | 0.5 | 0.5 | 1.9 | 1.9 | 0.5 |
| $A_{67}$~$A_{70}$ | 0.5 | 0.4 | 0.7 | 0.9 | 0.3 |
| $A_{71}$~$A_{72}$ | 0.7 | 0.6 | 1.6 | 1.3 | 0.7 |
| Weight (lb) | 400.66 | 387.94 | 1089.88 | 1069.79 | 388.94 |

**Table 3.26** Comparison of optimal designs for the 72-bar spatial truss structure (case 2)

| Variables (in$^2$) | Wu [24] | Li [26] PSO | Li [26] PSOPC | Li [26] HPSO |
|---|---|---|---|---|
| $A_1$~$A_4$ | 0.196 | 7.22 | 4.49 | 4.97 |
| $A_5$~$A_{12}$ | 0.602 | 1.80 | 1.457 | 1.228 |
| $A_{13}$~$A_{16}$ | 0.307 | 1.13 | 0.111 | 0.111 |
| $A_{17}$~$A_{18}$ | 0.766 | 0.196 | 0.111 | 0.111 |
| $A_{19}$~$A_{22}$ | 0.391 | 3.09 | 2.620 | 2.88 |
| $A_{23}$~$A_{30}$ | 0.391 | 0.785 | 1.130 | 1.457 |
| $A_{31}$~$A_{34}$ | 0.141 | 0.563 | 0.196 | 0.141 |
| $A_{35}$~$A_{36}$ | 0.111 | 0.785 | 0.111 | 0.111 |
| $A_{37}$~$A_{40}$ | 1.800 | 3.09 | 1.266 | 1.563 |
| $A_{41}$~$A_{48}$ | 0.602 | 1.228 | 1.457 | 1.228 |
| $A_{49}$~$A_{52}$ | 0.141 | 0.111 | 0.111 | 0.111 |
| $A_{53}$~$A_{54}$ | 0.307 | 0.563 | 0.111 | 0.196 |
| $A_{55}$~$A_{58}$ | 1.563 | 1.990 | 0.442 | 0.391 |
| $A_{59}$~$A_{66}$ | 0.766 | 1.620 | 1.457 | 1.457 |
| $A_{67}$~$A_{70}$ | 0.141 | 1.563 | 1.228 | 0.766 |
| $A_{71}$~$A_{72}$ | 0.111 | 1.266 | 1.457 | 1.563 |
| Weight (lb) | 427.203 | 1209.48 | 941.82 | 933.09 |

The Fig. 3.28 and Fig. 3.29 are comparison of convergence rates for the 72-bar spatial truss structure in two load cases.

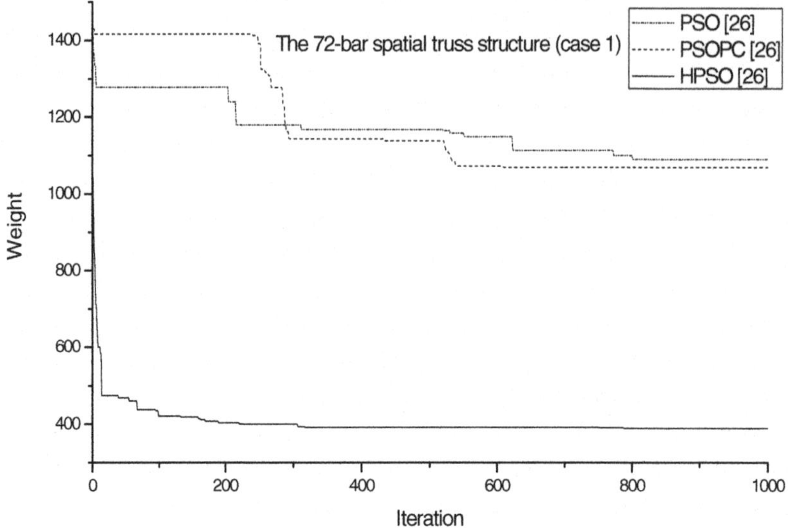


**Fig. 3.28** Comparison of convergence rates for the 72-bar spatial truss structure (case 1)

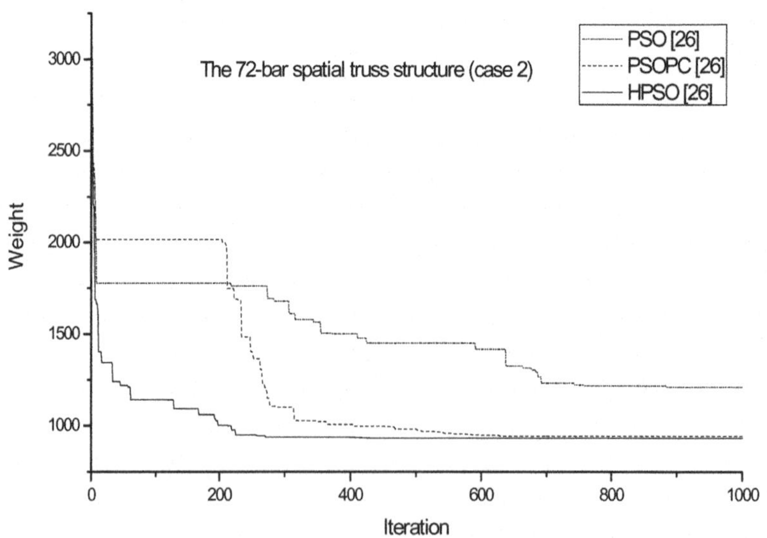


**Fig. 3.29** Comparison of convergence rates for the 72-bar spatial truss structure (case 2)

For both of the cases, it seems that Wu's results [24] achieve smaller weight. However, we discovered that both of these results do not satisfy the constraints. The results are unacceptable.

In case 1, the HPSO algorithm gets the optimal solution after 1000 iterations and shows a fast convergence rate, especially during the early iterations. For the PSO and PSOPC algorithms, they do not get optimal results when the maximum number of iterations is reached. In case 2, the HPSO algorithm gets best optimization result comparatively among three methods and shows a fast convergence rate.

## 3.6 The Weight Optimization of Grid Spherical Shell Structure

A double-layer grid steel shell structure with 83.6 m span, 14.0 m arc height and 1.5 shell thickness is shown in Fig. 3.30. The elastic module is 210 GPa and the density is 7850 kg/m$^3$. There are 6761 nodes and 1834 bars in this shell. The 1834 bars were divided into three groups, which were upper chord bars, lower chord bars and belly chord bars. All chords were thin circular tubes and their sections were limited to Chinese Criterion GB/T8162-1999 [31], which has 779 types of size to choose. The circumference nodes of lower chords are constrained. 50 kN vertical load is acted on each node of upper chords. The maximum permit displacement for all nodes is 1/400 of the length of span, that is ±0.209 m. The maximum permit stress for all chord bars is ±215 MPa. The stability of compressive chords is considered according to Chinese Standard GB50017-2003 [32]. The maximum slenderness ratio for compressive chords and tensile chords are 180 and 300 respectively.

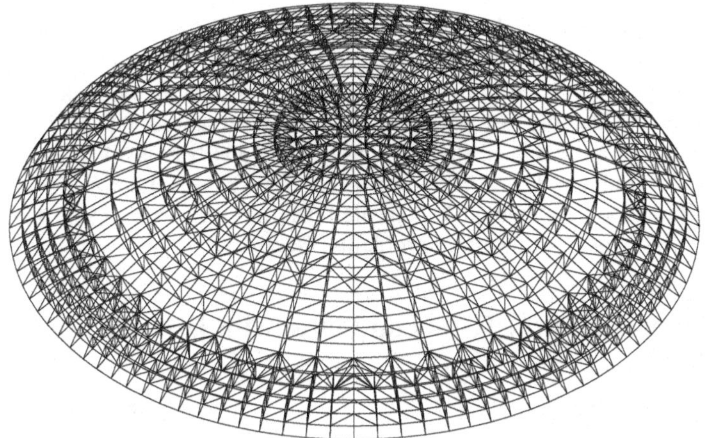

**Fig. 3.30** The double layer reticulated spherical shell structure

The optimization results [26] are shown in Table 3.27. The convergence velocity is shown in Fig. 3.31. It can be seen from Fig. 3.31 that HPSO can be used effectively to optimize the complicated engineering structures and can obtained a global optimization solution.

**Table 3.27** The optimal solution for the double layer reticulated spherical shell structure

| Upper chord bars | Lower chord bars | Belly chord bars | Weight (kg) |
|---|---|---|---|
| φ108×4 | φ83×3.5 | φ89×3.5 | 148811.71 |

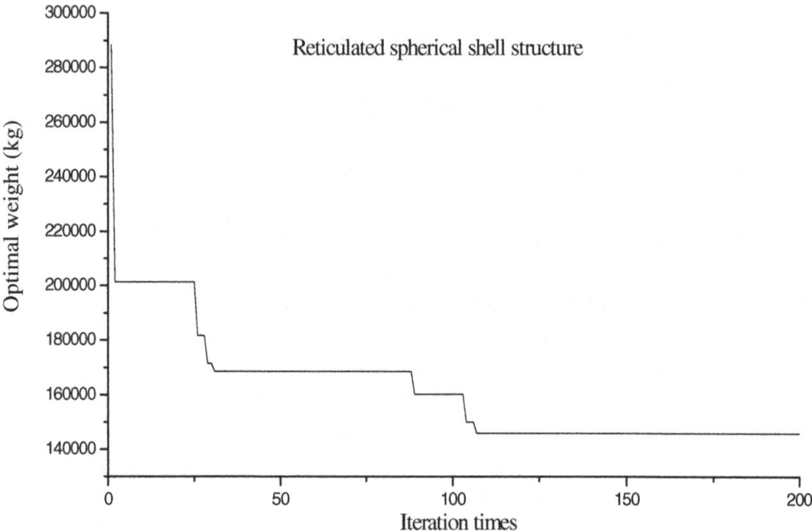


**Fig. 3.31** The convergence rate of the HPSO for the double layer grid spherical shell structure

As there are only three group optimal variables chosen in this example, the convergence rate is considerably fast, within only about 120 iterations. Anyway, the weight optimization for a reticulated shell structure with 1834 bars is a very complicated engineering problem. It is almost impossible to get a global optimal solution using a traditional optimal method. It is desired that the HPSO has an ability of handling complex structural optimization problems effectively.

## 3.7 The Application of the HPSO on Plates with Discrete Variables

### 3.7.1 Numerical Examples

#### (1) Quadrate plate

This example is taken from the literature [33]. The material parameters of this plate is as follows: Elastic modulus is 210 GPa, Poisson's ratio is 0.3, density is $7.85 \times 10^3$

kg/m$^3$, allowable stress is 120 MPa, allowable displacement is 0.0005 m, the four corners of plate are clamped, the plate is affected by uniform load of 200 kN/m$^2$, the size of plate is 1 m×1 m, and the calculation model is shown in Fig. 3.32.

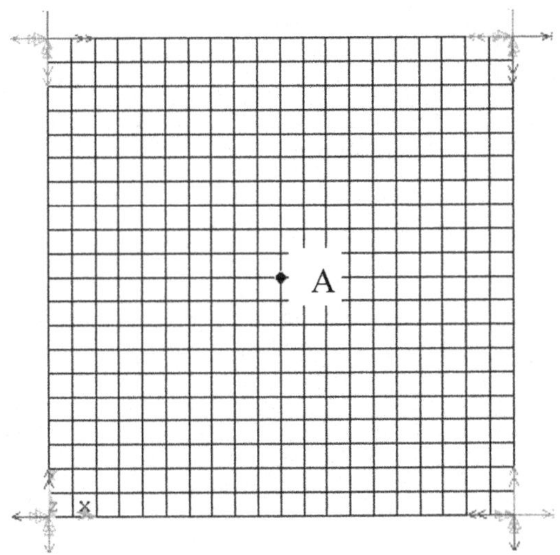


**Fig. 3.32** Quadrate plate

In optimization process, the selected range of discrete variables is based on the boundary of thickness in small deflection theory of plate: (1/80-1/100)≤h/l≤(1/5-1/8), whose maximum and minimum values is taken as boundary. 40 discrete values of plate thickness is as follows: [0.01, 0.016, 0.02, 0.026, 0.03, 0.034, 0.038, 0.042, 0.046, 0.052, 0.056, 0.06, 0.064, 0.07, 0.074, 0.078, 0.082, 0.086, 0.09, 0.094, 0.1, 0.104, 0.108, 0.112, 0.116, 0.122, 0.128, 0.134, 0.14, 0.146, 0.15, 0.156, 0.16, 0.166, 0.17, 0.176, 0.18, 0.186, 0.19, 0.2].

Consider the structural weight as objective function, thickness as design variables, and then the result of the plate thickness is 0.078 m by PSO, PSOPC and HPSO three optimization algorithms, and the weight convergence is 612.3 kg. Fig. 3.33 provide a comparison of the convergence rates of the three algorithms, where the vertical axis is the weight of the plate, while horizontal axis is the number of iterations.

In the literature [33]

$$\Delta = fql_r^4 / B_c \tag{3.7}$$

where $\Delta$ is the largest displacement under the uniform loads. In this case, f=0.02551, q is uniform load and $l_r$ is the length of square plate, respectively.

$$B_c = Eh^3 / \left[12\left(1-\mu^2\right)\right] \tag{3.8}$$

where E is elastic modulus, h is thickness of plate, $\mu$ is Poisson's ratio. Substitute equation (3.8) into equation (3.7), then

$$h = \left[12\left(1-\mu^2\right) fql_r^4 / E\Delta\right]^{1/3} \tag{3.9}$$

The theory solution of optimized thickness can be gotten according to equation (3.9). For the optimized square plate, the maximum stress of the plate is 110.35 MPa by Ansys (finite element analyze software), and the maximum displacement is 0.0003498 m. The optimized results of the plate are shown in Table 3.28.

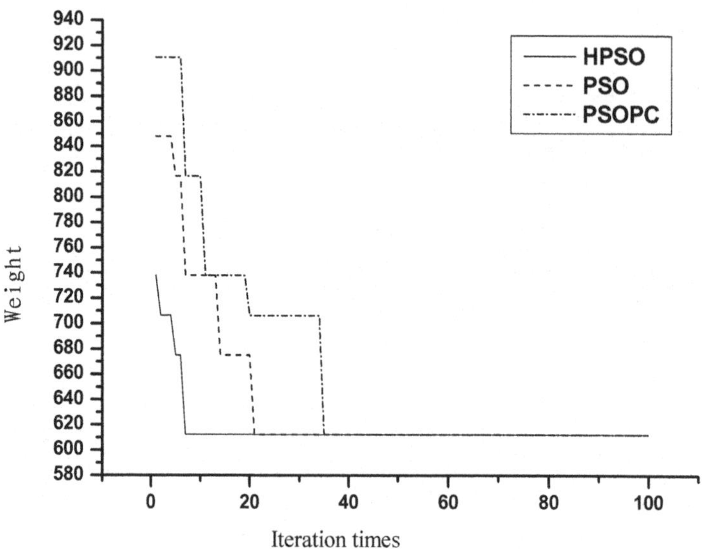


**Fig. 3.33** Comparison of convergence rates of three algorithms for the quadrate plate

**Table 3.27** The optimal result of the quadrate plate

| Constraints | Design variable constraints (Thickness) $0.01 \le h \le 0.2$ (m) | Allowable stress constraints 120 (MPa) | Allowable displacement constraints 0.0005(m) |
|---|---|---|---|
| Results | Optimized thickness (m) | Maximum stress (MPa) | Maximum dis-placement (m) |
| Optimization results | 0.078 | 110.35 | 0.0003498 |
| Theoretical solution | 0.081 | / | 0.0005 |

From Fig. 3.33 it can be seen that optimized results can be obtained by the three optimization algorithm with 40 iterations, however, the convergence rate of the HPSO algorithm is faster than other two algorithms. From Table 3.28 we can see that the feasibility and effectiveness of particle swarm optimization in plate optimization design. In the case of the plate optimization. The particle swarm optimization algorithm can get thinner and lighter plate, which can save more material.

### *(2) Combination of four plates*

The structural model of combination of four plates shown in Fig. 3.34, take the thickness of plate as the independent design variables, and the same material and material parameters as previous example are used, except that the allowable deflection of the plate is 0.005 m. Two adjacent sides of the plates are clamped, each plate is affected by uniform load of 50 kN/m$^2$,and the size of each plate is 0.5×0.5 m$^2$. The calculation model is shown in Fig. 3.34.

The selected discrete variables are the same as the previous example in optimization process. The boundary of plate thickness is $1/100 \le h/l \le 1/5$, 36 discrete values in total are as follows: [0.005, 0.006, 0.007, 0.008, 0.009, 0.01, 0.012, 0.014, 0.016, 0.018, 0.02, 0.022, 0.024, 0.026, 0.028, 0.03, 0.032, 0.034, 0.036, 0.038, 0.04, 0.044, 0.048, 0.052, 0.056, 0.060, 0.064, 0.068, 0.072, 0.076, 0.08, 0.084, 0.88, 0.092, 0.096, 0.1]

Fig. 3.35 provide a comparison of the convergence rates of the three algorithms, where the vertical axis is the weight of the plate, and the horizontal axis is the number of iterations.

Consider the structural weight as objective function, thickness as design variable. The thickness of each plate is $H_1$, $H_2$, $H_3$ and $H_4$, respectively. The three optimization algorithm (HPSO, PSOPC, PSO) were used separately with 3000 iterations and the optimal result is 247.28 kg, 262.975 kg and 298.3 kg respectively.

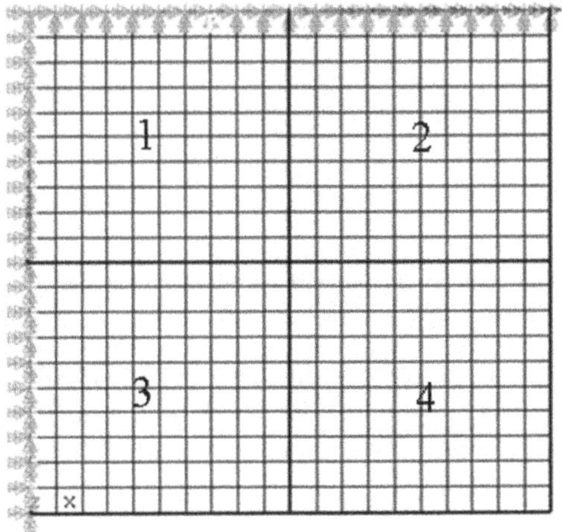


**Fig. 3.34** Combination of four plates

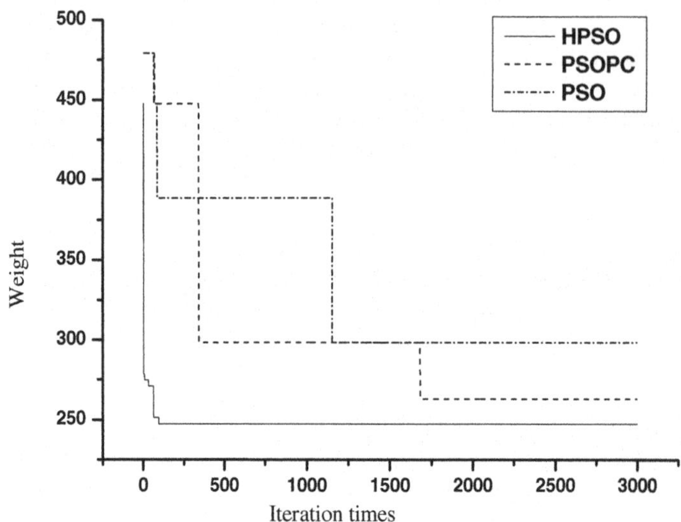


**Fig. 3.35** Comparison of convergence rates of three algorithms for a combination plate

The optimized results obtained by different optimization algorithm and Ansys are shown in Table 3.28.

**Table 3.28** The optimal result of the combination plate

| Constraints | | HPSO | | PSO | | PSOPC | |
|---|---|---|---|---|---|---|---|
| Variable constraints (Thickness) | Thickness | Optimized thickness (m) | Max. stress (MPa) | Optimized thickness (m) | Max. stress (MPa) | Optimized thickness (m) | Max Stress (MPa) |
| $0.005 \le h \le 0.1$ (m) | $H_1$ | 0.036 | 56.102 | 0.036 | 69.561 | 0.028 | 106.116 |
| | $H_2$ | 0.032 | 71.004 | 0.03 | 100.168 | 0.030 | 92.439 |
| | $H_3$ | 0.032 | 71.004 | 0.048 | 39.128 | 0.038 | 57.614 |
| | $H_4$ | 0.026 | 107.556 | 0.038 | 62.432 | 0.038 | 57.614 |
| Displacement constraints 0.005 (m) | Maximum displacement | 0.003731(m) | | 0.001092(m) | | 0.002803(m) | |

Table 3.28 and Fig. 3.35 illustrate that the result obtained by PSO and PSOPC algorithm with 3000 iterations is not converge, while the HPSO algorithm in 1000 iterations can get preferably results, which showed the superiority of HPSO algorithm in the calculation of complex plate issues.

### *(3) Combination of nine plates*

The structural model of combination of nine plates shown in Fig. 3.36. The opposite sides of the plate are clamped. The thickness of each plate is considered as the independent design variables. The material and geometrical parameters are the same as the previous combination of four plates.

The selected conditions of discrete variables are the same as the combination of four plates in optimization process. Fig. 3.37 provide a comparison of the convergence rates of the three algorithms, where the vertical axis is the weight of the plate, while horizontal axis is for the number of iterations.

Consider the structural weight as objective function, thickness as design variable. The thickness of each plate is $H_1$-$H_9$ respectively. The maximum iteration times is 3000. Convergence results for the three optimization algorithm is 537.73 kg, 580.09 kg and 573.05 kg respectively. The maximum stress and strain is shown in Table 3.29.

Fig. 3.37 illustrate that the convergence rate and optimized result of HPSO algorithm is obviously better than PSO algorithm and PSOPC algorithm.

The results in Table 3.29 show that HPSO algorithm is very effective when the structure with more independent variables.

The thickness for the combination plate optimized by HPSO algorithm is shown in Fig. 3.38.

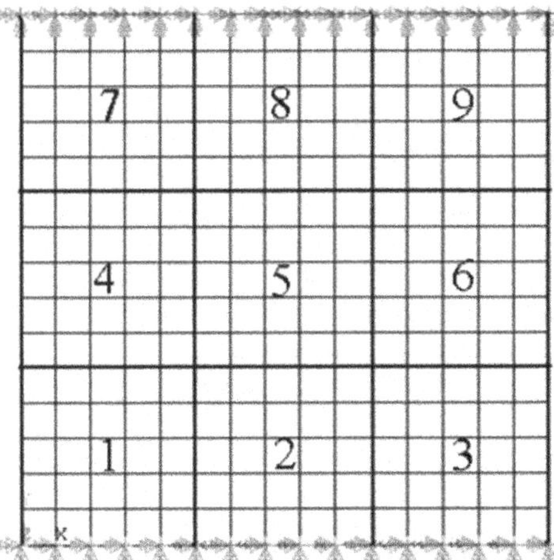


**Fig. 3.36** Combination of plate

**Table 3.29** The optimal result of the combination plate

| Constraints | | HPSO | | PSO | | PSOPC | |
|---|---|---|---|---|---|---|---|
| variable constraints (Thickness) | Thickness | Optimized thickness (m) | Max Stress (MPa) | Optimized thickness (m) | Max Stress (MPa) | Optimized thickness (m) | Max stress (MPa) |
| $0.005 \leq h \leq 0.1$ (m) | $H_1$ | 0.032 | 67.029 | 0.032 | 60.001 | 0.036 | 66.973 |
| | $H_2$ | 0.038 | 47.533 | 0.030 | 68.267 | 0.044 | 44.833 |
| | $H_3$ | 0.028 | 87.548 | 0.040 | 38.400 | 0.060 | 24.110 |
| | $H_4$ | 0.038 | 47.533 | 0.032 | 60.001 | 0.036 | 66.973 |
| | $H_5$ | 0.034 | 59.375 | 0.034 | 53.149 | 0.038 | 60.108 |
| | $H_6$ | 0.030 | 76.264 | 0.036 | 47.408 | 0.034 | 75.084 |
| | $H_7$ | 0.024 | 119.162 | 0.032 | 60.001 | 0.036 | 66.973 |
| | $H_8$ | 0.024 | 119.162 | 0.032 | 60.001 | 0.040 | 54.248 |
| | $H_9$ | 0.026 | 101.535 | 0.030 | 68.267 | 0.044 | 44.833 |
| Displacement constraints 0.005 (m) | Maximum displacement | 0.001604 (m) | | 0.001154 (m) | | 0.0007581 (m) | |

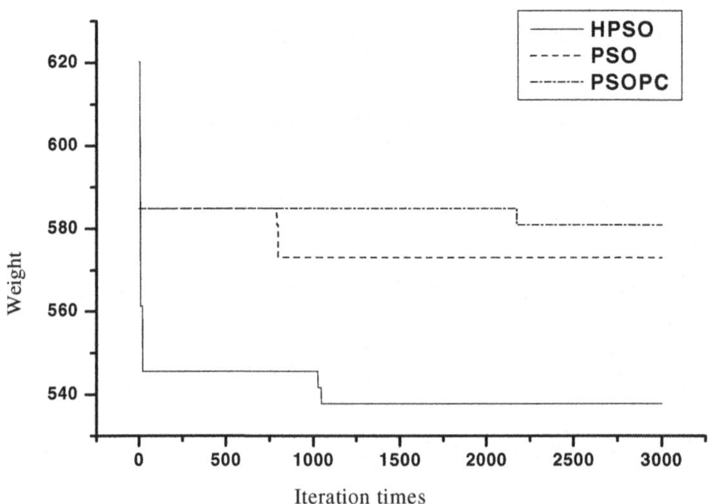


**Fig. 3.37** Convergence rates of three algorithms for the combination of nine plates

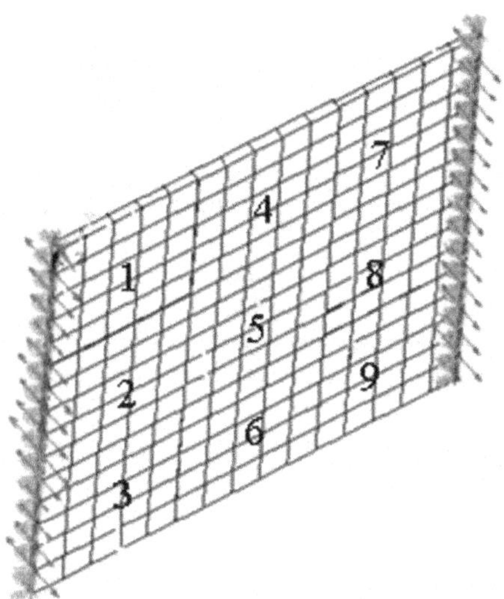


**Fig. 3.38** Thickness for the combination plate

## 3.8 Conclusions Remarks

In this chapter, a heuristic particle swarm optimizer (HPSO), based on the particle swarm optimizer with passive congregation (PSOPC), and the harmony search (HS) algorithm are presented. The HPSO algorithm handles the constraints of variables using the harmony search scheme in corporation with the 'fly-back mechanism' method used to deal with the problem-specific constraints. Compared with the PSO and the PSOPC algorithms, the HPSO algorithm does not allow any particles to fly outside the boundary of the variables and makes a full use of algorithm flying behaviour of each particle. Thus this algorithm performs more efficient than the others.

The efficiency of the HPSO algorithm presented is tested for optimum design of planar and spatial pin-connected structures with continuous and discrete variables, as well as plates. A double-layer grid shell structure was also used to test the HPSO. All the results show that the HPSO algorithm has better global/local search behaviour avoiding premature convergence while rapidly converging to the optimal solution. And the HPSO algorithm converges more quickly than the PSO and the PSOPC algorithms, in particular, in the early iterations.

A drawback of this HPSO algorithm at present is that its convergence rate will slow down when the number of the iterations increase. Further study is being conducted for improvement.

## References


1. Coello Coello, C.A.: Theoretical and numerical constraint-handling techniques used with evolutionary algorithms: a survey of the state of the art. Computer Methods in Applied Mechanics and Engineering 191(11-12), 1245–1287 (2002)
2. Nanakorn, P., Konlakarn, Meesomklin: An adaptive penalty function in genetic algorithms for structural design optimization. Computers and Structures 79(29-30), 2527–2539 (2001)
3. Deb, K., Gulati, S.: Design of truss-structures for minimum weight using genetic algorithms. Finite Elements in Analysis and Design 37(5), 447–465 (2001)
4. Ali, N., Behdinan, K., Fawaz, Z.: Applicability and viability of a GA based finite element analysis architecture for structural design optimization. Computers and Structures 81(22-23), 2259–2271 (2003)
5. Kennedy, J., Eberhart, R.C.: Particle swarm optimization. In: IEEE International Conference on Neural Networks, vol. 4, pp. 1942–1948. IEEE Press, Los Alamitos (1995)
6. Kennedy, J., Eberhart, R.C.: Swarm intelligence. Morgan Kaufmann, San Francisco (2001)
7. Angeline, P.: Evolutionary optimization versus particle swarm optimization: philosophy and performance difference. In: Proceeding of the Evolutionary Programming Conference, San Diago, USA (1998)
8. He, S., Wu, Q.H., Wen, J.Y., Saunders, J.R., Paton, R.C.: A particle swarm optimizer with passive congregation. BioSystem 78(1-3), 135–147 (2004)
9. He, S., Prempain, E., Wu, Q.H.: An improved particle swarm optimizer for mechanical design optimization problems. Engineering Optimization 36(5), 585–605 (2004)
10. Davis, L.: Genetic algorithms and simulated annealing. Pitman, London (1987)
11. Le Riche, R.G., Knopf-Lenoir, C., Haftka, R.T.: A segregated genetic algorithm for constrained structural optimization. In: Sixth International Conference on Genetic Algorithms, pp. 558–565. University of Pittsburgh, Morgan Kaufmann (1995)
12. Li, L.J., Ren, F.M., Liu, F., Wu, Q.H.: An improved particle swarm optimization method and its application in civil engineering. In: Proceedings of the Fifth International Conference on Engineering Computational Technology, Spain, paper 42 (2006)
13. Lee, K.S., Geem, Z.W., Lee, S.H., Bae, K.W.: The harmony search heuristic algorithm for discrete structural optimization. Engineering Optimization 37(7), 663–684 (2005)
14. Shi, Y., Eberhart, R.C.: A modified particle swarm optimizer. In: Proc. IEEE Inc. Conf. On Evolutionary Computation, pp. 303-308 (1997)
15. Geem, Z.W., Kim, J.H., Loganathan, G.V.: A new heuristic optimization algorithm: harmony search. Simulation 76(2), 60–68 (2001)
16. Lee, K.S., Geem, Z.W.: A new structural optimization method based on the harmony search algorithm. Computers and Structures 82(9-10), 781–798 (2004)
17. Schmit Jr, L.A., Farshi, B.: Some approximation concepts for structural synthesis. AIAA J. 12(5), 692–699 (1974)
18. Rizzi, P.: Optimization of multiconstrained structures based on optimality criteria. In: AIAA/ASME/SAE 17th Structures, Structural Dynamics and Materials Conference, King of Prussia, PA (1976)
19. Li, L.J., Huang, Z.B., Liu, F.: A heuristic particle swarm optimizer (HPSO) for optimization of pin connected structures. Computers and Structures 85(7-8), 340–349 (2007)

20. Khot, N.S., Berke, L.: Structural optimization using optimality criteria methods. In: Atrek, E., Gallagher, R.H., Ragsdell, K.M., Zienkiewicz, O.C. (eds.) New directions in optimum structural design. John Wiley, New York (1984)
21. Adeli, H., Kumar, S.: Distributed genetic algorithm for structural optimization. J. Aerospace Eng. ASCE 8(3), 156–163 (1995)
22. Sarma, K.C., Adeli, H.: Fuzzy genetic algorithm for optimization of steel structures. J. Struct. Eng. ASCE 126(5), 596–604 (2000)
23. He, D.K., Wang, F.L., Mao, Z.Z.: Study on application of genetic algorithm in discrete variables optimization. Journal of System Simulation 18(5), 1154–1156 (2006)
24. Wu, S.J., Chow, P.T.: Steady-state genetic algorithms for discrete optimization of trusses. Computers and Structures 56(6), 979–991 (1995)
25. Parsopoulos, K.E., Vrahatis, M.N.: Recent approaches to global optimization problems through particle swarm optimization. Natural Computing 12(1), 235–306 (2002)
26. Li, L.J., Huang, Z.B., Liu, F.: A heuristic particle swarm optimization method for truss structures with discrete variables. Computers and Structures 87(7-8), 435–443 (2009)
27. Li, L.J., Huang, Z.B., Liu, F.: An improved particle swarm optimizer for truss structure optimization. In: Wang, Y., Cheung, Y.-m., Liu, H. (eds.) CIS 2006. LNCS (LNAI), vol. 4456, pp. 1–10. Springer, Heidelberg (2007)
28. Rajeev, S., Krishnamoorthy, C.S.: Discrete optimization of structures using genetic algorithm. Journal of Structural Engineering, ASCE 118(5), 1123–1250 (1992)
29. Ringertz, U.T.: On methods for discrete structural constraints. Engineering Optimization 13(1), 47–64 (1988)
30. Zhang, Y.N., Liu, J.P., Liu, B., Zhu, C.Y., Li, Y.: Application of improved hybrid genetic algorithm to optimize. Journal of South China University of Technology 33(3), 69–72 (2003)
31. Seamless steel tubes for structural purposes, GB/T8162-1999. Standards Press of China, Beijing (1999)
32. Code for design of steel structures, GB50017-2003. China Architecture and Building Press, Beijing (2006)
33. Lin, Z.Q., Liu, C.C.: Static Calculation note Book. China Architecture and Building Press, Beijing (1976)

# Chapter 4
# Optimum Design of Structures with Group Search Optimizer Algorithm

**Abstract.** This chapter introduces a novel optimization algorithm, group search optimizer (GSO) algorithm. The implementation method of this algorithm is presented in detail. The GSO was used to investigate the truss structures with continuous variables and was tested by five planar and space truss optimization problems. The efficiency of GSO for frame structure with discrete variables was valued by three frame structures. The optimization results were compared with that of the particle swarm optimizer (PSO), the particle swarm optimizer with passive congregation (PSOPC) and the heuristic particle swarm optimizer (HPSO), ant colony optimization algorithm (ACO) and genetic algorithms (GA). Results from the tested cases illustrate the competitive ability of the GSO to find the optimal results.

## 4.1 Introduction

Over the past decade a family of group intelligence optimization algorithms, notably inspired by animal ethology, have been used extensively in various scientific and engineering problems such as logic circuit design, control design, power systems design and so on. Moreover, these group intelligence optimization algorithms have been gradually used in structural optimization problems. The most promising algorithms among this family are Ant Colony Optimizer (ACO) [1, 2] and Particle Swarm Optimizer (PSO) [3-9]. With respect to other traditional algorithms, a number of advantages make PSO and improved PSO ideal candidates to be used in structural optimization tasks. For examples, they can handle target functions and constraint functions without any specific operation. However, some disadvantages exist in these algorithms, such as the PSO, the particle swarm optimizer with passive congregation (PSOPC) and the heuristic particle swarm optimizer (HPSO) need to handle the constraints at each searching iteration. Such a technique leads to a long computation time when dealing with large and sophisticated structures. Thus researches concentrate on creating new algorithms, which can overcome such disadvantages while maintain the advantages. Group search optimizer (GSO) has a markedly superior search performance for complex structural optimization problems, which makes itself a promising optimization algorithm for further investigation [10, 11].

L. Li & F. Liu: Group Search Optimization for Applications in Structural Design, ALO 9, pp. 69–96.
springerlink.com 

This chapter firstly introduces the basic idea of GSO algorithm, then, describes the optimization settings, finally, five typical structural optimization problems were tested to value the probability of GSO for optimal structure design.

## 4.2 Group Search Optimizer (GSO)

The GSO algorithm was firstly proposed by He [12]. It is based on the biological model: Producer-Scrounger (PS) model [9], which assumes group members search either for 'finding' (producer) or for 'joining' (scrounger) opportunities. Animal scanning mechanisms (e.g., vision) [13, 14] are incorporated to develop the GSO algorithm. GSO also employs 'rangers' which perform random walks to avoid entrapment in local minima.

The population of the GSO algorithm is called a group and each individual in the population is called a member. In an n-dimensional search space, the $i_{th}$ member at the $k_{th}$ searching bout (iteration), has a current position $X_i^k \in R^n$, a head angle $\varphi_i^k = (\varphi_{i1}^k, ..., \varphi_{i(n-1)}^k) \in R^{n-1}$ and a head direction $D_i^k(\varphi_i^k) = (d_{i1}^k, ..., d_{in}^k) \in R^n$ which can be calculated from $\varphi_i^k$ via a Polar to Cartesian coordinates transformation:

$$
\begin{aligned}
d_{i1}^k &= \prod_{p=1}^{n-1} \cos(\varphi_{ip}^k) \\
d_{ij}^k &= \sin(\varphi_{i(j-1)}^k) \cdot \prod_{p=i}^{n-1} \cos(\varphi_{ip}^k) \\
d_{in}^k &= \sin(\varphi_{i(n-1)}^k)
\end{aligned}
\tag{4.1}
$$

In GSO, a group consists three kinds of members: producers and scroungers whose behaviors are based on the PS model, and rangers who perform random walk motions. Recently, Couzin [15] suggested that the larger the group, the smaller the proportion of informed individuals need to guide the group with better accuracy. Therefore, for accuracy and convenience of computation, we simplify the PS model by assuming that there is only one producer at each searching bout and the remaining members are scroungers and rangers. It is also assumed that the producer, the scroungers and the rangers do not differ in their relevant phenotypic characteristics. Therefore, they can switch between the three roles. At each iteration, a group member, located in the most promising area, conferring the best fitness value, is chosen as the producer. The producer's scanning field of vision is generalized to a $n$-dimensional space, which is characterized by maximum pursuit angle $\theta_{max} \in R^{n-1}$ and maximum pursuit distance $l_{max} \in R^1$ as illustrated in a 3D space [16] in Fig. 4.1.

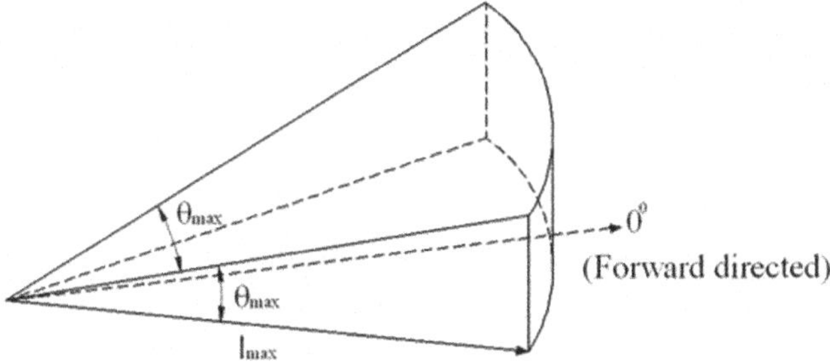


**Fig. 4.1** Scanning field in 3D space

In the GSO algorithm, at the $k_{th}$ iteration the producer $X_p$ behaves as follows:

(1) The producer will scan at zero degree and then scan laterally by randomly sampling three points in the scanning field: one point at zero degree:

$$X_z = X_p^k + r_1 l_{\max} D_p^k(\varphi^k) \tag{4.2}$$

one point in the right hand side hypercube:

$$X_l = X_p^k + r_1 l_{\max} D_p^k(\varphi^k - r_2\theta_{\max}/2) \tag{4.3}$$

and one point in the left hand side hypercube:

$$X_r = X_p^k + r_1 l_{\max} D_p^k(\varphi^k + r_2\theta_{\max}/2) \tag{4.4}$$

where $r_1 \in R^1$ is a normally distributed random number with mean 0 and standard deviation 1 and $r_2 \in R^{n-1}$ is a random sequence in the range (0, 1).

(2) The producer will then find the best point with the best resource (fitness value). If the best point has a better resource than its current position, then it will fly to this point. Or it will stay in its current position and turn its head to a new angle:

$$\varphi^{k+1} = \varphi^k + r_2\alpha_{\max} \tag{4.5}$$

where, $\alpha_{\max}$ is the maximum turning angle.

(3) If the producer cannot find a better area after $a$ iterations, it will turn its head back to zero degree:

$$\varphi^{k+a} = \varphi^k \tag{4.6}$$

where, $a$ is a constant.

At the $k_{th}$ iteration, the area copying behaviour of the $i_{th}$ scrounger can be modeled as a random walk towards the producer:

$$X_i^{k+1} = X_i^k + r_3(X_p^k - X_i^k) \tag{4.7}$$

where, $r_3 \in R^n$ is a uniform random sequence in the range (0, 1).

Besides the producer and the scroungers, a small number of rangers have been also introduced into our GSO algorithm. Random walks, which are thought to be the most efficient searching method for randomly distributed resources, are employed by rangers. If the $i_{th}$ group member is selected as a ranger, at the $k_{th}$ iteration, firstly, it generates a random head angle $\varphi_i$:

$$\varphi^{k+1} = \varphi^k + r_2\alpha_{\max} \tag{4.8}$$

where, $\alpha_{\max}$ is the maximum turning angle; and secondly, it chooses a random distance:

$$l_i = a \cdot r_1 l_{\max} \tag{4.9}$$

and move to the new point:

$$X_i^{k+1} = X_i^k + l_i D_i^k(\varphi^{k+1}) \tag{4.10}$$

To maximize their chances of finding resources, the GSO algorithm employs the fly-back mechanism [17] to handle the problem specified constraints: When the optimization process starts, the members of the group search the solution in the feasible space. If any member moves into the infeasible region, it will be forced to move back to the previous position to guarantee a feasible solution. The pseudocode for structural optimization by GSO is listed in Table 4.1.

**Table 4.1** Pseudocode for structural optimization by GSO algorithm

```
Set k = 0;
Randomly initialize positions X_i and head angles φ_i of all members;
FOR (each member i in the group)
WHILE (the constraints are violated)
  Randomly re-generate the current member Xi
END WHILE
END FOR
WHILE (the termination conditions are not met)
FOR (each members i in the group)
 Calculate fitness: Calculate the fitness value of current member: f(X_i)
 Choose producer: Find the producer X_p of the group;
```

**Table 4.1** (Contd.)

Perform producing: 1) The producer will scan at zero degree and then scan laterally by randomly sampling three points in the scanning field using equations (4.2) to (4.4).

2) If any point violates the constraints, it will be replaced by the producer's previous position.

3) Find the best point with the best resource (fitness value). If the best point has a better resource than its current position, then it will fly to this point. Otherwise it will stay in its current position and turn its head to a new angle using equation (4.5).

4) If the producer can not find a better area after a iterations, it will turn its head back to zero degree using equation (4.6);

Perform scrounging: Randomly select 80% from the rest members to perform scrounging using equation (4.7);

Perform ranging: For the rest members, they will perform ranging: firstly, generate a random head angle using equation (4.8); and secondly, choose a random distance $l_i$ from the Gauss distribution using equation (4.9) and move to the new point using equation (4.10);

Check feasibility: Check whether each member of the group violates the constraints. If it does, it will move back to the previous position to guarantee a feasible solution.

END FOR

Set k = k + 1;

END WHILE

## 4.3 The Application of the GSO to Truss Structure Optimal Design

In this item, five pin connected structures commonly used in literature are selected as benchmark structures to test the GSO. The examples given in the simulation studies include

(1) A 10-bar planar truss structure as shown in Fig. 4.2;
(2) A 17-bar planar truss structure as shown in Fig. 4.4;
(3) A 22-bar space truss structure as shown in Fig. 4.6;
(4) A 25-bar space truss structure as shown in Fig. 4.8;
(5) A 72-bar space truss structure as shown in Fig. 4.10.

Every structure has its own displacement constraints and stress constraints. All the stresses and displacements of these truss structures are analyzed by the finite element method (FEM). The PSO, PSOPC, HPSO and GSO schemes are applied, respectively, to all these examples and the results are compared in order to evaluate the performance of the new algorithm. For all these four algorithms, the maximum number of iterations is limited to 3000 and the population size is set to at 50; the inertia weight $\omega$ decrease linearly from 0.9 to 0.4; and the value of acceleration constants $c_1$ and $c_2$ are set to be the same and equal to 0.8. The passive congregation coefficient $c_3$ is given as 0.6 for the PSOPC and the HPSO algorithms. For the GSO algorithm, 20% of the population will be selected as rangers; the initial head angle $\varphi_0$ of each individual is set to be $\pi/4$. The constant $a$ is given by round$(\sqrt{n+1})$. The maximum pursuit angle $\theta_{max}$ is $\pi/a^2$. The maximum turning angle $\alpha_{max}$ is set to be $\pi/2a^2$. The maximum pursuit distance $l_{max}$ is calculated from:

$$l_{max} = |U_i - L_i| = \sqrt{\sum_{i=1}^{n} (U_i - L_i)^2} \tag{4.11}$$

Where $L_i$ and $U_i$ are the lower and upper bounds for the $i_{th}$ dimension respectively.

### 4.3.1 The 10-Bar Planar Truss Structure

The 10-bar planar truss structure, shown in Fig. 4.2, has previously been analyzed by many researchers.

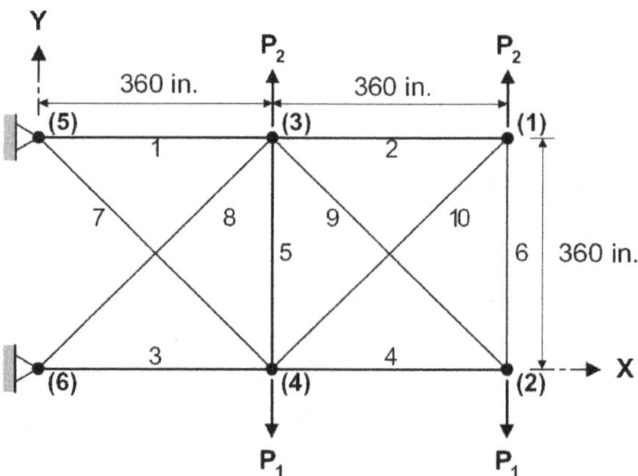


**Fig. 4.2** The 10-bar planar truss structure

The material density is 0.1 lb/in$^3$ and the modulus of elasticity is 10,000 ksi. The members are subjected to the stress limits of ±25 ksi. All nodes in both vertical and horizontal directions are subjected to the displacement limits of ±2.0 in. There are 10 design variables in this example and the minimum permitted cross-sectional area of each member is 0.1 in$^2$. The two loads are : $P_1 = 100$ kips, $P_2 = 0$ kips.

Table 4.2 shows the optimal design results and Fig. 4.3 provides a comparison of the convergence rates of the four algorithms. All these four algorithms achieve good solution after 3000 iterations. The HPSO and GSO algorithm shows better convergence rate than the others, especially at the early stage of iterations.

**Table 4.2** Comparison of optimal design results for the 10-bar truss structure

| Variables | | Optimal cross-sectional areas (in.$^2$) | | | |
|---|---|---|---|---|---|
| | | GSO [10] | PSO [8] | PSOPC [8] | HPSO [8] |
| 1 | $A_1$ | 31.289 | 31.560 | 30.569 | 30.704 |
| 2 | $A_2$ | 0.251 | 0.100 | 0.100 | 0.100 |
| 3 | $A_3$ | 22.606 | 20.974 | 22.974 | 23.167 |
| 4 | $A_4$ | 14.831 | 18.381 | 15.148 | 15.183 |
| 5 | $A_5$ | 0.166 | 0.836 | 0.100 | 0.100 |
| 6 | $A_6$ | 0.227 | 3.849 | 0.547 | 0.551 |
| 7 | $A_7$ | 8.775 | 6.085 | 7.493 | 7.460 |
| 8 | $A_8$ | 20.724 | 25.878 | 21.159 | 20.978 |
| 9 | $A_9$ | 21.700 | 19.804 | 21.556 | 21.508 |
| 10 | $A_{10}$ | 0.490 | 0.100 | 0.100 | 0.100 |
| Weight(lb) | | 5128.94 | 5365.79 | 5061.00 | 5060.92 |

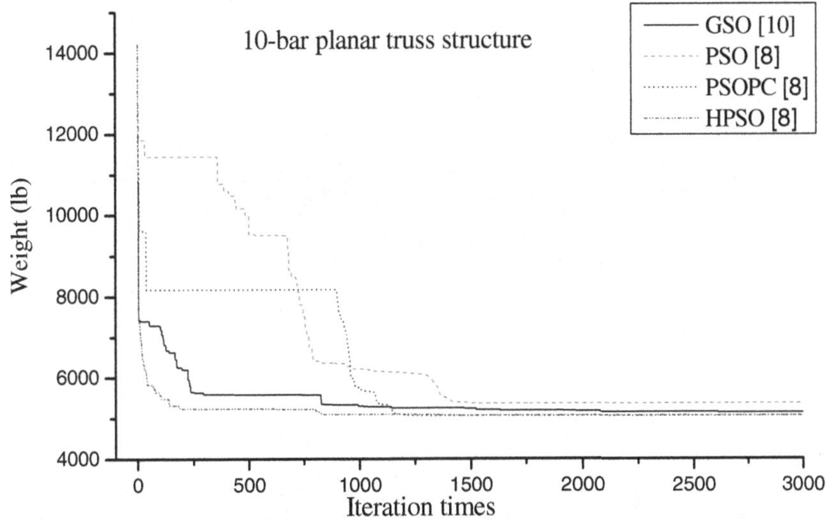


**Fig. 4.3** Comparison of convergence rates for the 10-bar truss structure

### *4.3.2 The 17-Bar Planar Truss Structure*

The 17-bar planar truss structure is shown in Fig. 4.4. The material density is 0.268 lb/in$^3$ and the modulus of elasticity is 30,000 ksi. The members are subjected to

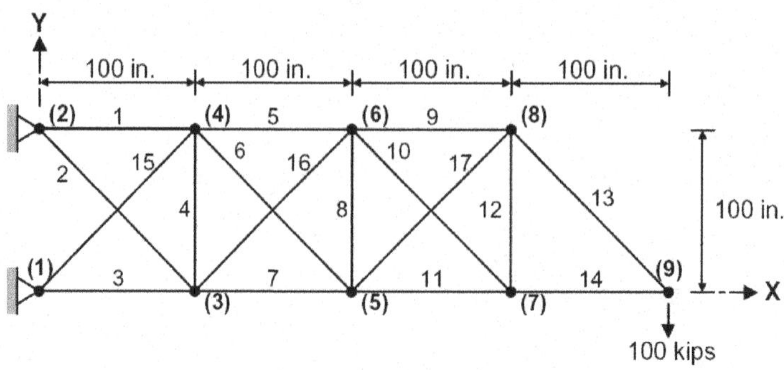


**Fig. 4.4** The 17-bar planar truss structure

the stress limits of ±50 ksi. All nodes in both directions are subjected to the displacement limits of ±2.0 in. There are 17 design variables in this example and the minimum permitted cross-sectional area of each member is 0.1 in$^2$. A single vertical downward load of 100 kips at node 9 is considered.

**Table 4.3** Comparison of optimal design results for the 17-bar truss structure

| Variables | | Optimal cross-sectional areas (in.$^2$) | | | |
|---|---|---|---|---|---|
| | | GSO [10] | PSO [8] | PSOPC [8] | HPSO [8] |
| 1 | $A_1$ | 15.940 | 14.915 | 15.981 | 15.929 |
| 2 | $A_2$ | 0.646 | 1.607 | 0.100 | 0.104 |
| 3 | $A_3$ | 12.541 | 12.738 | 12.142 | 12.016 |
| 4 | $A_4$ | 0.331 | 0.299 | 0.100 | 0.101 |
| 5 | $A_5$ | 7.361 | 9.458 | 8.098 | 8.091 |
| 6 | $A_6$ | 4.920 | 3.738 | 5.566 | 5.578 |
| 7 | $A_7$ | 11.072 | 11.558 | 11.732 | 11.883 |
| 8 | $A_8$ | 0.335 | 0.100 | 0.100 | 0.104 |
| 9 | $A_9$ | 8.535 | 6.065 | 7.982 | 7.955 |
| 10 | $A_{10}$ | 0.385 | 1.767 | 0.113 | 0.101 |
| 11 | $A_{11}$ | 4.525 | 5.844 | 4.074 | 4.094 |
| 12 | $A_{12}$ | 0.237 | 1.523 | 0.132 | 0.100 |
| 13 | $A_{13}$ | 6.034 | 5.449 | 5.667 | 5.677 |
| 14 | $A_{14}$ | 3.916 | 4.638 | 3.991 | 3.969 |
| 15 | $A_{15}$ | 5.149 | 4.184 | 5.555 | 5.552 |
| 16 | $A_{16}$ | 0.605 | 1.653 | 0.101 | 0.101 |
| 17 | $A_{17}$ | 5.416 | 4.264 | 5.555 | 5.590 |
| Weight(lb) | | 2613.96 | 2658.18 | 2582.85 | 2582.01 |

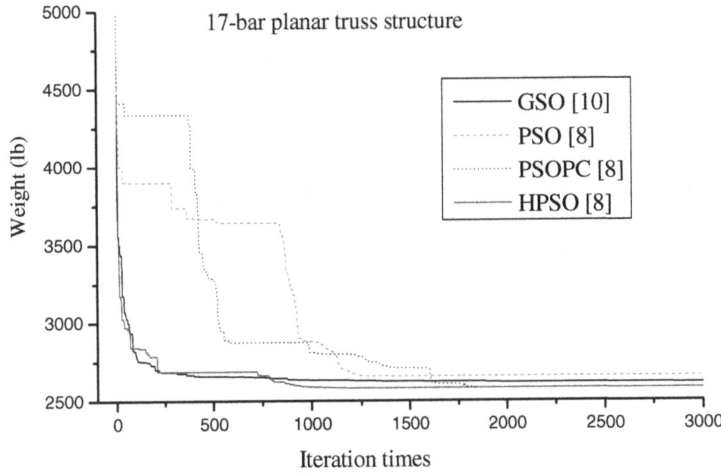


**Fig. 4.5** Comparison of convergence rates for the 17-bar truss structure

Table 4.3 shows the 17-bar planar truss structure optimization solutions and Fig. 4.5 compares the convergence rates of the above four algorithms. Fig. 4.5 shows both the PSOPC and HPSO algorithms achieve a good solution after 3000 iterations, their results are statistically better than that of GSO and PSOPC. However, the GSO and HPSO algorithm provide similar convergence rates, which are better than that of the PSO and PSOPC.

### *4.3.3 The 22-Bar Space Truss Structure*

The 22-bar space truss structure is shown in Fig. 4.6. The material density is 0.1 lb/in$^3$ and the modulus of elasticity is 10,000 ksi. The stress limits of the members are listed in Table 4.4. All nodes in all three directions are subjected to the

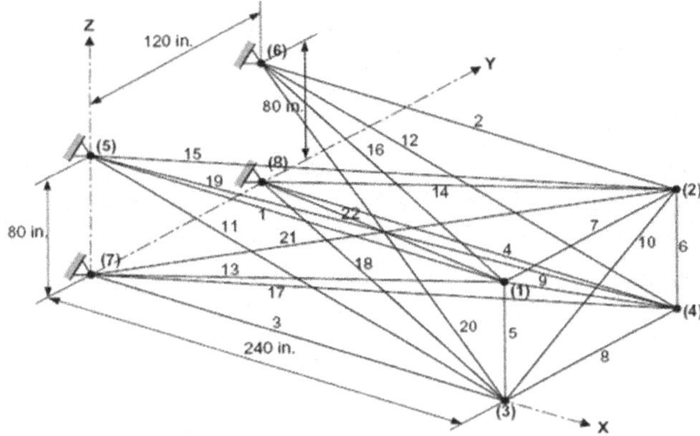


**Fig. 4.6** The 22-bar space truss structure

displacement limits of ±2.0 in. Three load cases are listed in Table 4.5. There are 22 members, which fall into 7 groups, as follows: (1) $A_1$–$A_4$, (2) $A_5$–$A_6$, (3) $A_7$–$A_8$, (4) $A_9$–$A_{10}$, (5)$A_{11}$–$A_{14}$, (6) $A_{15}$–$A_{18}$, and (7) $A_{19}$–$A_{22}$. The minimum permitted cross-sectional area of each member is 0.1 in$^2$.

**Table 4.4** Member stress limitations for the 22-bar space truss structure

| Variables | | Compressive stress limits (ksi) | Tensile stress limits (ksi) |
|---|---|---|---|
| 1 | $A_1$ | 24.0 | 36.0 |
| 2 | $A_2$ | 30.0 | 36.0 |
| 3 | $A_3$ | 28.0 | 36.0 |
| 4 | $A_4$ | 26.0 | 36.0 |
| 5 | $A_5$ | 22.0 | 36.0 |
| 6 | $A_6$ | 20.0 | 36.0 |
| 7 | $A_7$ | 18.0 | 36.0 |

**Table 4.5** Load cases for the 22-bar space truss structure

| Node | Case 1 (kips) | | | Case 2 (kips) | | | Case 3 (kips) | | |
|---|---|---|---|---|---|---|---|---|---|
| | $P_X$ | $P_Y$ | $P_Z$ | $P_X$ | $P_Y$ | $P_Z$ | $P_X$ | $P_Y$ | $P_Z$ |
| 1 | -20.0 | 0.0 | -5.0 | -20.0 | -5.0 | 0.0 | -20.0 | 0.0 | 35.0 |
| 2 | -20.0 | 0.0 | -5.0 | -20.0 | -50.0 | 0.0 | -20.0 | 0.0 | 0.0 |
| 3 | -20.0 | 0.0 | -30.0 | -20.0 | -5.0 | 0.0 | -20.0 | 0.0 | 0.0 |
| 4 | -20.0 | 0.0 | -30.0 | -20.0 | -50.0 | 0.0 | -20.0 | 0.0 | -35.0 |

**Table 4.6** Comparison of optimal design results for the 22-bar truss structure

| Variables | | Optimal cross-sectional areas (in$^2$) | | | |
|---|---|---|---|---|---|
| | | GSO [10] | PSO [8] | PSOPC [8] | HPSO[8] |
| 1 | $A_1$ | 2.803 | 2.702 | 2.623 | 2.613 |
| 2 | $A_2$ | 1.197 | 1.191 | 1.154 | 1.151 |
| 3 | $A_3$ | 0.332 | 0.354 | 0.355 | 0.346 |
| 4 | $A_4$ | 0.458 | 0.405 | 0.418 | 0.419 |
| 5 | $A_5$ | 2.634 | 2.679 | 2.792 | 2.797 |
| 6 | $A_6$ | 2.104 | 1.794 | 2.099 | 2.093 |
| 7 | $A_7$ | 2.003 | 2.357 | 2.010 | 2.022 |
| Weight (lb) | | 1026.02 | 1026.33 | 1023.91 | 1023.90 |

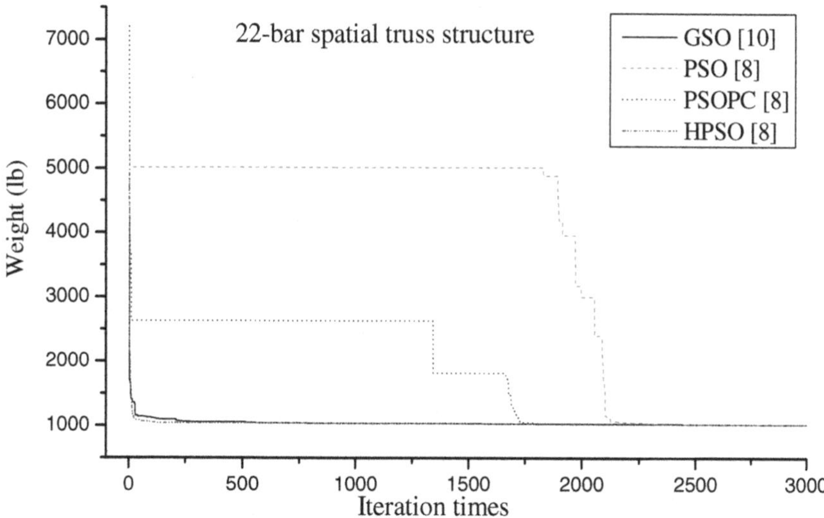


**Fig. 4.7** Convergence rates of four algorithms for the 22-bar truss structure

Table 4.6 shows the optimal solutions of the four algorithms and Fig. 4.7 provides the convergence rates of the four algorithms. In this example, the GSO and HPSO algorithm have converged after about 100 iterations, while the PSOPC and PSO algorithms need more than 1500 and 2000 iterations, respectively. But the optimum results obtained by using the HPSO and PSOPC algorithm are slightly better than that obtained by the GSO and the PSO algorithms. It can be seen form Fig. 4.7 that The GSO has much better convergence rate than PSO and PSOPC, while it has same convergence level with that of HPSO.

### *4.3.4 The 25-Bar Space Truss Structure*

The 25-bar space truss structure is shown in Fig. 4.8. The material density is 0.1 lb/in$^3$ and the modulus of elasticity is 10,000 ksi. The stress limits of the members are listed in Table 4.7. All nodes in all directions are subjected to the displacement limits of ±0.35 in. Two load cases listed in Table 4.8 are considered. There are 25 members, which are divided into 8 groups, as follows: (1) $A_1$, (2) $A_2$–$A_5$, (3) $A_6$–$A_9$, (4) $A_{10}$–$A_{11}$, (5) $A_{12}$–$A_{13}$, (6) $A_{14}$–$A_{17}$, (7) $A_{18}$–$A_{21}$ and (8) $A_{22}$–$A_{25}$. The minimum permitted cross-sectional area of each member is 0.01 in$^2$.

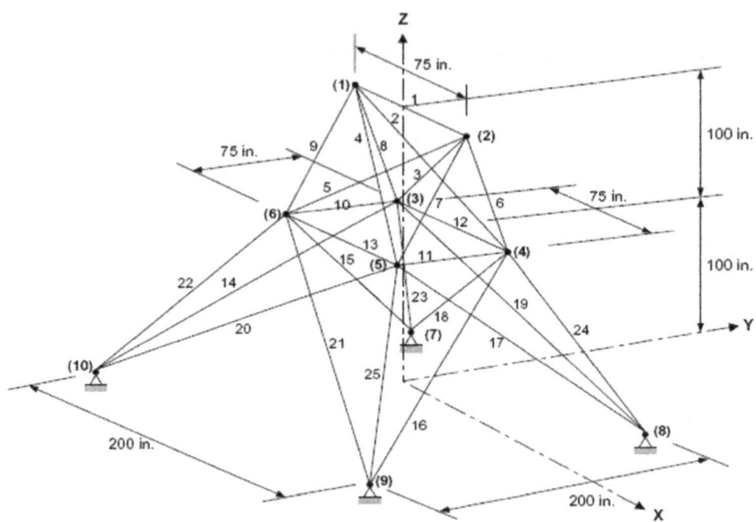


**Fig. 4.8** The 25-bar space truss structure

**Table 4.7** Member stress limitations for the 25-bar space truss structure

| Variables | | Compressive stress limits (ksi) | Tensile stress limits (ksi) |
|---|---|---|---|
| 1 | $A_1$ | 35.092 | 40.0 |
| 2 | $A_2$ | 11.590 | 40.0 |
| 3 | $A_3$ | 17.305 | 40.0 |
| 4 | $A_4$ | 35.092 | 40.0 |
| 5 | $A_5$ | 35.902 | 40.0 |
| 6 | $A_6$ | 6.759 | 40.0 |
| 7 | $A_7$ | 6.959 | 40.0 |
| 8 | $A_8$ | 11.802 | 40.0 |

**Table 4.8** Load cases for the 25-bar space truss structure

| Node | Case 1 | | | Case 2 | | |
|---|---|---|---|---|---|---|
| | $P_X$ (kips) | $P_Y$ (kips) | $P_Z$ (kips) | $P_X$ (kips) | $P_Y$ (kips) | $P_Z$ (kips) |
| 1 | 0.0 | 20.0 | -5.0 | 1.0 | 10.0 | -5.0 |
| 2 | 0.0 | -20.0 | -5.0 | 0.0 | 10.0 | -5.0 |
| 3 | 0.0 | 0.0 | 0.0 | 0.5 | 0.0 | 0.0 |
| 6 | 0.0 | 0.0 | 0.0 | 0.5 | 0.0 | 0.0 |

**Table 4.9** Comparison of optimal designs for the 25-bar truss structure

| Variables | | Optimal cross-sectional areas (in$^2$) | | | |
|---|---|---|---|---|---|
| | | GSO [10] | PSO [8] | PSOPC [8] | HPSO [8] |
| 1 | $A_1$ | 0.119 | 0.010 | 0.010 | 0.010 |
| 2 | $A_2$~$A_5$ | 1.838 | 1.704 | 1.948 | 1.970 |
| 3 | $A_6$~$A_9$ | 2.773 | 3.433 | 3.054 | 3.016 |
| 4 | $A_{10}$~$A_{11}$ | 0.017 | 0.010 | 0.010 | 0.010 |
| 5 | $A_{12}$~$A_{13}$ | 0.031 | 0.010 | 0.010 | 0.010 |
| 6 | $A_{14}$~$A_{17}$ | 0.729 | 0.921 | 0.684 | 0.694 |
| 7 | $A_{18}$~$A_{21}$ | 1.988 | 1.734 | 1.683 | 1.681 |
| 8 | $A_{22}$~$A_{25}$ | 2.610 | 2.333 | 2.644 | 2.643 |
| Weight (lb) | | 552.20 | 552.94 | 545.21 | 545.19 |

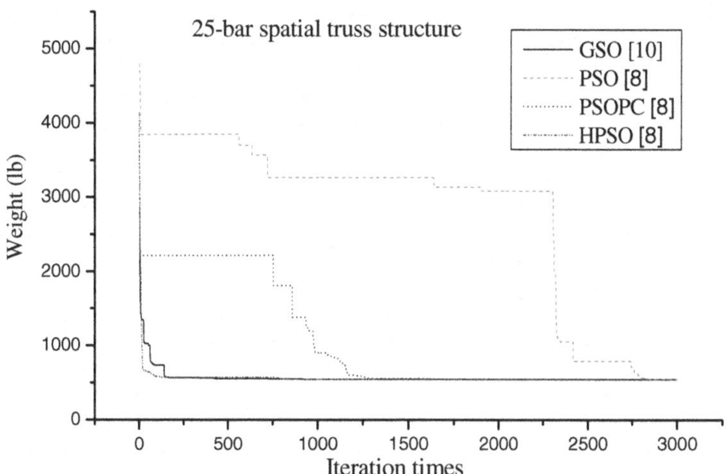


**Fig. 4.9** Convergence rates of four algorithms for the 25-bar truss structure

Table 4.9 shows the optimization solutions and Fig. 4.9 compares the convergence rate of the four algorithms. It can be seen from Fig. 4.9 that, for this space truss structure, it takes about 1200 and 2700 iterations for the PSOPC and the PSO algorithms to converge, respectively. However the HPSO and GSO algorithm take less than 500 iterations to converge. The optimum result obtained by using the GSO algorithm is slightly better than that obtained by the PSO algorithm, but has same level with that of the HPSO and PSOPC algorithms.

### *4.3.5 The 72-Bar Space Truss Structure*

The 25-bar space truss structure is shown in Fig. 4.10. The material density is 0.1 lb/in$^3$ and the modulus of elasticity is 10,000 ksi. The members are subjected to the stress limits of ±25 ksi. The uppermost nodes are subjected to the displacement limits of ±0.25 in. in both the X and Y directions. Two load cases are listed in Table 4.10. There are 72 members classified into 16 groups: (1) $A_1$–$A_4$, (2) $A_5$–$A_{12}$, (3) $A_{13}$–$A_{16}$, (4) $A_{17}$–$A_{18}$, (5) $A_{19}$–$A_{22}$, (6) $A_{23}$–$A_{30}$, (7) $A_{31}$–$A_{34}$, (8) $A_{35}$–$A_{36}$, (9) $A_{37}$–$A_{40}$, (10) $A_{41}$–$A_{48}$,(11) $A_{49}$–$A_{52}$, (12) $A_{53}$–$A_{54}$, (13) $A_{55}$–$A_{58}$, (14) $A_{59}$–$A_{66}$ (15) $A_{67}$–$A_{70}$, (16) $A_{71}$–$A_{72}$. For case 1, the minimum permitted cross-sectional area of each member is 0.1 in$^2$. For case 2, the minimum permitted cross-sectional area of each member is 0.01 in$^2$.

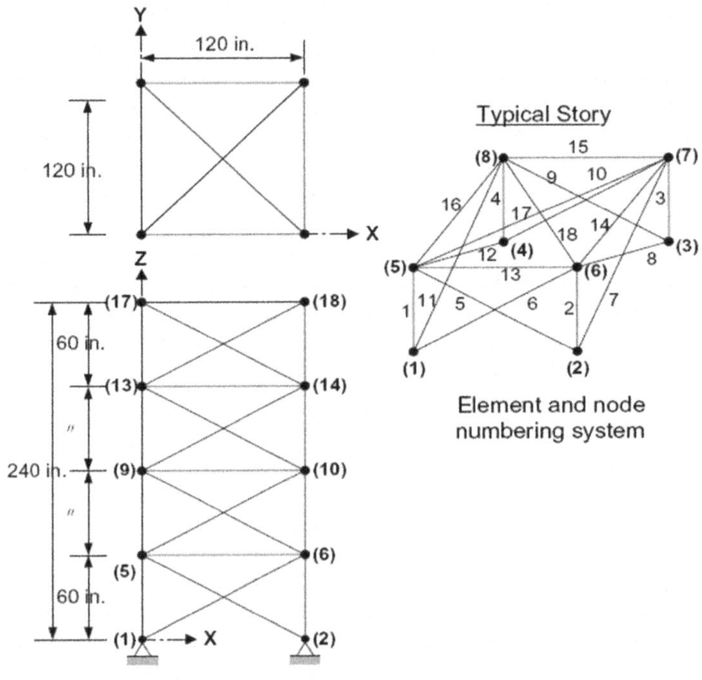


**Fig. 4.10** The 72-bar space truss structure

**Table 4.10** Load conditions for the 72-bar space truss structure

| Node | Case 1 | | | Case 2 | | |
|---|---|---|---|---|---|---|
| | $P_X$ (kips) | $P_Y$ (kips) | $P_Z$ (kips) | $P_X$ (kips) | $P_Y$ (kips) | $P_Z$ (kips) |
| 17 | 5.0 | 5.0 | -5.0 | 0.0 | 0.0 | -5.0 |
| 18 | 0.0 | 0.0 | 0.0 | 0.0 | 0.0 | -5.0 |
| 19 | 0.0 | 0.0 | 0.0 | 0.0 | 0.0 | -5.0 |
| 20 | 0.0 | 0.0 | 0.0 | 0.0 | 0.0 | -5.0 |

The optimization solutions for two loading cases are given in Tables 4.11 and 4.12, respectively. Fig. 4.11 and Fig. 4.12 give the convergence rates of the four algorithms for the two loading cases.

**Table 4.11** Comparison of optimal design results for the 72-bar truss structure (Case 1)

| Variables | | Optimal cross-sectional areas (in$^2$) | | | |
|---|---|---|---|---|---|
| | | GSO [10] | PSO [8] | PSOPC [8] | HPSO [8] |
| 1 | $A_1$~$A_4$ | 3.129 | 3.156 | 1.940 | 1.889 |
| 2 | $A_5$~$A_{12}$ | 0.539 | 6.835 | 0.508 | 0.510 |
| 3 | $A_{13}$~$A_{16}$ | 0.133 | 13.218 | 0.100 | 0.100 |
| 4 | $A_{17}$~$A_{18}$ | 0.152 | 26.802 | 0.100 | 0.100 |
| 5 | $A_{19}$~$A_{22}$ | 1.161 | 0.976 | 1.299 | 1.265 |
| 6 | $A_{23}$~$A_{30}$ | 0.410 | 1.893 | 0.524 | 0.510 |
| 7 | $A_{31}$~$A_{34}$ | 0.102 | 0.812 | 0.102 | 0.100 |
| 8 | $A_{35}$~$A_{36}$ | 0.101 | 1.344 | 0.100 | 0.100 |
| 9 | $A_{37}$~$A_{40}$ | 0.489 | 0.800 | 0.518 | 0.523 |
| 10 | $A_{41}$~$A_{48}$ | 0.366 | 0.228 | 0.510 | 0.519 |
| 11 | $A_{49}$~$A_{52}$ | 0.101 | 11.251 | 0.100 | 0.100 |
| 12 | $A_{53}$~$A_{54}$ | 0.100 | 13.261 | 0.103 | 0.100 |
| 13 | $A_{55}$~$A_{58}$ | 0.152 | 48.118 | 0.157 | 0.156 |
| 14 | $A_{59}$~$A_{66}$ | 0.676 | 0.255 | 0.549 | 0.548 |
| 15 | $A_{67}$~$A_{70}$ | 0.590 | 3.656 | 0.357 | 0.411 |
| 16 | $A_{71}$~$A_{72}$ | 0.633 | 24.707 | 0.580 | 0.568 |
| Weight (lb) | | 409.86 | 5894.87 | 380.10 | 379.63 |

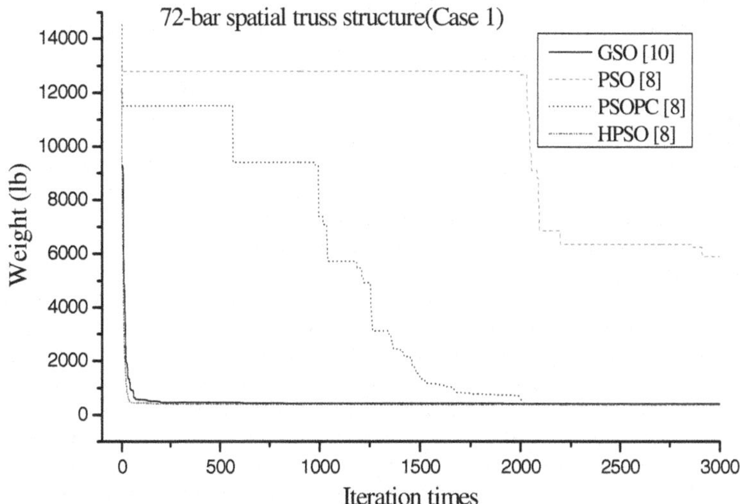


**Fig. 4.11** Comparison of convergence rates for the 72-bar truss structure (Case 1)

**Table 4.12** Comparison of optimal designs for the 72-bar truss structure (Case 2)

| Variables | | Optimal cross-sectional areas (in$^2$) | | | |
|---|---|---|---|---|---|
| | | GSO [10] | PSO [8] | PSOPC [8] | HPSO [8] |
| 1 | $A_1$~$A_4$ | 3.056 | 1.439 | 1.652 | 1.907 |
| 2 | $A_5$~$A_{12}$ | 0.356 | 3.137 | 0.547 | 0.524 |
| 3 | $A_{13}$~$A_{16}$ | 0.014 | 0.014 | 0.100 | 0.010 |
| 4 | $A_{17}$~$A_{18}$ | 0.083 | 21.029 | 0.101 | 0.010 |
| 5 | $A_{19}$~$A_{22}$ | 1.347 | 0.875 | 1.102 | 1.288 |
| 6 | $A_{23}$~$A_{30}$ | 0.432 | 0.450 | 0.589 | 0.523 |
| 7 | $A_{31}$~$A_{34}$ | 0.072 | 0.226 | 0.011 | 0.010 |
| 8 | $A_{35}$~$A_{36}$ | 0.040 | 0.022 | 0.010 | 0.010 |
| 9 | $A_{37}$~$A_{40}$ | 0.431 | 0.391 | 0.581 | 0.544 |
| 10 | $A_{41}$~$A_{48}$ | 0.488 | 0.468 | 0.458 | 0.528 |
| 11 | $A_{49}$~$A_{52}$ | 0.056 | 22.716 | 0.010 | 0.019 |
| 12 | $A_{53}$~$A_{54}$ | 0.096 | 49.589 | 0.152 | 0.020 |
| 13 | $A_{55}$~$A_{58}$ | 0.177 | 24.017 | 0.161 | 0.176 |
| 14 | $A_{59}$~$A_{66}$ | 0.726 | 0.606 | 0.555 | 0.535 |
| 15 | $A_{67}$~$A_{70}$ | 0.430 | 7.217 | 0.514 | 0.426 |
| 16 | $A_{71}$~$A_{72}$ | 1.015 | 0.184 | 0.648 | 0.612 |
| Weight (lb) | | 404.45 | 4993.69 | 368.45 | 364.86 |

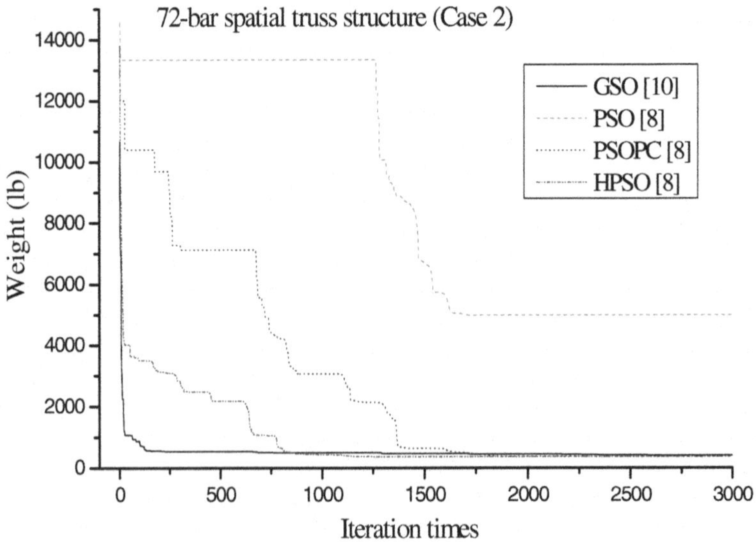


**Fig. 4.12** Comparison of convergence rates for the 72-bar truss structure (Case 2)

It can be seen from Fig. 4.11 and Fig. 4.12 that for both loading cases, the optimum results achieved by the PSOPC and the HPSO algorithms are better than that of the GSO algorithm. Indeed, in this example, the PSO algorithm did not fully converge when the maximum number of iterations is reached. However, in terms of the convergence rate, the GSO algorithm has the best performance especially in Case 2. Fig.4.12 also shows that the GSO algorithm may outperform the other three algorithms for large dimension problems.

## 4.4 The Application of GSO to Frame Cross-Section Optimization

Since the stochastic algorithms are introduced to the civil structural design, a lot of corresponding research results has been proposed. But the researches on the frame optimization study are less than others. Some researches, such as Saka and Kameshki, used genetic algorithm to do frame cross-section optimization work [18] and seismic cross-section optimum design work based on the semi-rigid assumption [19]. Camp [20] used ant colony algorithm to do steel frame structural optimal cross-section design. Chuang [21] used simulated annealing algorithm to do cross-section design to improve the particle swarm optimization algorithm. Zhu [22] improved genetic algorithm and used it for framework structural topology optimization. Group search optimization with the simple constraint handling (the penalty function processing) can avoid a large number of structural re-analysis, and had a good application to truss optimal design [10]. In this item, some frame optimization problems used by literatures are taken to prove that GSO is feasible and robust and has the obvious advantage over traditional algorithms. It is expected that GSO can work well in a more complex and bigger frame structure optimal design.

### *4.4.1 The Optimal Model of Frame Cross-Section Optimization*

The minimum-weight mathematical model of frame cross-section optimization is,

$$\begin{aligned} &\min W = \sum_{i=1}^{n} \rho A_i L_i \\ s.t.\quad & g_i^{\sigma} = [\sigma_i] - \sigma_i \ge 0 (i = 1, 2, \cdots, k) \\ & g_{jl}^{u} = [u_{jl}] - u_{jl} \ge 0 ( j = 1, 2, \cdots, m; l = 1, 2, \cdots, p) \\ & A_i \in D(i = 1, 2, \cdots, n) \\ & D = [A_1 \quad A_2 \quad \cdots \quad A_n]^T \end{aligned} \tag{4.12}$$

where $A_i$ and $L_i$ represent the cross-section and length of the $i$th element respectively. $u_{jl}$ represents the displacement of $l$ direction of $j$th node. $W$ is the weight of the structure. $g_i^{\sigma}$ and $g_{jl}^{u}$ represent the stress constraint and the displacement constraint respectively. $D$ is a set of discrete cross-section.

For frame structures, the working stress of component is mainly coming from the combination of moment stress and axial force. Ignoring the impact of shear stress, the maximum stress of the combination of moment stress and axial force is as follows:

$$\sigma_k = \left|\frac{N_k}{A_k}\right| + \left|\frac{(M_k)_x}{(W_k)_x}\right| + \left|\frac{(M_k)_y}{(W_k)_y}\right| \tag{4.13}$$

where $A_k$ represents the cross-section of the $k_{th}$ component; $(M_k)_x$ and $(M_k)_y$ represent the strong-axis and the weak-axis bending moment respectively; $(W_k)_x$ and $(W_k)_y$ represent the strong axis and the weak axis bending modulus respectively.

### 4.4.2 . Numerical Examples

In this item, three planar frame structures commonly used in literature are selected as benchmark problem to test the GSO. The proposed algorithm is coded in Matlab language. All results of optimization are modelled and checked by ANSYS to make sure that all of them are correct.

The examples given in the simulation studies include:

- a single-bay and 8-storey frame structure as shown in Fig. 4.13;
- a 2-bay and 5-storey frame structure as shown in Fig. 4.15;
- a 3-bay and 24-storey as shown in Fig. 4.18.

All these frame structures are optimized by GSO, HPSO (the previous chapter's method) respectively and analysed by the finite element method (FEM). The results are compared in order to evaluate the performance of the new algorithm. For the two algorithms, the maximum number of iterations is limited to 1000 and the population size is set to 50; the inertia weight $\omega$ decrease linearly from 0.9 to 0.4; and the value of acceleration constants c1 and $c_2$ are set to be the same and equal to 0.8. The passive congregation coefficient $c_3$ is given as 0.6 for the HPSO algorithms. For the GSO algorithm, 20% of the population is selected as rangers; the initial head angle $\varphi_0$ of each individual is set to be $\pi/4$. The constant $a$ is given by $\text{round}(\sqrt{n+1})$. The maximum pursuit angle $\theta_{\max}$ is $\pi/a^2$. The maximum turning angle $\alpha$ is set to be $\pi/2a^2$. The maximum pursuit distance $l_{\max}$ is calculated from:

$$l_{\max} = \| d_i^u - d_i^l \| = \sqrt{\sum_{i=1}^{n} (d_i^u - d_i^l)^2} \tag{4.14}$$

where $d_i^l$ and $d_i^u$ are the lower and upper bounds for the $i_{th}$ dimension.

#### *(1) A Single-Bay and 8-Storey Frame Structure Optimal Design*

The single-bay and 8-storey planar frame shown in Fig. 4.13 had been studied by several researchers, such as Chuang [21], Camp [23] and so on. The material

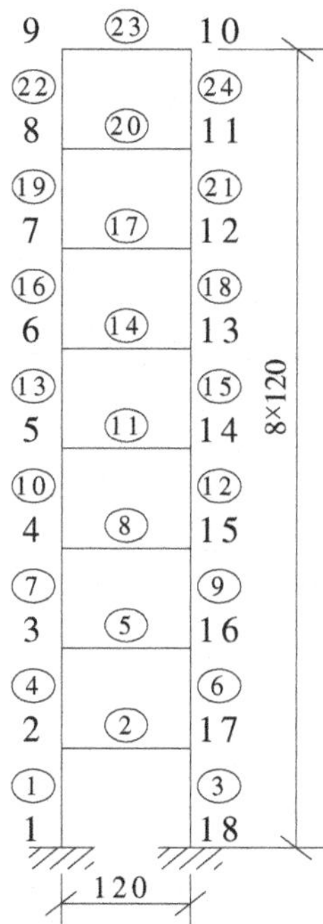


**Fig. 4.13** A single-bay and 8-storey planar frame

density is 0.283 lb/in.$^3$ and the modulus of elasticity is 10,000 ksi. Especially, there are not the stress limits of the members. But all nodes in horizontal directions are subjected to the displacement limits of ±2.0 in. There are 23 members, which fall into 8 groups, as follows: $A_1$ (1, 3, 4, 6), $A_2$ (7, 9, 10, 12), $A_3$ (13, 15, 16, 18), $A_4$ (19, 21, 22, 24), $A_5$ (2, 5), $A_6$ (8, 11), $A_7$ (14, 17), $A_8$ (20, 23). Discrete section set for the variables is selected from the W-beam of U.S. AISC, total of 268 groups (AISC 2001) [21]. The optimal model is only one load case: nodes 2-17, have the vertical direction loads of -100 kips; nodes 2-9, have the horizontal direction loads respectively, as follows, 0.272, 0.544, 0.816, 1.088, 1.361, 1.633, 1.905, 2.831 kips.

The optimization results of single-bay and 8-storey planar frame are listed in Table 4.13. And it is worth noticing that the optimal result of Chuang [21] is 7060.2840 lb with 3000 iterations (GSO with 1000 iterations), and the detail result of Camp [23] is not listed in table but with a optimal weight of 7380 lb. Compared with the optimal results of these optimizers, It is obvious that the GSO and HPSO are both good at this optimization problem. As Fig. 4.14 shows, the convergence rate of GSO is faster than HPSO's. Being of unique searching mechanism, GSO runs with the same iteration times as HPSO but with much fewer structural analysis times. In order to verify the correctness of optimal results of Matlab program, the optimization results of GSO and HPSO are all modeled and checked with

**Table 4.13** Comparison of optimal design for the single-bay and 8-storey frame structure

| Variables | Chuang [21] | Camp [23] | HPSO | GSO |
|---|---|---|---|---|
| $A_1$ | W18×40 | - | W21×44 | W18×40 |
| $A_2$ | W16×31 | - | W14×30 | W16×26 |
| $A_3$ | W16×26 | - | W16×26 | W16×26 |
| $A_4$ | W12×19 | - | W14×22 | W12×16 |
| $A_5$ | W18×35 | - | W18×35 | W18×35 |
| $A_6$ | W18×35 | - | W16×26 | W18×35 |
| $A_7$ | W18×35 | - | W16×26 | W18×35 |
| $A_8$ | W12×16 | - | W16×26 | W16×26 |
| Weight (lb) | 7060.2840 | 7380.0000 | 7157.4096 | 6949.5744 |

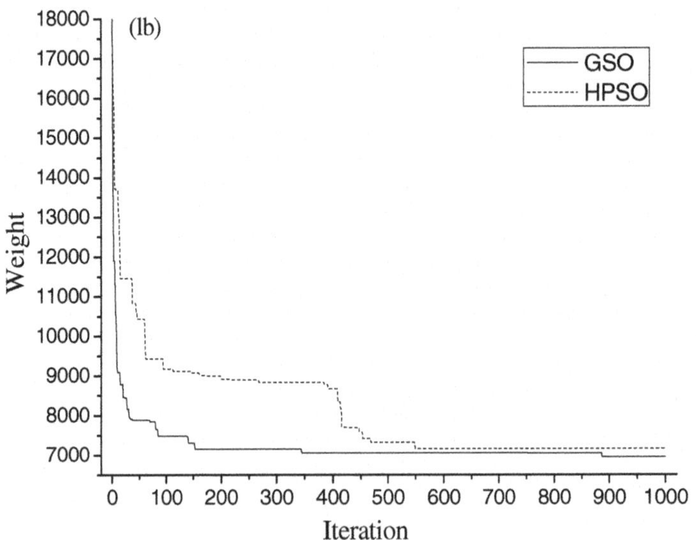


**Fig. 4.14** Convergence curves for the single-bay and 8-storey frame structure

commercial ANSYS program. The calculation results show that the biggest displacement 1.998 in. of nodes (the displacement of node 10 in x direction) meet the displacement constraints.

### *(2) A 2-Bay and 5-Storey Frame Structure Optimal Design*

The 2-bay and 5-storey frame structure, shown in Fig. 4.15, has previously been analysed by Chuang [21]. The material density is 7800 N/m$^3$ and the modulus of elasticity is $2.058\times10^{11}$ N/m$^2$. The members are subjected to stress limitations of $\pm$ $1.666\times10^8$ N/m$^2$. Node 1, 2, 3 in x directions are subjected to displacement limitations of $\pm$ 4.58 cm. There are 23 variables together and they are divided into 15 groups, as follows: $A_1$(1, 2, 3), $A_2$(4, 5, 6), $A_3$(7, 8, 9), $A_4$(10, 11, 12),

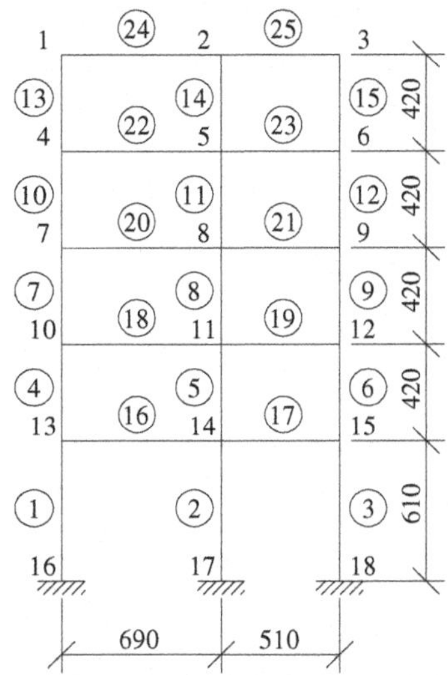


**Fig. 4.15** A 2-bay and 5-storey frame structure

**Table 4.14** Cross-section group for the 2-bay and 5-storey planar frame structure

| Section No. | Area ($cm^2$) | Bending modulus ($cm^3$) | Moment of inertia ($cm^4$) |
|---|---|---|---|
| 1 | 51.38 | 282.83 | 2545.50 |
| 2 | 57.66 | 356.08 | 3560.80 |
| 3 | 63.67 | 435.25 | 4787.70 |
| 4 | 69.81 | 537.46 | 6710.20 |
| 5 | 79.81 | 579.13 | 7239.10 |
| 6 | 80.04 | 678.13 | 9505.10 |
| 7 | 91.24 | 731.20 | 10236.80 |
| 8 | 97.00 | 938.83 | 15021.30 |
| 9 | 109.80 | 1007.10 | 16113.50 |
| 10 | 121.78 | 1319.35 | 23748.20 |
| 11 | 136.18 | 1405.75 | 25303.40 |
| 12 | 150.09 | 1757.77 | 35155.40 |
| 13 | 166.09 | 1864.44 | 37288.70 |
| 14 | 182.09 | 1971.10 | 39422.10 |
| 15 | / | / | / |

$A_5$(13, 14, 15), $A_6$(16), $A_7$(17), $A_8$(18), $A_9$(19), $A_{10}$(20), $A_{11}$(21), $A_{12}$(22), $A_{13}$(23), $A_{14}$(24), $A_{15}$(25). The discrete variables are selected from the set D listed in Table 4.3. Three load cases are considered: Case 1: G+0.9(Q+W); Case 2: G+W; Case 3: G+Q. Cross-section group are displayed in Table 4.14. The details of loads are listed in Table 4.15. A maximum number of 1000 iterations is imposed in calculation.

**Table 4.15** The details about the loads of 2-bay and 5-storey frame structure

| Load | Acting position | Type | Direction | Value |
|---|---|---|---|---|
| Dead load /G | Member 16-25 | Distributed load | -y | 11.76kN/m |
| | Node 1 ,3 | Concentrated load | -y | 19.6kN |
| | Node 4, 6, 7,8, 9, 12, 13, 15 | Concentrated load | -y | 40.2kN |
| Live load /Q | Member 16-25 | Distributed load | -y | 10.78kN/m |
| Wind load /W | Node 1 | Concentrated load | +x | 5.684kN |
| | Node 4 | Concentrated load | +x | 7.252kN |
| | Node 7 | Concentrated load | +x | 6.664kN |
| | Node 10 | Concentrated load | +x | 5.978kN |
| | Node 13 | Concentrated load | +x | 6.272kN |

Considering stress constraints and multi-load cases, the 2-bay and 5-storey frame optimization model is more complicated than the single-bay and 8-storey one. As Table 4.16 shows, the result of Chuang [21] seems to be the best one, but it violates the stress constraints when checked with ANSYS program. Table 4.16 and Fig. 4.16 give the comparison of optimal design results and convergence rates of 2-bay and 5-storey frame structure respectively. It can be seen that, after about 400 iterations, two algorithms have obtained good results.

**Table 4.16** Comparison of optimal designs 2-bay and 5-storey planar frame structure

| Variables | Chuang [21] | HPSO | GSO |
|---|---|---|---|
| $A_1$ | 6 | 6 | 6 |
| $A_2$ | 4 | 4 | 4 |
| $A_3$ | 4 | 4 | 4 |
| $A_4$ | 3 | 1 | 2 |
| $A_5$ | 1 | 2 | 2 |
| $A_6$ | 8 | 8 | 8 |
| $A_7$ | 8 | 6 | 6 |
| $A_8$ | 6 | 8 | 6 |
| $A_9$ | 4 | 3 | 6 |
| $A_{10}$ | 6 | 8 | 6 |
| $A_{11}$ | 3 | 2 | 4 |
| $A_{12}$ | 6 | 6 | 6 |
| $A_{13}$ | 1 | 2 | 2 |
| $A_{14}$ | 4 | 6 | 6 |
| $A_{15}$ | 1 | 3 | 3 |
| Weight (kg) | 71611.1370 | 72977.346 | 72903.4956 |
| Feasibility | No | Yes | Yes |

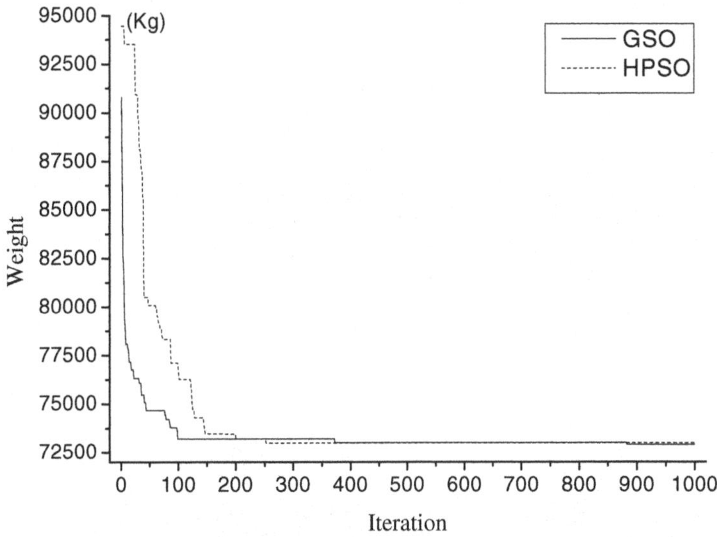


**Fig. 4.16** Convergence curves for the 2-bay and 5-storey planar frame structure

The Fig. 4.15 shows that the GSO algorithm has the faster convergence rate than HPSO, especially in the early iterations. The checking results of all the optimization are listed on Table 4.6 and the moment curve of Case 2 is shown in Fig. 4.17.

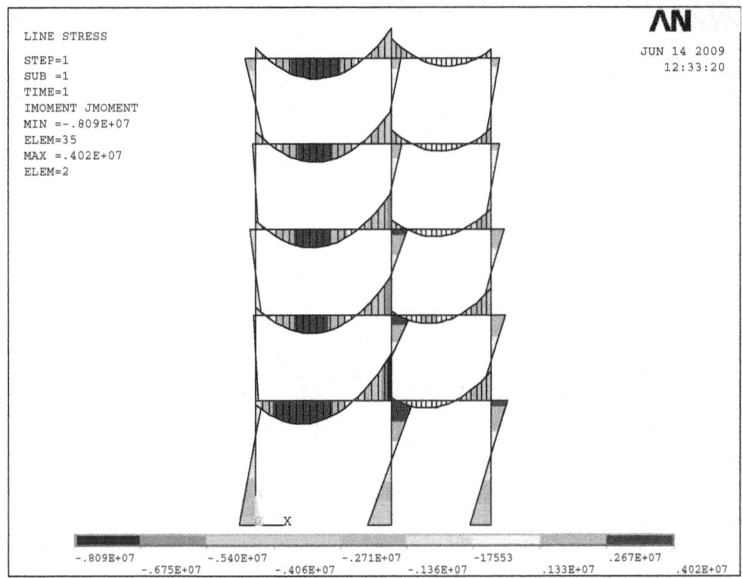


**Fig. 4.17** Moment curve for the 2-bay and 5-storey planar frame structure (In Case 2)

**Table 4.17** The checking detail of the optimal results of 2-bay and 5-storey planar frame structure

| Load case | Constraints | Details of the calculating results |
|---|---|---|
| 1 | Max stress | Member 18 $1.5409\times10^8$ N/m$^2$ |
| | 1, 2, 3 max displacement | Node 2 4.1775 cm |
| 2 | Max stress | Member 2 $1.1123\times10^8$ N/m$^2$ |
| | 1, 2, 3 max displacement | Node 1 4.5620 cm |
| 3 | Max stress | Member 25 $1.6213\times10^8$ N/m$^2$ |
| | 1, 2, 3 max displacement | Node 1 0.1887 cm |

***(3) A 3-Bay and 24-Storey Frame Structure Optimal Design***

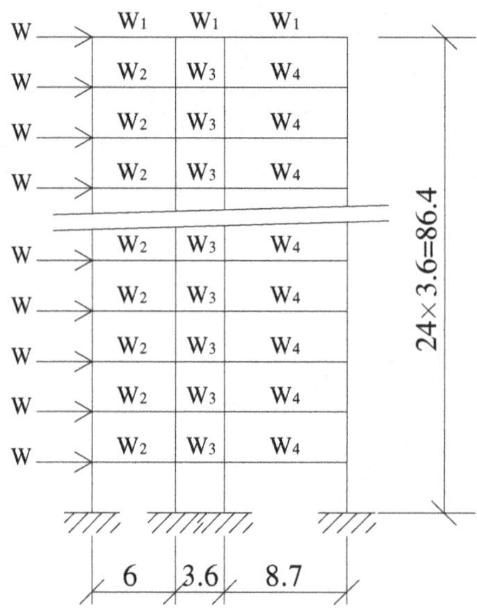


**Fig. 4.18** A 3-bay and 24-storey frame structure

A 3-bay and 24-storey frame structure, shown in Fig. 4.18, is an actual project. It has three spans of 6 m (20 ft), 3.6 m (12 ft), 8.7 m (28 ft) respectively. The optimal model is only one load case: the wind loads, w = 25629.86 N; and the distributed loads, $w_1$ = 4380 N/m; $w_2$ = 6360 N/m; $w_3$ = 10901 N/m; $w_4$ = 5950 N/m. Saka and Kameshki [18] optimized this frame structure with genetic algorithm and got the result (in weigh) of 100,000.72 kg in accordance with the code of BS5990 and AISC2001. Camp Charles V [20] did it with ant colony algorithm and got the result (in weigh) of 114,099.20 kg.

To make this optimization with practical significance of engineering application, this section does the optimization with GSO in accordance with the code of GB50017-2003 (Chinese code). The material density is 7850 kg/m$^3$, the modulus of elasticity is 206 GPa and Poisson's ratio is 0.27. The members are subjected to stress limitations of $[\sigma] = \pm$ 230 MPa. The model is subjected to the displacement constraints: the maximum horizontal displacement is 0.2448 m (1/500 the total floors), the maximum layer-relative displacement is 0.009 m (1/400 floors). 168 members are divided into 20 groups as shown in Table 4.18. Considering the property of H type steel (GB/T11263-200, Chinese code), HW type steel (group 29 in H type steel) is chosen as the steel frame columns, and HM type steel (group 62 in H type steel) is chosen as the steel frame beams. GSO is taken to make the optimization with these constraints, noted by GSO-2. Besides, The result of another constraint condition, with the maximum layer-relative displacement of 0.012 m (1/300 floors), is noted by GSO-1.

As the inflexible searching mechanism that HPSO requires all the initial population particles are workable and needs to calculate all the particles at each iteration. HPSO need too much computing time to work well in this 3-bay and 24-storey frame optimization. Thus, only GSO is taken to make this optimization. Besides, in order to make the optimization run well and save computing time, the upper boundary value of the variables is offered to the initialization particles. The optimal results with the different constraints are listed in Table 4.18. It is worthy noticing that the results of GSO-1 and GSO-2 seems to be quite different from the results of Saka [18] and Camp [20], the reason is the different constraints of different design code, such as the GB50017-2003 (Chinese), BS5990 (UK) and AISC 2001 (UAS). Another reason for the different results of GSO is the different selection of cross-section. Overall, as Fig. 4.8 shows, the GSO algorithm has a good convergence rate in this 3-bay and 24-storey frame optimization and can find the optimal solution in 100 iterations.

**Table 4.18** Comparison of optimal designs for the 3-bay and 24-storey frame structure

| Variables | Members | Variables | Members |
|---|---|---|---|
| $A_1$ | 1~23$^{th}$ storey & 1,3$^{rd}$ bay beams | $A_{11}$ | 19-21$^{st}$ storey side column |
| $A_2$ | 24$^{th}$ storey & 1,3$^{rd}$ bay beams | $A_{12}$ | 22-24$^{th}$ storey side column |
| $A_3$ | 1~23$^{th}$ storey & 2$^{nd}$ bay beams | $A_{13}$ | 1-3$^{rd}$ storey interior column |
| $A_4$ | 24$^{th}$ storey & 2$^{nd}$ bay beams | $A_{14}$ | 4-6$^{th}$ storey interior column |
| $A_5$ | 1-3$^{rd}$ storey side column | $A_{15}$ | 7-9$^{h}$ storey interior column |
| $A_6$ | 4-6$^{th}$ storey side column | $A_{16}$ | 10-12$^{th}$ storey interior column |
| $A_7$ | 7-9$^{h}$ storey side column | $A_{17}$ | 13-15$^{th}$ storey interior column |
| $A_8$ | 10-12$^{th}$ storey side column | $A_{18}$ | 16-18$^{th}$ storey interior column |
| $A_9$ | 13-15$^{th}$ storey side column | $A_{19}$ | 19-21$^{st}$ storey interior column |
| $A_{10}$ | 16-18$^{th}$ storey side column | $A_{20}$ | 22-24$^{th}$ storey interior column |

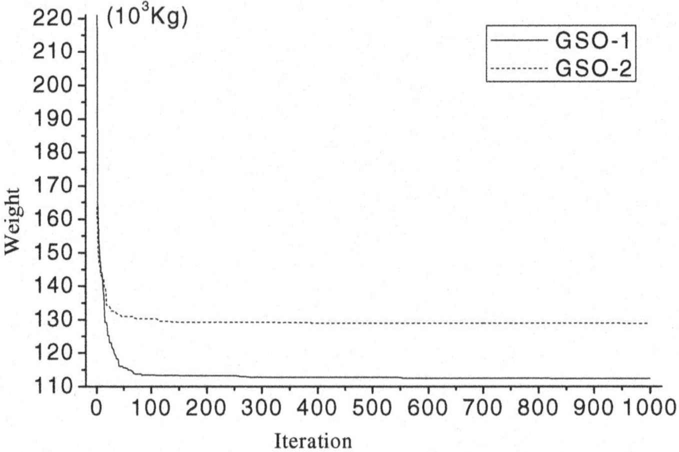


**Fig. 4.19** Convergence rates for the 3-bay and 24-storey frame structure (case 2)

**Table 4.19** Comparison of optimal designs for the 3-bay and 24-storey frame structure

| Variables | Camp [20] (ACO) | Saka [18] (GA) | GSO-1 | GSO-2 |
|---|---|---|---|---|
| $A_1$ | W30×90 | 838×292×194 UC | 850×300a | 850×300a |
| $A_2$ | W8×8 | 305×102×25 UC | 200×100b | 300×150a |
| $A_3$ | W25×250 | 457×191×82 UC | 850×300a | 800×300b |
| $A_4$ | W8×21 | 305×102×25 UC | 100×50 | 125×60 |
| $A_5$ | W14×145 | 305×305×198 UC | 400×400d | 500×500c |
| $A_6$ | W14×132 | 356×368×129 UC | 350×350d | 400×400b |
| $A_7$ | W14×132 | 305×305×97 UC | 350×350a | 400×400b |
| $A_8$ | W14×132 | 356×368×129 UC | 350×350a | 400×400c |
| $A_9$ | W14×68 | 305×305×97 UC | 400×400a | 350×350b |
| $A_{10}$ | W14×53 | 203×203×71 UC | 350×350a | 350×350b |
| $A_{11}$ | W14×43 | 305×305×118 UC | 300×300a | 300×300b |
| $A_{12}$ | W14×43 | 152×152×23 UC | 300×300a | 350×350b |
| $A_{13}$ | W14×145 | 305×305×137 UC | 400×400b | 500×500a |
| $A_{14}$ | W14×145 | 305×305×198 UC | 350×350b | 500×500a |
| $A_{15}$ | W14×145 | 356×368×202 UC | 350×350d | 500×500a |
| $A_{16}$ | W14×120 | 356×368×129 UC | 350×350b | 400×400b |
| $A_{17}$ | W14×90 | 356×368×129 UC | 300×300b | 400×400b |
| $A_{18}$ | W14×61 | 356×368×153 UC | 300×300a | 400×400a |
| $A_{19}$ | W14×30 | 203×203×60 UC | 250×250b | 400×400a |
| $A_{20}$ | W14×26 | 254×254×89 UC | 100×100 | 125×125 |
| Weight (kg) | 100 000.72 | 114 099.20 | 112404.65 | 128923.63 |

## 4.5 Conclusions

GSO is a novel optimization algorithm, based on animal searching behavior and group living theory. In this chapter, the principle of the GSO algorithm and its implementation method on truss structure and frame structure design is presented in detail.

The efficiency of the GSO algorithm presented in this paper is tested for optimum design of five pin connected structures and three frame structures. The truss structure optimal results show that the GSO algorithm converges much quickly than the PSO and the PSOPC while its convergence rate is at the same level with that of HPSO. The frame structure results show that the GSO spends less times than HPSO to get the optimal solutions as GSO need less analysis times. Which means GSO may be effectively used to optimal structural design if possibly improved. In particular, the GSO algorithm may outperform the other three algorithms when the variable dimension is bigger.

The GSO algorithm is conceptually simple and easy to implement than other three algorithms. In most of tested cases, the performance of GSO is not sensitive to some parameters such as maximum pursuit angle. These features make it particularly attractive for sophisticate practical engineering problems.

The optimum designs of truss structures of this chapter are based on continuous variables, while the practical engineering tasks are discrete variables. The efficiency of GSO needs to be proved on discrete variable structures. Three frame structures are used to test the efficiency and robust property of GSO. In addition, the improvement of the performance of GSO algorithm will also be the subject of forward investigation work.

## References

1. Kaveh, A., Shojaee, S.: Optimal design of scissor-link foldable structures using ant colony optimization algorithm. Computer-Aided Civil and Infrastructure Engineering 22(1), 56–64 (2007)
2. Dorigo, M., Di Caro, G., Gambardella, L.: Ant algorithms for discrete optimization. Artificial Life 5(3), 137–172 (1999)
3. Kenndy, J., Eberhart, R.C.: Particle swarm optimization. In: Proceedings of the 1995 IEEE International Conference on Neural Networks, Piscataway, NJ, USA, pp. 1942–1948 (1995)
4. Li, L.J., Ren, F.M., Liu, F., Wu, Q.H.: An improved particle swarm optimization method and its application in civil engineering. In: The 8th International Conference on Com-putation and Structures Technology, Las Palmas de Gran Canaria, Spain (2006) paper 42
5. He, S., Prempain, Wu, Q.H.: An improved particle swarm optimizer for mechanical design optimization problems. Engineering Optimization 36(5), 585–605 (2004)
6. Perez, R.E., Behdinan, K.: Particle swarm approach for structural design optimization. Computers and Structures 85(19-20), 1579–1588 (2007)

7. Li, L.J., Ren, F.M., Liu, F.: Structure optimization with discrete variables based on heuristic particle swarm optimal algorithm. In: Proceedings of the 3rd International Conference on Steel and Composite Structures, pp. 1017–1021. Taylor & Francis, Manchester (2007)
8. Li, L.J., Huang, Z.B., Liu, F., Wu, Q.H.: A heuristic particle swarm optimizer for optimization of pin connected structures. Computers and Structures 85(7-8), 340–349 (2007)
9. Barnard, C.J., Sibly, R.M.: Producers and scroungers: a general model and its application to captive flocks of house sparrows. Animal Behavior 29(2), 543–550 (1981)
10. Li, L.J., Liu, F., Xu, X.T., Liu, F.: The group search optimizer and its application to truss structure design. Advances in Structural Engineering 13(1), 43–51 (2010)
11. Liu, F., Xu, X.T., Li, L.J.: The group search optimizer and its application on truss structure design. In: The 4th International Conference on Natural Computation, Jinan, China, pp. 688–692 (2008)
12. He, S., Wu, Q.H., Saunder, J.R.: A novel group search optimizer inspired by animal behaviour ecology. In: 2006 IEEE Congress on Evolutionary Computation, Vancouver, BC, Canada, pp. 4415–4421 (2006)
13. Carpenter, R.H.S.: Eye movements. Macmilan, London (1991)
14. Liversedge, S.P., Findley, J.M.: Saccadic eye movements and cognition. Trends in Cognitive Sciences 4(1), 6–14 (2000)
15. Couzin, I.D., Krause, J., Franks, N.R., Levin, S.A.: Effective leadership and decision-making in animal groups on the move. Nature 433(7025), 513–516 (2005)
16. O'Brien, W.J., Evans, B.I., Howick, G.L.: A new view of the predation cycle of a planktivorous fish, white crappie (pomoxis annularis). Canadian Journal of Fisheries and Aquatic Sciences 43, 1894–1899 (1986)
17. Dixon, A.F.G.: An experimental study of the searching behavior of the predatory coccinellid beetle adalia decempunctata. Journal of Animal Ecology 28, 259–281 (1959)
18. Saka, M.P., Kameshki, E.S.: Optimum design of unbraced rigid frames. Computer & Structures 69, 433–442 (1998)
19. Kameshki, E.S., Saka, M.P.: Optimum design of nonlinear steel frames with semirigid connections using a genetic algorithm. Computer and Structures 79, 1593–1604 (2001)
20. Charles, V.C., Barron, J.B., Scott, P.S.: Design of steel frames using ant colony optimization. Structure Engineering ASCE 131(3), 369–379 (2005)
21. Chuang, W.S.: A PSO-SA hybrid searching algorithm for optimization of structures. Master's Thesis of National Taiwan University, 116–126 (2007)
22. Zhu, C.Y., Guo, P.F., et al.: Improving of genetic algorithm and its application to discrete topology optimization for frames. Journal of Mechanical Strength 27(1), 61–65 (2005)
23. Camp, C., Pezeshk, S., Cao, G.: Optimized design of two-dimensional structures using a genetic algorithm. Journal of Structural Engineering ASCE 124(5), 551–559 (1998)

# Chapter 5
# Improvements and Applications of Group Search Optimizer in Structural Optimal Design

**Abstract.** This chapter introduces the improvement and application of swarm algorithm named Group Search Optimizer (GSO) in civil structure optimization design. GSO is a new and robust stochastic searching optimizer, as it is based on PS (produce and scrounger) model and the animal scanning mechanisms. An improved group search optimizer named IGSO was presented based on harmony search mechanism and GSO. The implementation of IGSO for different optimal purposes is presented in detail, including the application of IGSO to truss structure shape optimal design, to truss structure dynamic optimal design, to truss structure topology optimization design. Different truss structures are used to test the GSO and IGSO in structural shape optimization, dynamic optimization and topology optimization.

## 5.1 Introduction

In the last 30 years, a great attention has been paid to structural optimization, since material consumption is one of the most important factors influencing building construction. The optimal design and research on structures is always the core of the researchers' and engineers' work. There is great and considerable economic significance to propose a new optimal structure, a new structure type, or to apply optimal design in the practical project successfully. However, many practical engineering optimal problems are very complex and hard to solve by the traditional optimal algorithms [1-3]. Thanks to the development of computer technology and random algorithms (random researching algorithms), there is a researching trend that this newborn random algorithms has been taking place in the traditional optimization theory. Since 1990s, evolutionary algorithms (EA) [4], such as genetic algorithms (GA) [5], evolutionary programming (EP) [6] and evolution strategies (ES) [7] have become more attractive, because they do not require conventional mathematical assumptions and thus possess better global search abilities than the conventional optimization algorithms. For example, GA has been applied for structural optimization problems [8]. Recently, swarm intelligence algorithm is considered one of the most important random algorithms [9-11], and group search optimizer (GSO) [12] is considered the most robust one of them. GSO is widely applied to structural optimal design which is still the most common optimization

L. Li & F. Liu: Group Search Optimization for Applications in Structural Design, ALO 9, pp. 97–159.
springerlink.com 

problem, such as the cross-section optimal design [13]. Based on the research results that we have achieved about swarm intelligence optimization algorithm in the past five years, we try to introduce the GSO and its effectiveness. In this chapter, the authors focus on improving the algorithm and developing its application in the complex structural optimal design, such as topology optimization, dynamic optimal design and shape optimization.

## 5.2 Group Search Optimizer

The group search optimizer algorithm was firstly proposed by He et al [12]. It is based on the biological Producer-Scrounger (PS) model [14], which assumes group members search either for 'finding' (producer) or for 'joining' (scrounger) opportunities. Animal scanning mechanisms (e.g., vision) are incorporated to develop the GSO algorithm. GSO also employs 'rangers' which perform random walks to avoid entrapment in local minima.

The population of the GSO algorithm is called a group and each individual in the population is called a member. In an n-dimensional search space, the $i_{th}$ member at the $k_{th}$ searching bout (iteration), has a current position $X_i^k \in R^n$, a head angle $\varphi_i^k = (\varphi_{i1}^k, \ldots, \varphi_{i(n-1)}^k) \in R^{n-1}$ and a head direction $D_i^k(\varphi_i^k) = (d_{i1}^k, \ldots, d_{in}^k) \in R^n$ which can be calculated from $\varphi_i^k$ via a Polar to Cartesian coordinates transformation:

$$
\begin{aligned}
d_{i1}^k &= \prod_{p=1}^{n-1} \cos(\varphi_{ip}^k) \\
d_{ij}^k &= \sin(\varphi_{i(j-1)}^k) \cdot \prod_{p=i}^{n-1} \cos(\varphi_{ip}^k) \\
d_{in}^k &= \sin(\varphi_{i(n-1)}^k)
\end{aligned}
\tag{5.1}
$$

In GSO, a group consists of three kinds of members: producers, scroungers whose behaviors are based on the PS model, and rangers who perform random walk motions. Recently, Couzin [15] suggested that the larger the group, the smaller the proportion of informed individuals needed to guide the group with better accuracy. Therefore, for accuracy and convenience of computation, the PS model is simplified by assuming that there is only one producer at each searching bout and the remaining members are scroungers and rangers. It is also assumed that the producer, the scroungers and the rangers do not differ in their relevant phenotypic characteristics. Therefore, they can switch between the three roles. At each iteration, a group member, located in the most promising area, conferring the best fitness value, is chosen as the producer. The producer's scanning field of vision is generalized to a n-dimensional space, which is characterized by maximum pursuit angle $\theta_{max} \in R^{n-1}$ and maximum pursuit distance $l_{max} \in R^1$ as illustrated in a 3D space [16] in Fig. 5.1.

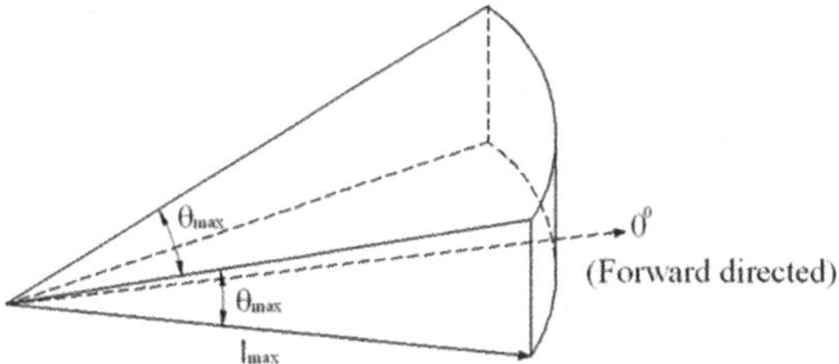


**Fig. 5.1** Scanning field in 3D space

In the GSO algorithm, at the $k_{th}$ iteration the producer $X_p$ behaves as follows:

(1) The producer will scan at zero degree and then scan laterally by randomly sampling three points in the scanning field: one point at zero degree:

$$X_z = X_p^k + r_1 l_{\max} D_p^k(\varphi^k) \tag{5.2}$$

one point in the left hand side hypercube:

$$X_l = X_p^k + r_1 l_{\max} D_p^k(\varphi^k - r_2\theta_{\max}/2) \tag{5.3}$$

and one point in the right hand side hypercube:

$$X_r = X_p^k + r_1 l_{\max} D_p^k(\varphi^k + r_2\theta_{\max}/2) \tag{5.4}$$

where $r_1 \in R^1$ is a normally distributed random number with mean 0 and standard deviation 1 and $r_2 \in R^{n-1}$ is a random sequence in the range (0, 1).

(2) The producer will then find the best point with the best resource (fitness value). If the best point has a better resource than its current position, then it will fly to this point. Otherwise it will stay in its current position and turn its head to a new angle:

$$\varphi^{k+1} = \varphi^k + r_2\alpha_{\max} \tag{5.5}$$

where $\alpha_{\max}$ is the maximum turning angle.

(3) If the producer cannot find a better area after $a$ iterations, it will turn its head back to zero degree:

$$\varphi^{k+a} = \varphi^{k} \tag{5.6}$$

where $a$ is a constant.

At the $k_{th}$ iteration, the area copying behavior of the $i_{th}$ scrounger can be modeled as a random walk towards the producer:

$$X_i^{k+1} = X_i^k + r_3(X_p^k - X_i^k) \tag{5.7}$$

where $r_3 \in R^n$ is a uniform random sequence in the range (0, 1).

Besides the producer and the scroungers, a small number of rangers have been also introduced into our GSO algorithm. Random walks, which are thought to be the most efficient searching method for randomly distributed resources, are employed by rangers. If the $i_{th}$ group member is selected as a ranger, at the $k_{th}$ iteration, firstly, it generates a random head angle $\varphi_i$:

$$\varphi^{k+1} = \varphi^{k} + r_2\alpha_{\max} \tag{5.8}$$

where $\alpha_{\max}$ is the maximum turning angle; and then, it chooses a random distance:

$$l_i = a \cdot r_1 l_{\max} \tag{5.9}$$

and move to the new point:

$$X_i^{k+1} = X_i^k + l_i D_i^k(\varphi^{k+1}) \tag{5.10}$$

To maximize their chances of finding resources, the GSO algorithm employs the fly-back mechanism [11] to handle the problem specified constraints: When the optimization process starts, the members of the group search the solution in the feasible. If any member moves into the infeasible region, it will be forced to move back to the previous position to guarantee a feasible solution. The pseudocode for GSO is listed in Table 5.1.

**Table 5.1** Pseudocode for the GSO algorithm

```
Set k = 0;
Randomly initialize positions X_i and head angles φ_i of all members;
FOR (each member i in the group)
  WHILE (the constraints are violated)
    Randomly re-generate the current member Xi
  END WHILE
END FOR
WHILE (the termination conditions are not met)
  FOR (each members i in the group)
  Calculate fitness: Calculate the fitness value of current member: f(X_i)
  Choose producer: Find the producer X_p of the group;
    Perform producing: 1) The producer will scan at zero degree and then scan laterally by
                          randomly sampling three points in the scanning field using
                          equations (5.2) to (5.4).
                       2) If any point violates the constraints, it will be replaced by the
                          producer's previous position.
                       3) Find the best point with the best resource (fitness value). If the
                          best point has a better resource than its current position, then it
                          will fly to this point. Otherwise it will stay in its current position
                          and turn its head to a new angle using equation (5.5).
                       4) If the producer can not find a better one area after a iterations, it
                          will turn its head back to zero degree using equation (5.6);
    Perform scrounging: Randomly select 80% from the rest members to perform
                          scrounging using equation (5.7);
    Perform ranging:    For the rest members, they will perform ranging:
                       1). Generate a random head angle using equation (5.8);
                       2). Choose a random distance l_i from the Gauss distribution
                       using equation (5.9) and move to the new point using equation
                       (5.10);
    Check feasibility: Check whether each member of the group violates the constraints.
                          If it does, it will move back to the previous position to guarantee
                          a feasible solution.
  END FOR
  Set k = k + 1;
END WHILE
```

## 5.3 The Improvement of GSO (IGSO)

The GSO algorithm just needs to find the best member as producer who is followed by the other members except the rangers. We can sort the members by the fitness without constraints, and then check the constraints by the sequence. For this, it is not needed to check all the members' constraints to find the producer, so GSO will save much computational time. More details about GSO can be found in reference [12] and [13]. Besides, there are two improvements for GSO as follow:

(1) Based on the main idea of harmony search algorithm [11-17], a harmony memory, the storage of best position of the members in the iteration history, is incorporated into the GSO algorithm. When the *i*th dimension of the *S* is out of the boundary of the design space, the algorithm will choose a member randomly from the harmony memory, and use the corresponding dimension to replace the *i*th dimension of the member. This mechanism can make full use of the personal best optimum information.

(2) Many studies about structure optimization show that the optimum (local and global) often locate in or nearly to the boundary of the design space, however, the members always fly out of the design space frequently during the optimization. In order to avoid the members' flying out, a new searching mechanism, called *adhering to the boundary*, is utilized to improve the failed searching behavior in this paper. With this searching mechanism, the members which fly out of the design space will have an opportunity (10% in improved GSO) to adhere to the boundary, so they can search from the boundary to the inside design space in the next iteration. Apparently, the mechanism can raise convergence speed. Even if the optimum locates in the boundary, the algorithm will converge to the optimum quickly. The mechanism can be expressed as:

for $\{X\}=[x_1,x_2,\ldots,x_i,\ldots,x_n]^T$

if $x_i>x_{imax}$ (or $x_i<x_{imin}$)

then $x_i=x_{imax}$ (or $x_i=x_{imin}$)

that is $\{X\}=[x_1,x_2,\ldots,x_{imax}$ (or $X_{imin}$)$,\ldots,x_n]^T$

where $\{X\}$is a member's position, $x_i$ is the *i*th dimension of the member, the $x_{imax}$ is the upper boundary of *i*th dimension of the design space.

The previous improved methods for GSO is named IGSO algorithm. A pseudo code for IGSO algorithm is listed in Table 5.2.

**Table 5.2** Pseudo code for IGSO algorithm

```
Set k = 0, iter=0;
WHILE (iter=0 or all the members violate the constraints)
        Randomly initialize positions and head angles of all members;
        Modify variables with;
        Calculate fitness with constraints, iter=1;
        Build the harmony searching population;
END WHILE
Choose producer: Find the producer of the group;
WHILE (the termination conditions are not met)
        FOR (each members i in the group)
            Perform producing:
                    1) The producer will scan at zero degree and then scan laterally by randomly
                       sampling three points in the scanning field, and modify topology
                       variables with one of the two methods.
                    2) If any point violates the constraints, it will be replaced by the producer's
                       previous position.
                    3) Find the best point with the best resource (fitness value). If the best point
                       has a better resource than its current position, then it will fly to this point.
                       Otherwise it will stay in its current position and turn its head to a new
                       angle.
                    4) If the producer can not find a better area after a iterations, it will turn its
                       head back to zero degree;
            Perform scrounging:
                       Randomly select 80% from the rest members to perform scrounging, and
                       modify variables with one of the two methods;
            Perform ranging:
                       For the rest members, they will perform ranging, and modify variables with
                       one of the two methods;
            Check feasibility:
                    Check whether each member of the group violates the constraints. If it does, use
                    the two new searching mechanisms to refresh the members' position.
        END FOR
        Calculate fitness without constraints and Sort: Calculate the fitness value of current
        member:
        Check the constraints by the sequence and Choose producer: Find the producer;
        Update the harmony searching population;
        Set k = k + 1;
    END WHILE
```

## 5.4 The Application of the IGSO on Truss Shape Optimization

With the development of social economy, people demand more of the civil structures, such as the internal space of building, the outer shape and so on. However, what kind of structure or shape is best and how to propose a optimal shape for a civil structure always troubles structure designers. Compared with truss cross-section optimization design, truss shape optimization adds shape variables and makes the optimization problem more complicated. With the discrete cross-sectional variables and the shape variables, the optimization problem is essentially a nonlinear optimization problem mixed various discrete variables. The traditional optimal theory, such as mathematical programming and the full stress criteria, can't handle with this highly non-linear discrete optimization well. In this section, IGSO is taken to make truss shape optimizations, and the result of several numerical examples show the algorithm is robust to solve this kind of optimization problem.

### 5.4.1 Mathematical Models

The mathematical models for shape optimization of truss structures generally have two kinds of variables, cross-sectional variables of bar and shape variables of nodes' coordinate. In this section, the objective of truss shape optimization is still the lightest weight of the structure. It is worth mentioning that there are many optimization models contain only shape variables, when bars' cross section has been assumed. According to the literatures [18], the mathematical model of truss shape optimization is as follows,

$$
\begin{aligned}
&\min . Weight\left(A_i, C_j\right)=\sum_{i=1}^{N} \rho_i A_i L_i \\
&\qquad L_i=L_i\left(C_j\right) \\
&s.t. \quad g_i^{\sigma}=\left[\sigma_i\right]-\sigma_i \geq 0(i=1,2,\cdots,n) \\
&\qquad g_{jl}^{u}=\left[u_{jl}\right]-u_{jl} \geq 0(i=1,2,\cdots,n ; l=1,2,\cdots N) \\
&\qquad A_i \in s(i=1,2,\cdots,n) \text{ 或 } A_i \in\left[A_{\min}, A_{\max}\right]
\end{aligned} \tag{5.11}
$$

where $A_i$ represents the cross-section of the $i$th bar. $\rho_i$ represents the density of the $i$th bar. $C_j$ represents the coordination of $j$th node. [σ] and [δ] represent the allowable stress and the allowable displacement respectively. S is a set of discrete cross-section of bar. It is worthy noticing that it is needed to check geometric variability of the optimal structures. Because a larger range of shape variables' value is set, the optimization process will encounter some infeasible structures.

### 5.4.2 Numerical Examples

In this section, five structures commonly used in literature are selected as benchmark problems to test the GSO, HPSO and IGSO. All these trusses are analysed by the finite element method (FEM), and the whole calculating programmes are coded with Matlab, and all the optimal trusses are modelled and checked by Ansys program.

The examples given in the simulation studies include:

- a 52-bar spatial truss;
- a 10-bar planar truss;
- a 15-bar planar truss A;
- a 15-bar planar truss B;
- a 18-bar planar truss;
- a 40-bar planar truss;
- a 25-bar spatial truss structure A;
- a 25-bar spatial truss structure B;
- a 39-bar spatial truss.

All these frame structures are analysed by the finite element method (FEM). The GSO, HPSO are applied, respectively, to all these examples and the results are compared in order to evaluate the performance of the new algorithm IGSO. For all these four algorithms, the maximum number of iterations is limited to 1000 and the population size is set to at 50; the inertia weight $\omega$ decrease linearly from 0.9 to 0.4; and the value of acceleration constants $c_1$ and $c_2$ are set to be the same and equal to 0.8. The passive congregation coefficient $c_3$ is given as 0.6 for the HPSO algorithms. For the GSO algorithm and IGSO, 20% of the population is selected as rangers; the initial head angle $\varphi_0$ of each individual is set to be $\pi/4$. The constant a is given by round$(\sqrt{n+1})$. The maximum pursuit angle $\theta_{max}$ is $\pi/a^2$. The maximum turning angle $\alpha$ is set to be $\pi/2a^2$. The maximum pursuit distance $l_{max}$ is calculated from:

$$l_{max} = \| d_i^u - d_i^l \| = \sqrt{\sum_{i=1}^{n} (d_i^u - d_i^l)^2}$$

where $d_i^l$ and $d_i^u$ are the lower and upper bounds for the $i_{th}$ dimension.

### *(1) A 52-Bar Spatial Truss*

The 52-bar spatial truss structure, shown in Fig. 5.2, has previously been analysed by Chuang [20]. The material density is 7850 kg/m$^3$ and the modulus of elasticity is $2.1\times10^5$ MPa. All the members' area is assumed as 10 cm$^2$. Node 1 in z direction is subjected to displacement limitations of ± 0.01m. The shaping variables grouping and the relative boundary are given in this section as: 9 m $\le z_1 \le$14 m, 1 m $\le x_2 \le$7 m, 7 m $\le z_2 \le$10 m, 7 m $\le x_6 \le$13 m, 4 m $\le z_6 \le$6 m. When node 2 moves, nodes 3, 4 and 5 must move to make sure that the structure is symmetric in xy and yz plane. And so when node 6 moves, nodes 7 to 13 must move in the same time. Four work cases are considered:

Case 1, node 1 is acted by $P_1$, -3.0×105 N in y direction;
Case 2, node 1 to 13 are acted by $P_1$, $-0.3\times10^5$ N in y direction;

Case 3, node 1, $P_1$, -1.5×10$^5$ N and node 4, 5 are acted by $P_2$, 1.0×105 N in y direction;

Case 4, node 1 is acted by $P_1$,- 1.5×105 N, node 2, 3 and 4 are acted by $P_2$, -0.7×10$^5$ N in y direction.

The connection nodes of each bar, described as *Bar number(node 1, node 2)* , is given as follows: 1(1, 2), 2(1, 3), 3(1, 4), 4(1, 5), 5(2, 3), 6(3, 4), 7(4, 5), 8(5, 2), 9(2, 6), 10(3, 8), 11(4, 10), 12(5, 12), 13(2, 7), 14(3, 7), 15(3, 9), 16(4, 9), 17(4, 11), 18(5, 11), 19(5, 13), 20(2, 13), 21(6, 7), 22(7, 8), 23(8, 9), 24(9, 10), 25(10, 11), 26(11, 12), 27(12, 13), 28(13, 6), 29(6, 21), 30(7, 20), 31(8, 19), 32 (9, 18), 33(10, 17), 34 (11, 16), 35(12, 15), 36(13, 14), 37(6, 14), 38(6, 20), 39(8, 20), 40(8, 18), 41(10, 18), 42(10, 16), 43(12, 16), 44(12, 14), 45(7, 21), 46(7, 19), 47(9, 19), 48(9, 17), 49(11, 17), 50(11, 15), 51(13, 15), 52(13, 21).

The 52-bar spatial truss optimization problem is subjected to four work cases, with the fixed cross-sectional variables. As Table 5.3 listed, the IGSO and HPSO get the same optimal result. The GSO gets a worse optimal result. All the three algorithms can get the better optimal result than the one of literatures. As Fig. 5.10 shows, the convergence rate of GSO and IGSO is both faster than the one of HPSO in early iterations, and the one of IGSO is best. All of them can find the final optimal result with about 250 iterations. It is worthy noticing that the IGSO and GSO are not needed to check the constraints of all particles, but HPSO is needed. Because of this, the IGSO and GSO can save a lot of computing time. The convergence rate of IGSO for 52-bar truss is shown in Fig. 5.3 and the optimal structure with IGSO is shown in Fig. 5.3.

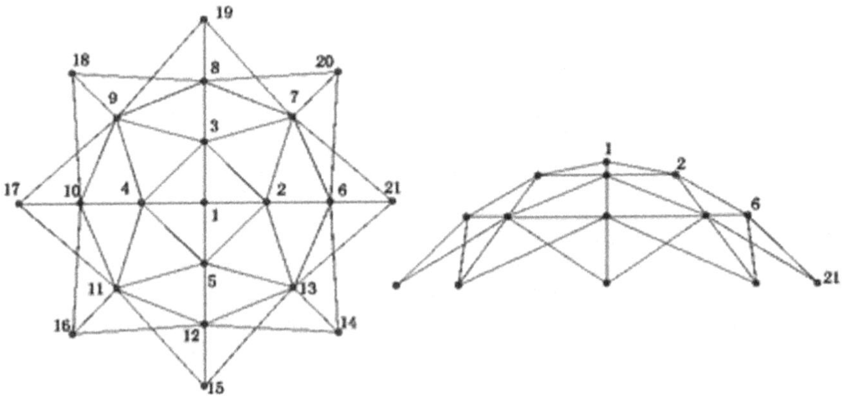


**Fig. 5.2** A 52-bar spatial truss

**Table 5.3** Comparison of the designs for the 52-bar spatial truss

| Variables | Wang [19] | PSO-SA [20] | HPSO | GSO | IGSO |
|---|---|---|---|---|---|
| $z_1$ | 9.620 | 9.000 | 9.000 | 9.007 | 9.000 |
| $x_2$ | 2.100 | 1.706 | 1.693 | 1.711 | 1.693 |
| $z_2$ | 7.410 | 7.403 | 7.407 | 7.401 | 7.407 |
| $x_6$ | 7.210 | 7.000 | 7.000 | 7.001 | 7.000 |
| $z_6$ | 4.080 | 4.000 | 4.000 | 4.009 | 4.000 |
| Weight (kg) | 3195.7559 | 3168.0756 | 3167.6756 | 3167.7056 | 3167.6756 |

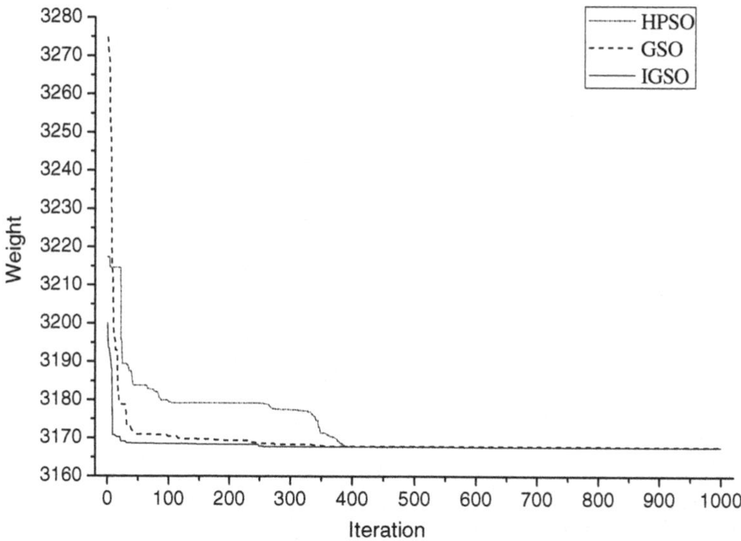


**Fig. 5.3** Convergence curves for the 52-bar spatial truss I

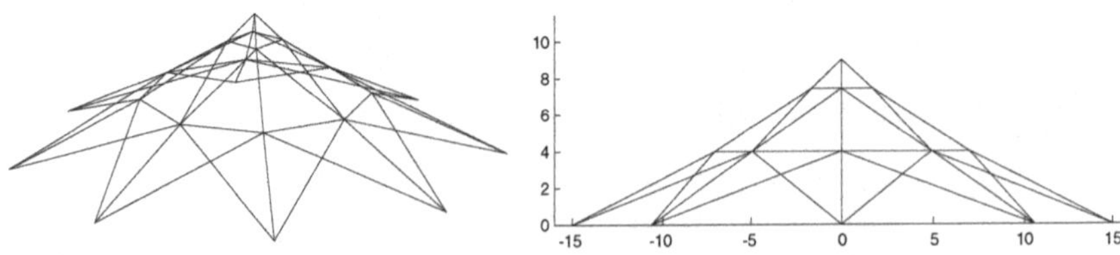

**Fig. 5.4** The optimized 52-bar spatial truss

### *(2) A 10-Bar Planar Truss*

The 10-bar truss structure, shown in Fig. 5.5, had been analyzed by Luo [20] as a shape optimization with dynamic constraints. But it is taken to check the algorithms as shape optimization without the dynamic constraints in this section. The material density is 7680 kg/m.$^3$ and the modulus of elasticity is 210 GPa. The members are subjected to the stress limits of ±100 MPa. There are 10 size variables and 5 shape variables. Discrete values considered for this example are taken from the set D[0.001, 0.01] (m$^2$) with the interval of 0.0005 m$^2$. The shape variables group and the relative boundary are given in this section as: $-2.5 \le y_1 \le 2.5$, $0 \le x_2 \le 2.5$, $-2.5 \le y_2 \le 2.5$, $2.5 \le x_3 \le 5$, $-2.5 \le y_3 \le 2.5$ (m). The vertical downward load of 100 kips on node 4 and 9 are considered.

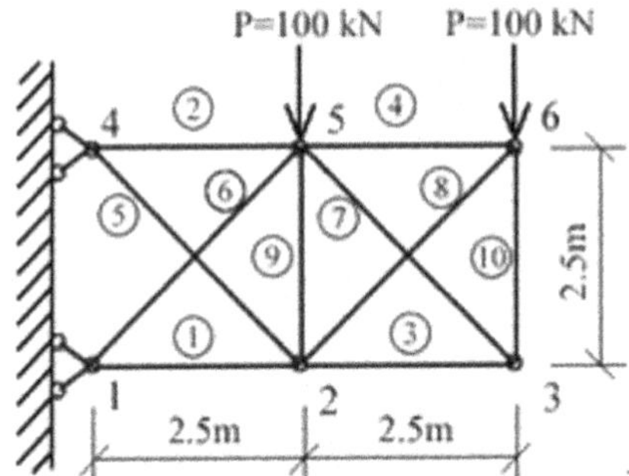


**Fig. 5.5** A 10-bar planar truss

Table 5.4 lists the optimal results. Fig. 5.6 shows the convergence rate. It is shown from table 5.11 that the optimization results of GSO and IGSO is superior to HPSO's. And as Fig. 5.6 shown, GSO and IGSO both have a good convergence rate in the early 100 iterations, and approach the global optimal result with 500 iterations. But the HPSO can't escape from the local optimal solution, which means HPSO is poorer in the ability of global searching than GSO and IGSO. Compared the GSO and IGSO in convergence rate, GSO is good as IGSO; but in the optimal result, GSO is better than IGSO. The optimal structure with GSO is shown in Fig. 5.7.

**Table 5.4** Comparison of the optimal results for the10-bar planar truss

| Variables | HPSO | GSO | IGSO |
|---|---|---|---|
| $A_1$ | 0.0055 | 0.0020 | 0.0020 |
| $A_2$ | 0.0025 | 0.0020 | 0.0020 |
| $A_3$ | 0.0010 | 0.0010 | 0.0010 |
| $A_4$ | 0.0010 | 0.0010 | 0.0010 |
| $A_5$ | 0.0010 | 0.0010 | 0.0010 |
| $A_6$ | 0.0010 | 0.0010 | 0.0010 |
| $A_7$ | 0.0010 | 0.0010 | 0.0010 |
| $A_8$ | 0.0010 | 0.0010 | 0.0010 |
| $A_9$ | 0.0010 | 0.0010 | 0.0010 |
| $A_{10}$ | 0.0070 | 0.0010 | 0.0010 |
| y1 | -0.846 | -0.438 | -0.438 |
| x2 | 0.100 | 2.476 | 2.421 |
| y2 | -0.746 | 0.332 | 0.293 |
| x3 | 4.996 | 3.663 | 3.713 |
| y3 | 2.400 | 0.979 | 1.034 |
| Weight (kg) | 275.5959 | 235.1748 | 235.2951 |

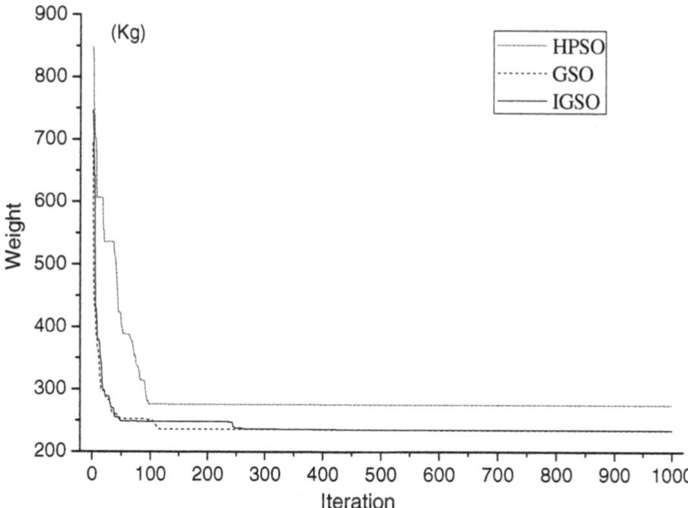


**Fig. 5.6** Convergence curves for the 10-bar planar truss

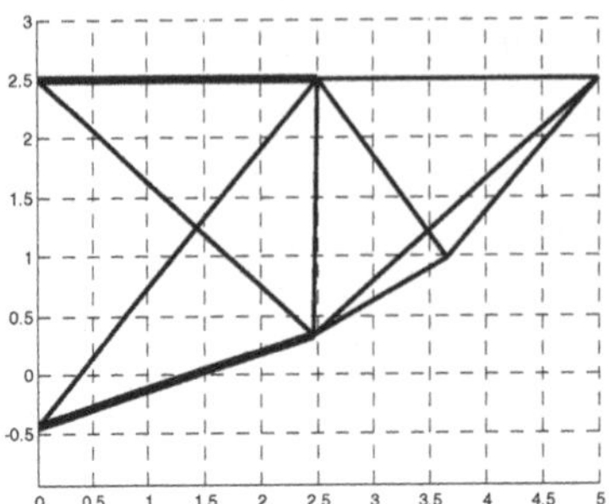

Fig. 5.7 The optimized 10-bar planar truss

### *(3) A 15-Bar Planar Truss A*

The 15-bar truss structure, shown in Fig. 5.8, has previously been analysed by many researchers, such as Wu [18] and Tang [21]. The material density is $2.768\times10^3$ kg/m$^3$ and the modulus of elasticity is 6.89×104 MPa. The members are subjected to the stress limits of ±172.25 MPa. There are 15 size variables and 8 shape variables ($x_2 = x_6$, $x_3 = x_7$, $y_2$, $y_3$, $y_4$, $y_6$, $y_7$, $y_8$) in this example. Discrete values considered for this example are taken from the set D={0.072, 0.091, 0.112, 0.142, 0.174, 0.185, 0.224, 0.284, 0.348, 0.615, 0.697, 0.757, 0.860, 0.960, 1.138, 1.382, 1.740, 1.806, 2.020, 2.300, 2.460, 3.100, 3.840, 4.240, 4.640, 5.500, 6.000, 7.000, 8.600, 9.219, 11.077, 12.374} ($\times10^{-3}$ m$^2$). The shape variables group and the relative boundary are $2.54\le x_2\le3.556$, $5.588\le x_3\le6.60$, $2.54\le y_2\le3.556$, $2.54\le y_2\le 3.556$, $1.27\le y_4\le2.286$, $-0.508\le y_6\le0.508$, $-0.508\le y_7\le 0.508$, $-0.508 \le y_8\le 1.524$ (m). Only one case is considered: Case 1, node 8 is acted by P, -44.45 kN in y direction. Table 5.5 give the optimal result. Fig. 5.9 shows the convergence of the optimal algorithm.

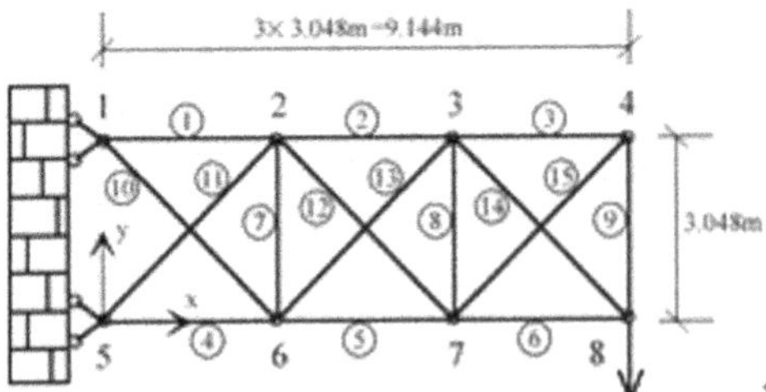


Fig. 5.8 A 15-bar planar truss structure A

As Table 5.5 listed, the GSO, HPSO and IGSO achieve the almost same optimal results as Tang [21], and in which the one of GSO is better than the IGSO's and HPSO's.

**Table 5.5** Comparison of the optimal results for the15-bar planar truss A

| Variables | Wu [18] | Tang [21] | HPSO | GSO | IGSO |
|---|---|---|---|---|---|
| $A_1$ | 1.174 | 0.954 | 1.081 | 1.081 | 1.081 |
| $A_2$ | 0.954 | 0.954 | 0.539 | 0.954 | 0.954 |
| $A_3$ | 0.440 | 0.287 | 0.270 | 0.141 | 0.220 |
| $A_4$ | 1.333 | 1.081 | 1.081 | 1.081 | 1.081 |
| $A_5$ | 0.954 | 0.440 | 0.954 | 0.539 | 0.539 |
| $A_6$ | 0.174 | 0.141 | 0.141 | 0.347 | 0.270 |
| $A_7$ | 0.440 | 0.111 | 0.270 | 0.111 | 0.111 |
| $A_8$ | 0.440 | 0.111 | 0.111 | 0.174 | 0.220 |
| $A_9$ | 1.081 | 1.764 | 1.333 | 0.141 | 0.220 |
| $A_{10}$ | 1.333 | 0.539 | 0.287 | 0.174 | 0.220 |
| $A_{11}$ | 0.174 | 0.220 | 0.539 | 0.220 | 0.220 |
| $A_{12}$ | 0.174 | 0.111 | 0.347 | 0.141 | 0.111 |
| $A_{13}$ | 0.347 | 0.440 | 0.270 | 0.539 | 0.539 |
| $A_{14}$ | 0.347 | 0.141 | 0.174 | 0.347 | 0.270 |
| $A_{15}$ | 0.440 | 0.287 | 0.270 | 0.141 | 0.220 |
| $x_2$ | 123.189 | 119.186 | 100.000 | 107.371 | 112.822 |
| $x_3$ | 231.595 | 251.751 | 235.4 | 234.566 | 231.804 |
| $y_2$ | 107.189 | 132.931 | 124.3 | 121.529 | 121.762 |
| $y_3$ | 119.175 | 122.394 | 139.1 | 111.338 | 112.118 |
| $y_4$ | 60.462 | 50.815 | 50.700 | 50.699 | 52.863 |
| $y_6$ | -16.728 | 4.380 | -8.300 | 14.417 | 15.658 |
| $y_7$ | 15.565 | 1.059 | -3.500 | -4.714 | -3.175 |
| $y_8$ | 36.645 | 51.030 | 50.600 | 47.350 | 52.860 |
| Weight (kg) | 120.528 | 79.078 | 84.6106 | 76.5114 | 77.0853 |

As Fig. 5.9 shows, all the three algorithms can converge to the global optimal solution with 1000 iterations. The optimal structures obtained by GSO and IGSO are shown in Fig. 5.10 respectively.

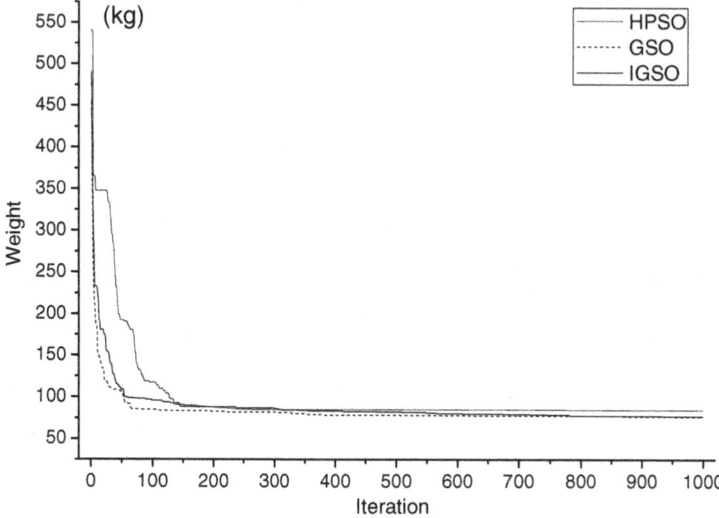


**Fig. 5.9** Convergence curves for the 15-bar planar truss A

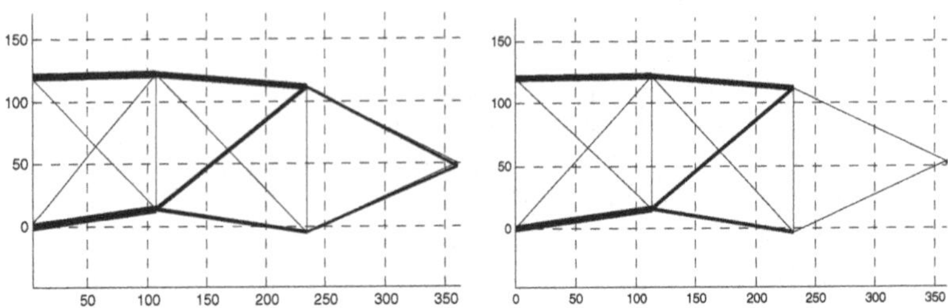

Fig. 5.10 The optimized 15-bar planar truss A

### (4) A 15-Bar Planar Truss B

The 15-bar truss structure B, shown in Fig. 5.11, has previously been analysed by Sun [21]. Young's modulus is specified as $1.379\times10^7$ N/cm$^2$, and the material density as 0.2768 N/cm$^3$. The stress limit is ±17243.5 N/cm$^2$ for all members, and the displacement limit is ±5.08 cm in y direction of node 5. Considering of the symmetry of 15-bar, there are 7 groups of size variable, as follows: $A_1$(11, 12, 13, 14), $A_2$(1, 2), $A_3$(5, 6), $A_4$(7, 8), $A_5$(9, 10), $A_6$(3, 4), $A_7$(15). The shape variable group and the relative boundary are given in this section as: $-1245 \leq x_3=-x_7\leq -25$, $-1245\leq x_4=-x_8\leq-25$, $100\leq y_4=y_8\leq1600$, $200\leq y_6\leq2200$(cm). Discrete values considered for this example are taken from the set D= {0.0645,0.645, 0.968, 1.29, 1.613, 1.935, 3.266, 4.194, 5.161, 6.452, 22.581, 25.806, 29.032, 32.258, 35.484, 38.71, 41.935, 45.161, 48.387, 51.613, 54.839, 58.064, 61.29, 64.516, 67.742, 70.968, 71.613, 72.258, 72.903, 73.548, 74.193, 77.419, 80.645, 83.871, 87.097, 90.322, 93.548, 96.774, 103.226, 116.129, 129.032, 141.935, 154.838, 167.742, 180.645} (cm$^2$). This model has only one load case: Case 1, node 4, 6, 8 all have a $-4.45\times10^5$ N vertical load. Table 5.6 give the optimal result. Fig. 5.12 shows the convergence of the optimal algorithm.

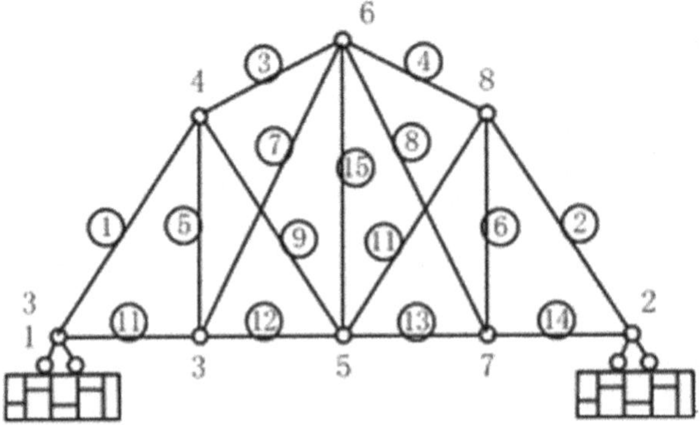

Fig. 5.11 A 15-bar planar truss B

**Table 5.6** Comparison of the designs for the 22-bar spatial truss structure

| Variables | Sun [21] | HPSO | GSO | IGSO |
|---|---|---|---|---|
| $A_1$ | 22.581 | 0.0645 | 0.968 | 0.0645 |
| $A_2$ | 45.161 | 45.161 | 45.161 | 45.161 |
| $A_3$ | 0.645 | 0.0645 | 0.645 | 0.645 |
| $A_4$ | 0.645 | 0.0645 | 0.645 | 0.0645 |
| $A_5$ | 0.645 | 0.968 | 0.0645 | 0.0645 |
| $A_6$ | 25.806 | 22.581 | 22.581 | 22.581 |
| $A_7$ | 0.645 | 0.645 | 0.0645 | 0.0645 |
| $x_3$ | -829.08 | -1174.396 | -1049.826 | -1175.06 |
| $x_4$ | -1234.44 | -1222.54 | -1212.417 | -1220.748 |
| $y_4$ | 312.49 | 100.029 | 122.05 | 102.879 |
| $y_6$ | 773.91 | 945.877 | 929.511 | 948.897 |
| Weight (N) | 4798.5 | 2228.2 | 2289.8 | 2161.9 |

The result of Sun [21] is obtained with traditional optimization method. There are not the boundary values of shape variables in the original optimization model. In this section, a suitable boundary of shape variables is given to make the optimization run well. As Table 5.6 lists, the result 2161.9 N of IGSO is the best of the four algorithms. As the convergence curve in Fig. 5.12 shows, IGSO is superior to the GSO and HPSO in convergence rate. And all the three algorithms can approach the global optimal solution well with 1000 iterations. The optimal truss obtained by IGSO is shown in Fig. 5.13.

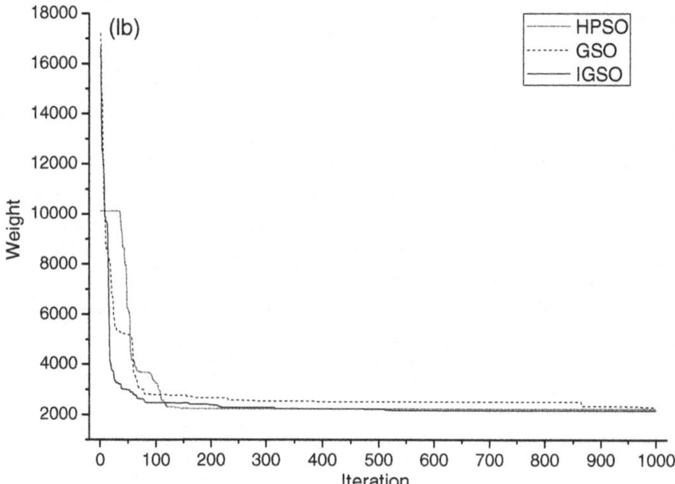


**Fig. 5.12** Convergence curves for the 15-bar B planar truss

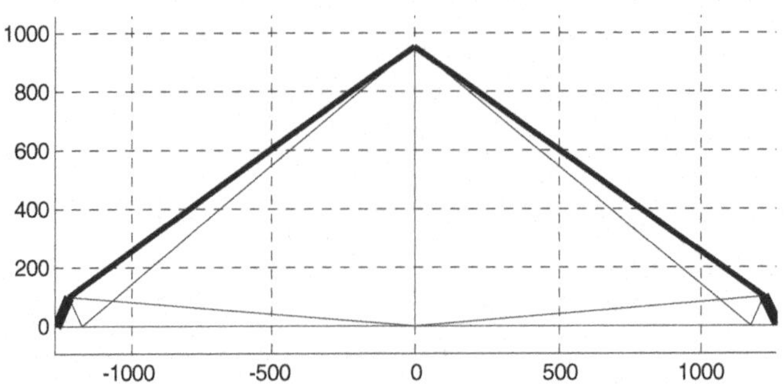


**Fig. 5.13** The optimized 15-bar planar truss B

### *(5) A 18-Bar Planar Truss*

The 18-bar truss structure, shown in Fig. 5.14, has previously been analysed by many researchers, such as Rajeev [23], Hasan [24] and Kaveh [25]. The material density is 0.1 lb/in$^3$ and the modulus of elasticity is 10,000 ksi. The members are subjected to the stress limits of ±20 ksi. There are 25 members, which are divided into 8 groups, as follows: ($A_1$) 1, 2, 3, 4; ($A_2$) 6, 7, 8, 9; ($A_3$) 11, 12, 13, 14; ($A_4$) 15, 16, 17, 18. Discrete values considered for this example are taken from the set D= [2, 20] (in$^2$) with the interval of 0.25 in$^2$. The shape variable group and the relative boundary are given in this section as: $-225 \leqslant y_3=y_5=y_7=y_9 \leqslant 245$, $775 \leqslant x_5 \leqslant 1225$, $525 \leqslant x_5 \leqslant 975$, $275 \leqslant x_7 \leqslant 725$, $25 \leqslant x_9 \leqslant 475$ (in). Only one case is considered: Case 1, node 1, 2, 4, 6, 8 are acted by P, -20 kips in y direction. This model has not any displacement limit of node, but considers the Euler stress constraint. The formula of calculating the Euler stress is given, as follows:

$$\sigma_i = \frac{\alpha E A_i}{L_i^2} \text{ (ksi)} \tag{5.16}$$

Where α represents Euler buckling coefficient. $L_i$ represents the length of i$^{th}$ bar. E represents the modulus of the material. $A_i$ represents the area of i$^{th}$ bar. The value of α is 4 in this 18-bar planar truss optimization.

Table 5.7 give the optimal result. Fig. 5.15 shows the convergence of the optimal algorithm.

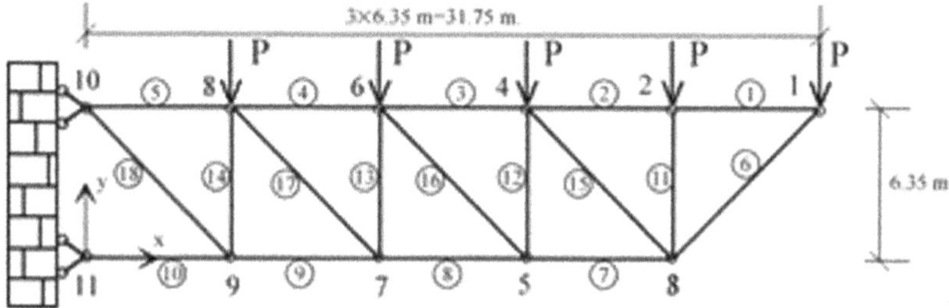


**Fig. 5.14** A 18-bar planar truss

With lots of research to the 18-bar truss optimization, we know its solution space is complicated and full of the infeasible solution area. With the rigor searching rules, HPSO can't work well in this optimization, because it takes a lot of time to find 50 initial particles which all meet the constraints. Only the GSO and IGSO are taken to make this optimization. As Table 5.7 lists, both the GSO and IGSO have a good accuracy in this 18-bar truss optimization. It can be seen from Fig. 5.15 both the two algorithms have a good convergence rate. The rate of IGSO is faster in early iterations, but the final result of it is inferior to GSO's with 1000 iterations. The optimal structures obtained by GSO and IGSO are shown in Fig. 5.16 respectively.

**Table 5.7** Comparison of the optimal results for the18-bar planar truss

| Variables | Rajeev [23] | Hasan [24] | Kaveh [25] | GSO | IGSO |
|---|---|---|---|---|---|
| $A_1$ | 12.50 | 12.50 | 13.00 | 12.25 | 12.25 |
| $A_2$ | 16.25 | 18.25 | 18.25 | 18.25 | 18.25 |
| $A_3$ | 8.00 | 5.50 | 5.50 | 4.75 | 4.75 |
| $A_4$ | 4.00 | 3.75 | 3.00 | 4.25 | 4.25 |
| $X_3$ | 891.90 | 933 | 913 | 916.9 | 920.812 |
| $Y_3$ | 145.30 | 188 | 182 | 191.971 | 170.912 |
| $X_5$ | 610.60 | 658 | 648 | 654.224 | 640.506 |
| $Y_5$ | 118.20 | 148 | 152 | 156.1 | 139.87 |
| $X_7$ | 385.40 | 422 | 417 | 423.5 | 409.416 |
| $Y_7$ | 72.50 | 100 | 103 | 102.571 | 91.774 |
| $X_9$ | 184.40 | 205 | 204 | 207.519 | 198.775 |
| $Y_9$ | 23.40 | 32 | 39 | 28.579 | 29.504 |
| Weight (lb) | 4616.800 | 4574.280 | 4566.210 | 4538.7676 | 4553.116 |

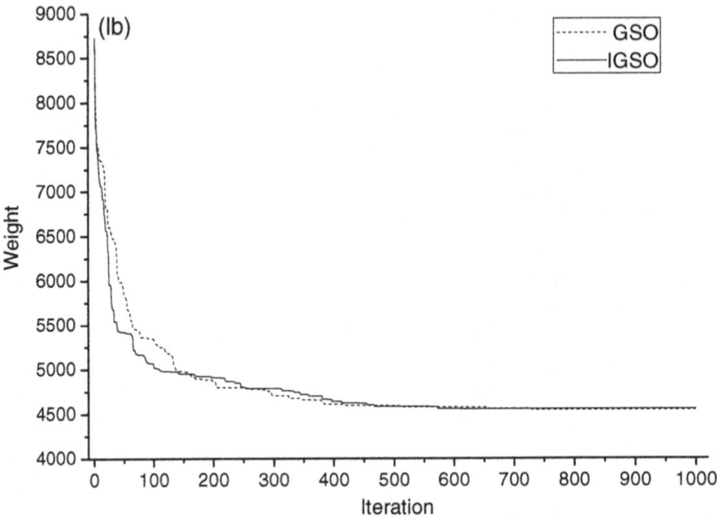


**Fig. 5.15** Convergence curves for the 18-bar planar truss

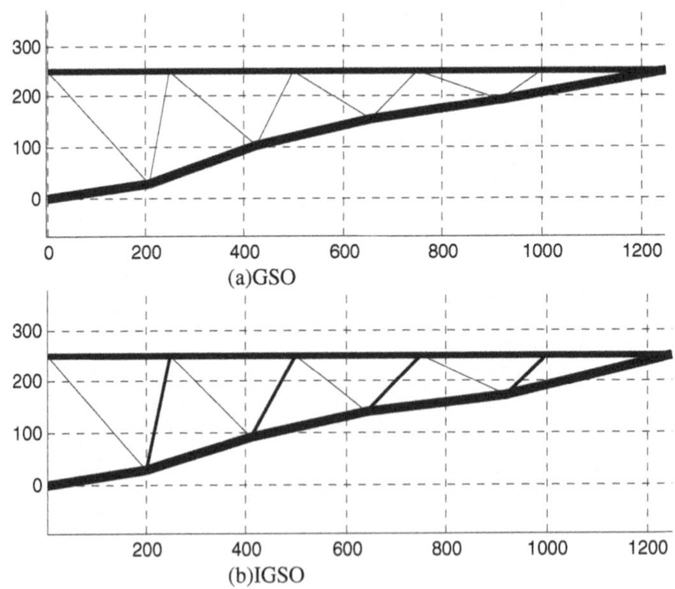


**Fig. 5.16** The optimized 18-bar planar truss

### *(6) A 40-Bar Planar Truss*

The 40-bar planar truss structure [26], shown in Fig. 5.17, is taken to test the three algorithms. The material density is 7800 kg/m$^3$ (7.8 t/m$^3$) and the modulus of elasticity is 196.13 GPa (2×10$^7$ t/m$^2$). The stress limits of the members are subjected to ±156.91 MPa (16000 t/m$^2$). Node 4 and 5 in y directions are subjected to the displacement limits of ±0.035 m (1/600 span). There are 40 members, which fall into 19 groups, as follows: ($A_1$) 1, 7; ($A_2$) 2, 6; ($A_3$) 3, 5; ($A_4$) 4; ($A_5$) 8, 14; ($A_6$) 9, 13; ($A_7$) 10, 12; ($A_{10}$) 16, 21; ($A_{11}$) 17, 20; ($A_{12}$) 18, 19; ($A_{13}$) 23, 36; ($A_{14}$) 24, 35; ($A_{15}$) 25, 34; ($A_{16}$) 26, 33; ($A_{17}$) 30, 29; ($A_{18}$) 31, 28; ($A_{19}$) 32, 27. Discrete values considered for this example are taken from the set D=[0.001, 0.7] (m$^2$) with the interval of 0.001 m$^2$. Considering the symmetry, the shape variable group and the relative boundary are given in this paper as, $1 \le y_9 = y_{16} \le 5$, $1 \le y_{10} = y_{15} \le 5$, $1 \le y_{11} = y_{14} \le 5$, $1 \le y_{12} = y_{13} \le 5$ (m). Only one case is considered: Case 1, node 2, 3, 4, 5, 6 and 7 are acted by P=10t in y direction.

Table 5.8 gives the optimal result and Fig. 5.18 shows the convergence rate.

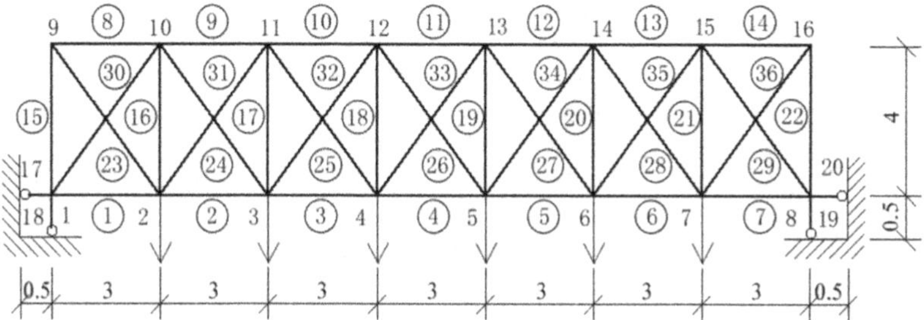

**Fig. 5.17** A 40-bar planar truss

**Table 5.8** Comparison of the optimal results for the 40-bar planar truss

| Variables | HPSO | GSO | IGSO |
|---|---|---|---|
| $A_1$ | 0.0055 | 0.0015 | 0.001 |
| $A_2$ | 0.001 | 0.001 | 0.001 |
| $A_3$ | 0.0105 | 0.001 | 0.001 |
| $A_4$ | 0.001 | 0.001 | 0.001 |
| $A_5$ | 0.001 | 0.001 | 0.0025 |
| $A_6$ | 0.0025 | 0.003 | 0.0035 |
| $A_7$ | 0.003 | 0.0035 | 0.004 |
| $A_8$ | 0.0245 | 0.0035 | 0.004 |
| $A_9$ | 0.0025 | 0.001 | 0.001 |
| $A_{10}$ | 0.001 | 0.001 | 0.001 |
| $A_{11}$ | 0.001 | 0.001 | 0.001 |
| $A_{12}$ | 0.001 | 0.001 | 0.001 |
| $A_{13}$ | 0.001 | 0.001 | 0.001 |
| $A_{14}$ | 0.001 | 0.001 | 0.001 |
| $A_{15}$ | 0.0015 | 0.001 | 0.001 |
| $A_{16}$ | 0.005 | 0.001 | 0.001 |
| $A_{17}$ | 0.004 | 0.0025 | 0.003 |
| $A_{18}$ | 0.001 | 0.001 | 0.001 |
| $A_{19}$ | 0.001 | 0.001 | 0.001 |
| $y_9$ | 1.006 | 1.069 | 1.021 |
| $Y_{10}$ | 2.791 | 2.307 | 1.894 |
| $Y_{11}$ | 3.541 | 2.851 | 2.355 |
| $Y_{12}$ | 3.396 | 3.287 | 2.954 |
| Weight (kg) | 3653.0103 | 2080.6733 | 2165.3412 |

The original model of this 40-bar truss comes from Qian [26]. As table 8 lists, the optimal result of GSO and IGSO is superior to the one of HPSO, and as Fig. 5.25 shows, GSO and IGSO is almost good at the global searching, and can find the final solution with less 200 iterations. Compared with convergence rate and result, IGSO is inferior to the GSO and HPSO. The optimal truss obtained by IGSO is shown in Fig. 5.19.

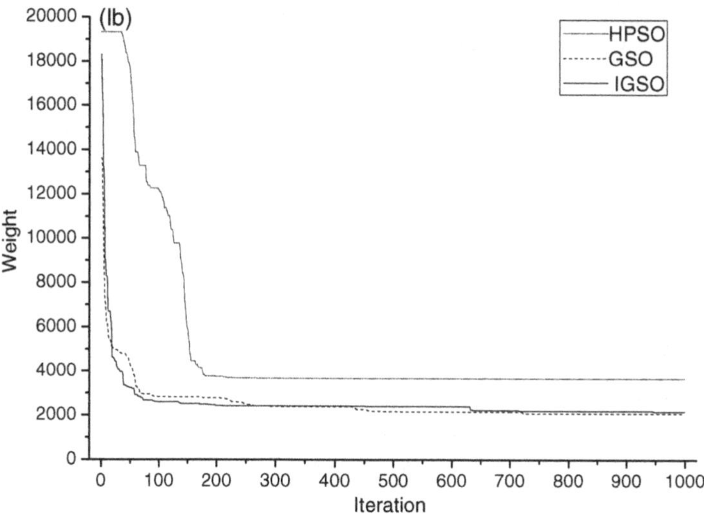


**Fig. 5.18** Convergence curves for the 40-bar planar truss

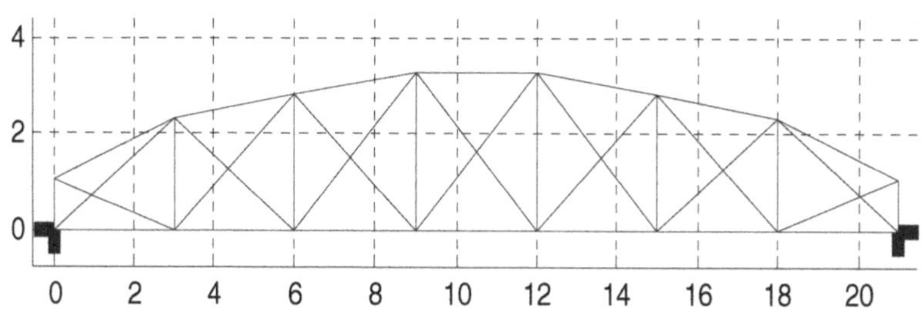


**Fig. 5.19** The optimized 40-bar planar truss

### *(7) A 25-Bar Spatial Truss Structure A*

This 25-bar spatial truss example is a sizing and geometry optimization model in Wu [18] and Kaveh [25], and the geometry is shown in Fig. 5.20. Young's modulus is specified as 10000 ksi, and the material density as 0.1 lb/in$^3$. The stress limit is ±40 ksi for all members and the displacement limit of node 1~6 in all three directions are ±0.35 in. Because of the symmetry of the truss, there are 8 sizing variables as table 5.16 listed and 5 shape variables. Discrete values considered for this example are taken from the set D={ 0.1, 0.2, 0.3, 0.4, 0.5, 0.6, 0.7, 0.8, 0.9, 1.0, 1.1, 1.2, 1.3, 1.4, 1.5, 1.6, 1.7, 1.8, 1.9, 2.0, 2.1, 2.2, 2.3, 2.4, 2.5, 2.6, 2.8, 3.0, 3.2, 3.4} (in$^2$). With the symmetry, the shaping variables grouping and the relative boundary are given in this paper as, $20 \leqslant x_4=x_5=-x_3=-x_6 \leqslant 60$, $40 \leqslant y_3=y_4=-y_5=-y_6 \leqslant 80$, $90 \leqslant z_3=z_4=z_5=z_6 \leqslant 130$, $40 \leqslant x_8=x_9=-x_7=-x_{10} \leqslant 80$, $100 \leqslant y_7=y_8=-y_9=-y_{10} \leqslant 140$ (in.). The bar grouping detail and load case is listed on Table 5.9 and Table 5.10

respectively.. The optimal results are shown in Table 5.11. The convergence rate is shown in Fig. 5.21. The optimal structure is shown in Fig. 5.22.

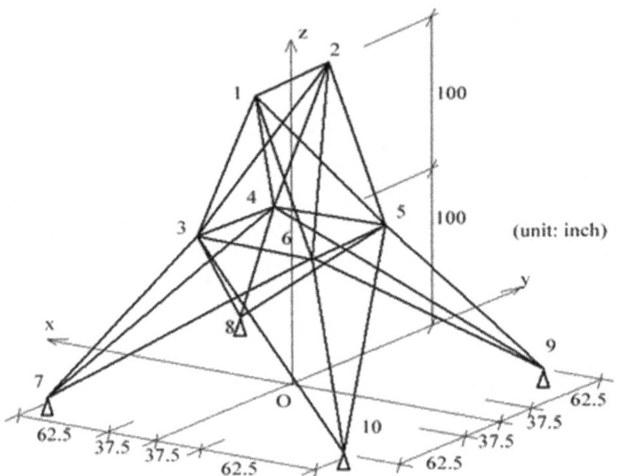


**Fig. 5.20** A 25-bar spatial truss structure

**Table 5.9** The bar grouping detail of 25-bar Spatial Truss A

| Variables | Members | Connective of Node |
|---|---|---|
| $A_1$ | 1 | (1,2) |
| $A_2$ | 2,3,4,5 | (1,4),(2,3),(1,5),(2,6) |
| $A_3$ | 6,7,8,9 | (2,5),(2,4),(1,3),(1,6) |
| $A_4$ | 10,11 | (3,6),(4,5) |
| $A_5$ | 12,13 | (3,4),(5,6) |
| $A_6$ | 14,15,16,17 | (3,10),(6,7),(4,9),(5,8) |
| $A_7$ | 18,19,20,21 | (3,8),(4,7),(6,9),(5,10) |
| $A_8$ | 22,23,24,25 | (3,7),(4,8),(5,9),(6,10) |

**Table 5.10** Load cases for the 25-bar spatial truss structure

| Variables | $F_x$ (kips) | $F_y$ (kips) | $F_z$ (kips) |
|---|---|---|---|
| 1 | 1.0 | -10.0 | -10.0 |
| 2 | 0.0 | -10.0 | -10.0 |
| 3 | 0.5 | 0.0 | 0.0 |
| 6 | 0.6 | 0.0 | 0.0 |

**Table 5.11** Comparison of the optimal results for the 25-bar spatial truss A

| Variables | Wu [18] | Kaveh [25] | HPSO | GSO | IGSO |
|---|---|---|---|---|---|
| $A_1$ | 0.1 | 0.1 | 0.1 | 0.1 | 0.1 |
| $A_2$ | 0.2 | 0.1 | 0.2 | 0.1 | 0.1 |
| $A_3$ | 1.1 | 1.1 | 1.0 | 1.0 | 1.0 |
| $A_4$ | 0.2 | 0.1 | 0.1 | 0.1 | 0.1 |
| $A_5$ | 0.3 | 0.1 | 0.1 | 0.1 | 0.1 |
| $A_6$ | 0.1 | 0.1 | 0.1 | 0.1 | 0.1 |
| $A_7$ | 0.2 | 0.1 | 0.1 | 0.2 | 0.2 |
| $A_8$ | 0.9 | 1.0 | 1.0 | 0.9 | 0.9 |
| $z_1$ | 41.070 | 36.230 | 34.084 | 32.149 | 31.754 |
| $x_2$ | 53.470 | 58.560 | 50.650 | 52.742 | 53.335 |
| $z_2$ | 124.600 | 115.590 | 129.978 | 128.230 | 127.058 |
| $x_6$ | 50.800 | 46.460 | 47.838 | 42.401 | 42.485 |
| $z_6$ | 131.480 | 127.950 | 129.584 | 132.603 | 132.979 |
| Weight (lb) | 136.1977 | 124.0015 | 124.6025 | 121.3684 | 121.4642 |

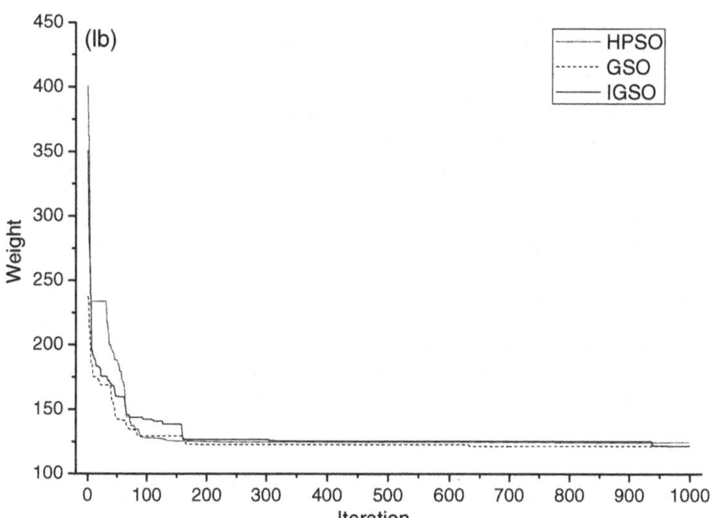


**Fig. 5.21** Convergence curves for the 25-bar spatial truss A

As can be seen from Table 5.11, all the three optimization algorithms used in this section have achieved results better than Wu [18] and Kaveh [25] with 1000 iterations. From which the 121.3684 lb result of GSO is the best. As Fig. 5.22 shows, all the three algorithms have a good convergence speed. They can get the global optimal solution with less than 200 iterations. The optimal structure obtained by GSO is shown in Fig. 5.22.

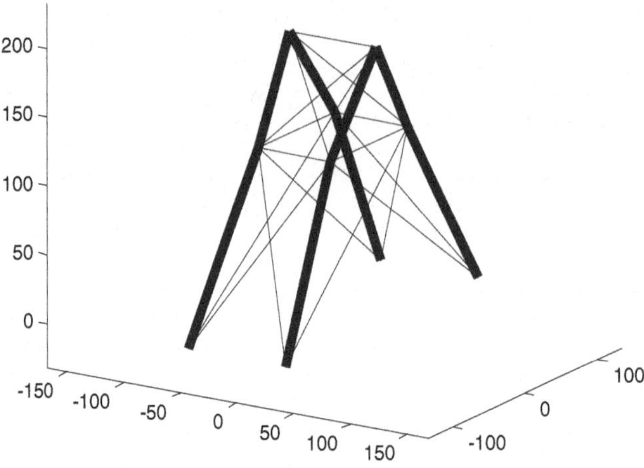


**Fig. 5.22** The optimized 25-bar spatial truss A

### *(8) The 25-Bar Spatial Truss Structure B*

The 25-bar spatial truss B is the continuation of 25-bar spatial truss A with additional Euler buckling stress constraint. The Euler buckling stress is calculated from equation (5.16), and its value of α is 8 in this example. Considering the Euler buckling stress constraint, parts of the original solution space become infeasible, so that the optimization is nonlinear and complicated than the truss model A.

**Table 5.12** Comparison of the optimal results for the 25-bar spatial truss B

| Variables | Wu [18] | Chuang [20] | HPSO | GSO | IGSO |
|---|---|---|---|---|---|
| $A_1$ | 0.9 | 0.1 | 0.5 | 0.1 | 0.1 |
| $A_2$ | 0.8 | 0.9 | 0.9 | 0.9 | 0.9 |
| $A_3$ | 1.3 | 1.2 | 1.1 | 1.3 | 1.2 |
| $A_4$ | 0.5 | 0.1 | 0.1 | 0.1 | 0.1 |
| $A_5$ | 0.3 | 0.2 | 0.2 | 0.2 | 0.2 |
| $A_6$ | 0.6 | 0.3 | 0.4 | 0.3 | 0.3 |
| $A_7$ | 1.2 | 0.9 | 1 | 0.9 | 0.9 |
| $A_8$ | 1.6 | 1.2 | 1.4 | 1.1 | 1.2 |
| $z_1$ | 22.22 | 20.143 | 20 | 20.46 | 20.881 |
| $x_2$ | 49.01 | 52.235 | 47.526 | 53.437 | 51.852 |
| $z_2$ | 106.98 | 97.152 | 105.186 | 93.601 | 96.705 |
| $x_6$ | 44.60 | 40.000 | 40 | 23.451 | 40.048 |
| $z_6$ | 102.44 | 100.00 | 100 | 100.93 | 100.011 |
| Weight (lb) | 301.5968 | 226.0832 | 246.7083 | 223.5674 | 226.2017 |

As Table 5.12 lists, the GSO and IGSO achieve almost the same optimal results as Chuang [20]. As Fig. 5.23 shows, the GSO and IGSO both can converge to the global optimal solution with 500 iterations, but the HPSO converges to the local optimal solution. It is noticed that the HPSO need to calculate the constraints of every particle, but the GSO and IGSO does not need. The optimal structure obtained by GSO is shown in Fig. 5.24.

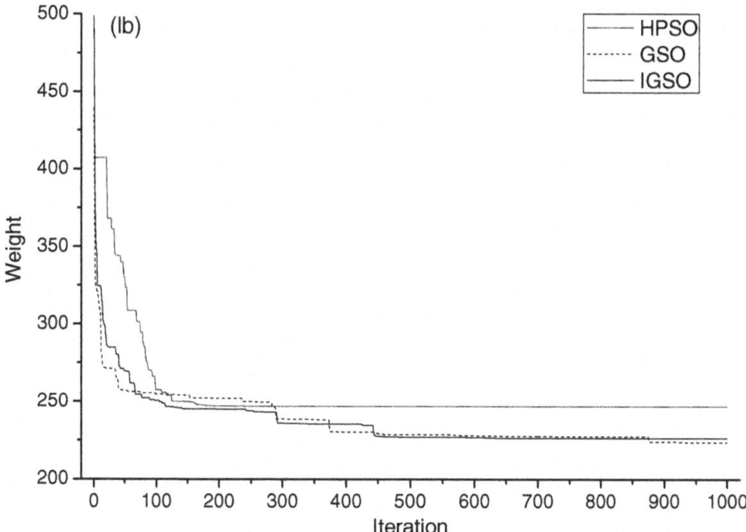


**Fig. 5.23** Convergence curves for the 25-bar spatial truss B

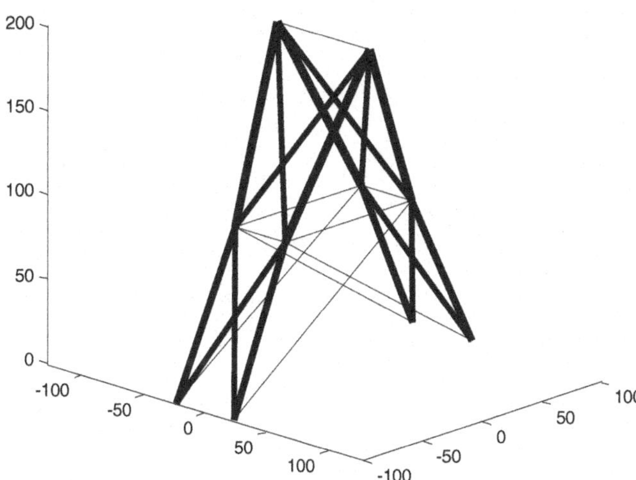


**Fig. 5.24** The optimized 25-bar spatial truss B

***(9) A 39-Bar Spatial Truss***

The 39-bar spatial truss structure shown in Fig. 5.25 had been studied by Thi [27] and PSO-SA [20]. The material density is $7.85\times10^{-6}$ kg/mm$^3$ and the modulus of elasticity is $2.1\times10^6$ N/mm$^2$. The stress limits of the members are subjected to ±150 N/mm$^2$. All nodes' coordination can be changed except node 1~3 and node 13~15. Considering the symmetry, the shape variable group and the relative boundary in Fig. 5.25 are given as: $500\le x_1\le 4000$, $1000\le x_2\le 5000$, $2000\le x_3\le 6000$, $1000\le x_{4\text{-}6}\le 4000$ (mm). Node 14 and 15 in y direction are subjected to the displacement limit of ±3 mm. 39 members, which are divided into 5 groups considering the symmetry are listed in Table 5.13. The initial coordinate of nodes is shown in Table 5.14. Discrete values of cross-section considered for this example are taken from the set D={112.0, 142.0, 174.0, 185.0, 185.0, 227.0, 267.0, 308.0, 328.0, 349.0, 379.0, 430.0, 480.0, 569.0, 582.0, 656.0, 691.0, 870.0, 903.0, 935.0, 940.0, 1010.0, 1150.0, 1190.0, 1220.0, 1230.0, 1510.0, 1550.0, 1920.0, 2120.0, 2270.0, 2320.0} (mm$^2$).

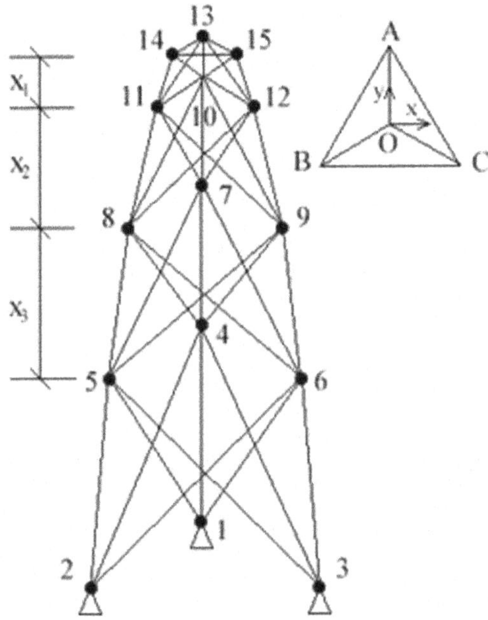


**Fig. 5.25** A 39-bar spatial truss structure

**Table 5.13** The category of 39-bar spatial truss's members

| Member | Connection nodes |
|---|---|
| 1~3 | (1, 4), (2, 5), (3, 6) |
| 4~6 | (4, 7), (5, 8), (6, 9) |
| 7~9 | (7, 10), (11, 14), (12, 15) |
| 10~12 | (10,13), (11,14), (12,15) |
| 13~39 | (2, 6), (3, 5), (5, 9), (6, 8), (8, 12), (9, 11), (11, 15), (12, 14), (3, 4), (1, 6), (6, 7), (4, 9), (9, 10), (7, 12), (12, 13), (10, 15), (2, 4), (1, 5), (5, 7), (4, 8), (8, 10), (7, 11), (11, 13), (10, 14), (13, 14), (14, 15), (13, 15) |

**Table 5.14** Node's coordinates of 39-bar spatial truss

| Node | Coordination (mm) | | | Node | Coordination (mm) | | |
|---|---|---|---|---|---|---|---|
| | x | y | z | | x | y | z |
| 1 | 0.0 | 4000.0 | 0.0 | 9 | 2361.3 | -1363.3 | 10877.9 |
| 2 | -3464.1 | -2000.0 | 0.0 | 10 | 0.0 | 1720.3 | 14441.4 |
| 3 | 3464.1 | -2000.0 | 0.0 | 11 | -1489.9 | -860.2 | 14441.4 |
| 4 | 0.0 | 3386.4 | 6345.3 | 12 | 1489.9 | -860.2 | 14441.4 |
| 5 | -2932.7 | -1693.0 | 6345.3 | 13 | 0.0 | 1120.0 | 16000.0 |
| 6 | 2932.7 | -1693.0 | 6345.3 | 14 | -970.0 | -560.0 | 16000.0 |
| 7 | 0.0 | 2726.5 | 10877.9 | 15 | 970.0 | -560.0 | 16000.0 |
| 8 | -2361.3 | -1363.3 | 10877.9 | / | / | / | / |

The optimal results are shown in Table 5.15. The convergence rate if the IGSO algorithm is shown in Fig. 5.26.

**Table 5.15** Comparison of the optimal results for the 39-bar spatial truss

| Variables | Thi [27] | PSO-SA [20] | HPSO | GSO | IGSO |
|---|---|---|---|---|---|
| $A_1$ | 1920 | 1550 | 1550 | 1550 | 1550 |
| $A_2$ | 1010 | 1230 | 1230 | 1230 | 1230 |
| $A_3$ | 656 | 870 | 903 | 870 | 903 |
| $A_4$ | 227 | 308 | 267 | 328 | 308 |
| $A_5$ | 227 | 227 | 227 | 227 | 227 |
| $X_1$ | 1705.71 | 1821.364 | 1814.5 | 1915.336 | 1901.615 |
| $X_2$ | 3563.70 | 3410.253 | 3539.9 | 3354.98 | 3444.49 |
| $X_3$ | 4524.76 | 4212.206 | 4194.2 | 4083.028 | 4251.895 |
| $X_4$ | 3361.76 | 3313.114 | 3337.7 | 3283.978 | 3322.667 |
| $X_5$ | 2727.62 | 2681.664 | 2714.9 | 2687.627 | 2690.444 |
| $X_6$ | 1689.01 | 1593.758 | 1590.9 | 1636.006 | 1628.618 |
| Weight (kg) | 725.7238 | 716.4761 | 716.5523 | 716.7290 | 716.5285 |

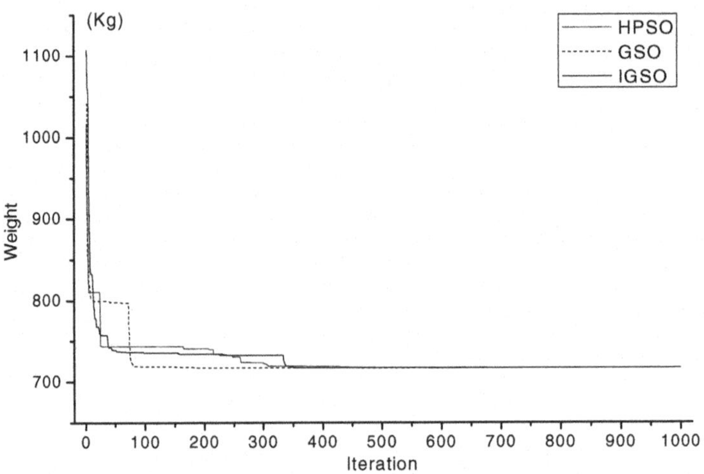


**Fig. 5.26** Convergence curves for the 39-bar spatial truss

As can be seen from Table 5.15, all the GSO, HPSO and IGSO can find the similar optimization results as the literatures. The 716.5285 kg result of IGSO is almost the same as the literature results of 716.4761 kg [20]. As Fig. 5.26 shows, all the three algorithms find the final optimal structure with 400 iterations, and IGSO is the best and fastest one. The optimal structure obtained by IGSO is shown in Fig. 5.27.

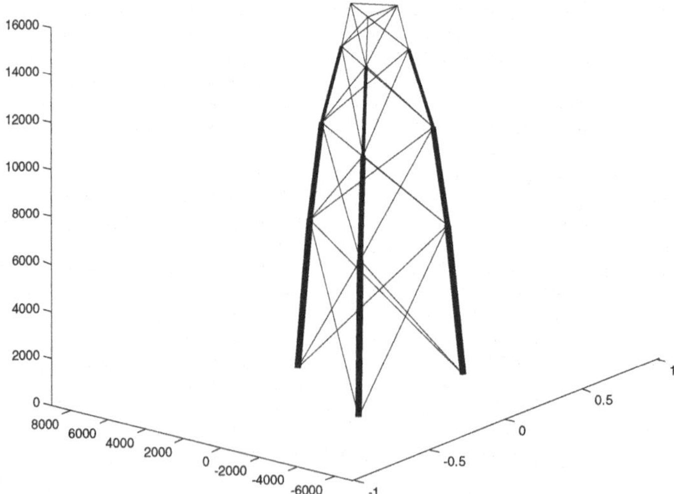


**Fig. 5.27** The optimized 39-bar spatial truss

## 5.5 The Application of IGSO in Truss Shape Optimization with Dynamic Constraint

Dynamic properties optimal design is an important research topic of structural seismic design. However, with the highly nonlinear objective function or constraint function for dynamic optimization model, the traditional optimizers, such as the distributed-parameter method [28], full stress criterion and mathematical programming [29-31], can't work well in the complex structure optimal design now. Thanks to the stochastic searching algorithms, people have found the shortcut to make this kind dynamic optimal design.

In this section, two types of optimization models, based on the natural frequencies constraints and the natural frequencies objective, were developed to deal with truss structure geometry optimization problems. A number of truss structure examples were calculated based on GSO or IGSO.

### *5.5.1 Two Kinds of Mathematical Model for Truss Dynamic Shape Optimization Problems*

In engineering practice, the main requirement of structure dynamic properties is a suitable natural frequency which can't too small and must far away to the frequency of outer force sources (avoiding the resonate action). And some other requirements are specifically to the vibration mode, structure damping, the mass distribution and stiffness distribution etc. Generally, the optimal models are established according two major purposes. Firstly, the minimum weight objective with frequency constraints. Secondly, the natural frequency objective with weight constraints. Both optimal models have some constraints of static analysis, such as the corresponding displacement, stress constraints and so on.

***(1) The Minimum Weight Optimal Model with Frequency Constraints (Model I)***

Generally, the minimum weight optimal model with frequency constraints can be defined as:

$$
\begin{aligned}
&\min Weight\left(A_i, C_j\right) = \sum_{i=1}^{N} \rho_i A_i L_i \\
&\omega_N\left(A_i, C_j\right), \qquad L_i = L_i\left(C_j\right) \\
&s.t.\begin{cases}
\qquad \omega_N \ge freqlimit \\
Or\ \ \omega_N \in \left[freq\min, freq\max\right] \\
Or\ \ \omega_N \notin \left[freq\min, freq\max\right] \\
\qquad \left|\sigma_i\right| \le [\sigma], i = 1, \cdots, n \\
\qquad \left|\delta_j\right| \le [\delta], j = 1, \cdots, m \\
\qquad A_i \in S = \left\{S_1 \quad \cdots \quad S_k\right\}
\end{cases}
\end{aligned}
\tag{5.17}
$$

where $A_i$ represents the cross-section of the ith bar. $C_j$ represents the coordination of jth node. $freq_{limit}$ is the minimum of $N$th natural frequency. $[freq_{min}, freq_{max}]$ is the allowable frequency interval of $N$th natural frequency. [σ] and [δ] represent the allowable stress and the allowable displacement respectively. $S$ is a set of discrete cross-section of bar.

***(2) The Maximum Natural Frequency with Maximum Weight Limit (Model II)***

Generally, the optimal model of the maximum natural frequency with weight constraints can be defined as:

$$\max .\omega_N\left(A_i, C_j\right)$$

$$weight = \sum_{i=1}^{N} \rho_i A_i L_i \ , \qquad L_i = L_i\left(C_j\right)$$

$$s.t.\begin{cases} weight \le \max Weightlimit \\ \left|\sigma_i\right| \le [\sigma] \, , i = 1, \cdots, n \\ \left|\delta_j\right| \le [\delta] \, , j = 1, \cdots, m \\ A_i \in S = \left\{S_1 \quad \cdots \quad S_k\right\} \end{cases} \tag{3.18}$$

where $A_i$ represents the cross-section of the ith bar. $C_j$ represents the coordination of jth node. *weightlimit* is the maximum of the whole structure weight. [σ] and [δ] represent the allowable stress and the allowable displacement respectively. S is a set of discrete cross-section of bar.

## *5.5.2 Numerical Examples*

In this section, five pin-connected structures commonly will be designed (modified the optimization problem of liberation) as benchmark problem to test the GSO and IGSO. The proposed algorithm is coded in Matlab language and ANSYS APDI language.

The examples given in the simulation studies include:

- a 40-bar planar truss structure in Model I as shown in Fig. 5.28;
- a 15-bar spatial truss structure in Model I as shown in Fig. 5.31;
- a 25-bar spatial truss structure in Model I as shown in Fig. 5.34;
- a 10-bar planar truss A structure in Model I as shown in Fig. 5.37;
- a 10-bar planar truss B structure in Model II .

All these frame structures are analysed by the finite element method (FEM). The GSO, IGSO are applied to all these examples respectively and the results are compared in order to evaluate the performance of the new algorithm. For the two algorithms, the maximum number of iterations is limited to 1000 and the population size is set to at 50. 20% of the population was selected as rangers. The initial head angle $\varphi_0$ of each individual is set to be $\pi/4$ . The constant $a$ is given by round$(\sqrt{n+1})$. The maximum pursuit angle $\theta_{max}$ is $\pi / a^2$. The maximum turning angle $\alpha$ is set to be $\pi / 2a^2$. The maximum pursuit distance $l_{max}$ is calculated from equation (5.14).

### *(1) A 40-Bar Planar Truss Structure (Using Model I)*

The 40-bar planar truss [18] (Model I) is shown in Fig. 5.28. The material density is 7800 kg/m$^3$ (7.8 t/m$^3$) and the modulus of elasticity is 196.13 GPa (2×10$^7$ t/m$^2$). The stress limits of the members are subjected to ±156.91 MPa (16000 t/m$^2$). Node 4 and 5 in y directions are subjected to the displacement limits of ±0.035 m (1/600 span).

The 1st natural frequency $\omega_1 \geq 200$ rad/s is taken to as the dynamic constraint of the optimization.

Table 5.16 gives the optimal results. Fig. 5.29 shows the convergence curves.

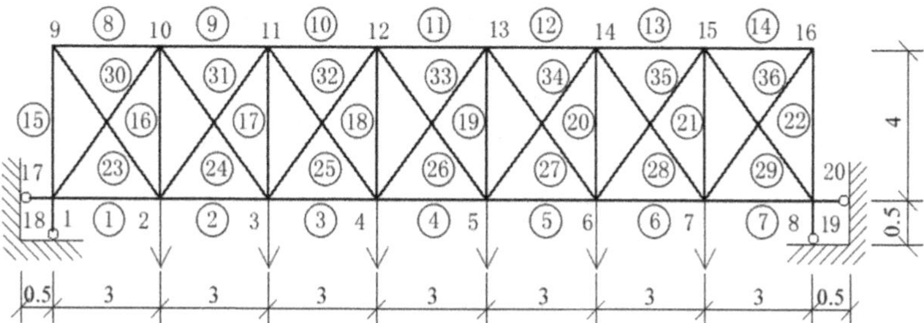

**Fig. 5.28** A 40-bar planar truss

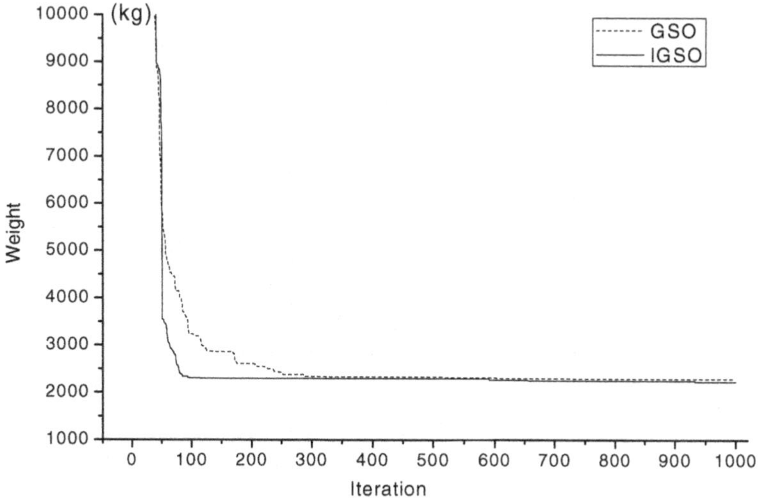


**Fig. 5.29** Convergence curves for the 40-bar planar truss

As Table 5.16 shows, the weight of the optimal structure with shape optimization for GSO is 2080.6733 kg. When the 1st frequency is taken as constraint, an optimal structure obtained by GSO weighs 2292.6556 kg, and the one by IGSO weighs 2228.7329 kg. By comparing three trusses with the corresponding 1st frequency, we can know the optimal truss with shape optimization is lightest but violated the frequency constraint. Thus the results show that parts of the solution space of the original shape optimization become infeasible because of the involving of frequency constraint. And the whole solution space becomes non-continuous and non-linear. It can be seen from Fig. 5.29 that both the GSO and IGSO algorithm can work well in this optimal problem, and can get the optimal solution with about 350 and 100 iterations respectively. The optimal structure with IGSO is shown in Fig. 5.30.

**Table 5.16** Comparison of the optimal results for the 40-bar planar truss

| Variables | Shape optimization | GSO | IGSO |
|---|---|---|---|
| $A_1$ | 0.0015 | 0.002 | 0.0015 |
| $A_2$ | 0.0010 | 0.001 | 0.001 |
| $A_3$ | 0.0010 | 0.001 | 0.001 |
| $A_4$ | 0.0010 | 0.001 | 0.001 |
| $A_5$ | 0.0010 | 0.001 | 0.001 |
| $A_6$ | 0.0030 | 0.003 | 0.003 |
| $A_7$ | 0.0035 | 0.003 | 0.003 |
| $A_8$ | 0.0035 | 0.003 | 0.003 |
| $A_9$ | 0.0010 | 0.001 | 0.001 |
| $A_{10}$ | 0.0010 | 0.001 | 0.001 |
| $A_{11}$ | 0.0010 | 0.001 | 0.001 |
| $A_{12}$ | 0.0010 | 0.001 | 0.001 |
| $A_{13}$ | 0.0010 | 0.001 | 0.001 |
| $A_{14}$ | 0.0010 | 0.001 | 0.001 |
| $A_{15}$ | 0.0010 | 0.001 | 0.001 |
| $A_{16}$ | 0.0010 | 0.001 | 0.001 |
| $A_{17}$ | 0.0025 | 0.005 | 0.004 |
| $A_{18}$ | 0.0010 | 0.001 | 0.001 |
| $A_{19}$ | 0.0010 | 0.001 | 0.001 |
| $y_9$ | 1.069 | 1.003 | 1.009 |
| $Y_{10}$ | 2.307 | 2.297 | 2.408 |
| $Y_{11}$ | 2.851 | 3.659 | 3.728 |
| $Y_{12}$ | 3.287 | 4.165 | 4.35 |
| Weight (kg) | 2080.6733 | 2292.6556 | 2228.7329 |
| $\omega_1$(rad/s) | 164.733 | 200.002 | 200.454 |

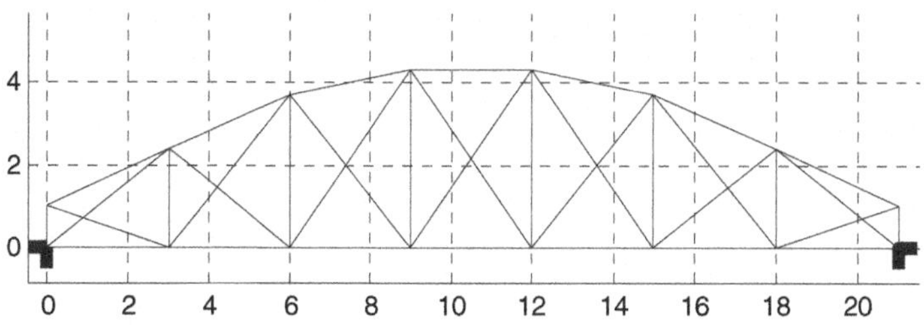


**Fig. 5.30** The optimized 40-bar planar truss

### *(2) A 15-Bar Spatial Truss Structure (Using Model I)*

The 15-bar spatial truss structure, shown in Fig. 5.31, has been analysed by Sadek [32] and Kang [33] Wang [34]. Its Young's modulus is specified as 68.9 GPa, and the material density is 2778 kg/m$^3$. Because of the symmetry of the truss, there are 7 shape variables. The initial coordination of nodes is listed as, 1(-0.635, -0.635, 2.54), 2(-0.635, 0.635, 2.54), 3(0.635, 0.00, 2.54), 4(-0.635, -0.635, 0.00), 5(-0.635, 0.635, 0.00), 6(0.635, 0.635, 0.00), 7(0.635, -0.635, 0.00)(m). All the members' area is fixed: the areas of bar 1~7 are 12.90 cm$^2$, and the areas of bar 8~15 are 6.45 cm$^2$. All the nodes' z coordination is fixed. And every node is allowed to moves among the range of ± 12.7 cm in x and y directions. Besides, the connection of bars is given as follows: 1(1, 2), 2(2, 3), 3(1, 3), 4(1, 4), 5(2, 5), 6(3, 7), 7(3, 6), 8 (2, 4), 9(1, 5), 10(3, 4), 11(2, 6), 12(2, 7), 13(3, 5), 14(1, 7), 15(1, 6).

Table 5.17 and Table 5.18 give the optimal results with different frequency constraints. Fig. 5.32 shows the convergence curves.

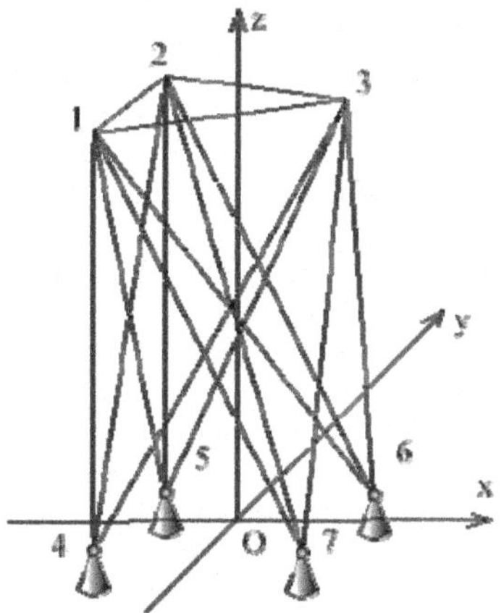


**Fig. 5.31** A 15-bar spatial truss

**Table 5.17** Comparisons of optimal designs of node coordinates for ($\omega_1 \geq 520$Rad/s)

| Variables | $\omega_1 \geq 520$ rad/s | | | |
|---|---|---|---|---|
| | Sadek [32] | Wang [34] | GSO | IGSO |
| $x_2$ | -0.508 | -0.508 | -0.506 | -0.506 |
| $y_2$ | 0.508 | 0.508 | 0.506 | 0.506 |
| $x_3$ | 0.508 | 0.508 | 0.506 | 0.506 |
| $x_5$ | -0.611 | -0.633 | -0.622 | -0.629 |
| $y_5$ | 0.620 | 0.618 | 0.641 | 0.643 |
| $x_6$ | 0.531 | 0.508 | 0.515 | 0.508 |
| $y_6$ | 0.558 | 0.544 | 0.510 | 0.506 |
| $\omega_1$ | 521.135 | 520.034 | 520.151 | 520.038 |
| Weight (kg) | 89.23 | 89.14 | 89.045 | 89.035 |

The optimal weight of 15-bar spatial truss with two different frequency constraints of 520 rad/s and 580 rad/s respectively was compared with that of Sadek [32] and Wang [34]. As Table 5.17 and Table 5.18 list, both optimizer algorithms have achieved similar optimization results to the literature. The best optimization weighs is 89.035 kg with the 1st minimum frequency constraint of 520 rad/s. Although GSO and IGSO have found the similar optimal results with different 1st frequency constraints, the difference between the design variables is obvious. Compared with two results, IGSO is better than GSO in the weight of structure. As is shown in Fig. 5.32, the convergence speed of two optimizers is the same well. The optimal structures with two optimizer algorithms are shown in Fig. 5.33.

**Table 5.18** Comparisons of optimal designs of node coordinates for ($\omega_1 \geq 580$Rad/s)

| Variables | $\omega_1 \geq 580$ rad/s | | | |
|---|---|---|---|---|
| | Sadek [32] | Wang [34] | GSO | IGSO |
| $x_2$ | -0.508 | -0.508 | -0.507 | -0.506 |
| $y_2$ | 0.508 | 0.508 | 0.506 | 0.506 |
| $x_3$ | 0.508 | 0.508 | 0.506 | 0.506 |
| $x_5$ | -0.740 | -0.762 | -0.725 | -0.735 |
| $y_5$ | 0.744 | 0.762 | 0.746 | 0.728 |
| $x_6$ | 0.574 | 0.540 | 0.571 | 0.560 |
| $y_6$ | 0.612 | 0.540 | 0.562 | 0.586 |
| $\omega_1$ | 585.448 | 580.006 | 580.029 | 580.017 |
| Weight (kg) | 90.14 | 89.92 | 89.88 | 89.88 |

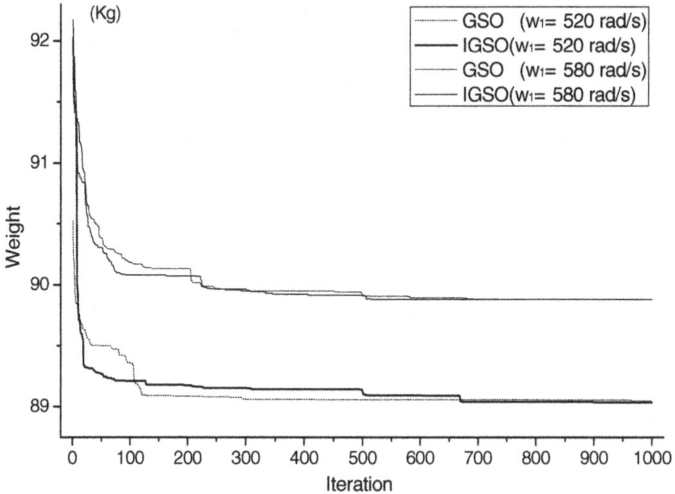


**Fig. 5.32** Convergence curves for the 15-bar spatial truss

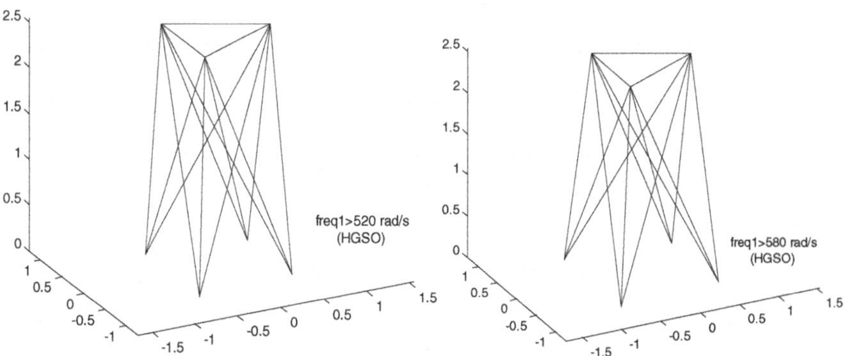


**Fig. 5.33** The optimized 15 bar spatial truss

### *(3) A 25-Bar Spatial Truss Structure (Using Model I)*

This 25-bar spatial truss example is a sizing and geometry optimization model in Wu [18] and Kaveh [25], and the geometry is shown in Fig. 5.34. Young's modulus is specified as $6.895\times10^4$ MPa, and the material density as 2767.99 kg/m$^3$. The stress limit is ±172.25MPa for all members and the displacement limit of node 1~6 in all three directions are ±0.0089 m. Because of the symmetry of the truss, there are 8 sizing variables as table 5.16 listed and 5 shape variables. Discrete values considered for this example are taken from the set D={0.645, 1.290, 1.936, 2.580, 3.226, 3.871, 4.516, 5.161, 5.806, 6.452, 7.097, 7.742, 8.387, 9.032, 9.677, 10.323, 10.968, 11.613, 12.258, 12.903, 13.548, 14.194, 14.839, 15.484, 16.129, 16.774, 17.419, 18.065, 19.355, 20.645} ($\times10^{-4}$ m$^2$). Considering the symmetry, the shaping variable group and the relative boundary are given in this paper as: 2.54

$\le x_4=x_5=-x_3=-x_6\le 3.556$, $5.588 \le y_3=y_4=-y_5=-y_6\le 6.604$, $2.54 \le z_3=z_4=z_5=z_6\le 3.556$, $2.54 \le x_8=x_9=-x_7=-x_{10}\le 3.556$, $1.27 \le y_7=y_8=-y_9=-y_{10}\le 2.286$ (m). The load case is listed on Table 5.10. This 25-bar spatial truss is from the 25-bar spatial truss A in previous items, and just adopts international system of units. And the dynamic constraint is the minimum 1st natural frequency of 250 rad/s.

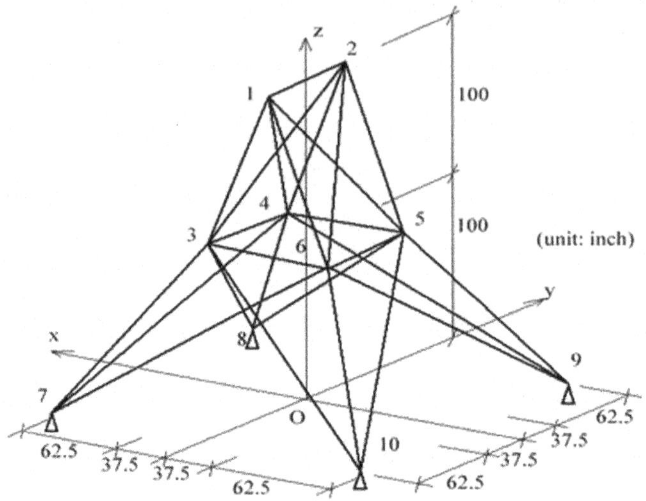


**Fig. 5.34** A 25-bar spatial truss

**Table 5.19** Comparison of the optimal results for the 25-bar spatial truss

| Variables | Shape optimization | GSO | IGSO |
|---|---|---|---|
| $A_1$ | 1.290 | 0.645 | 0.645 |
| $A_2$ | 0.645 | 1.290 | 1.936 |
| $A_3$ | 7.097 | 6.452 | 5.806 |
| $A_4$ | 0.645 | 0.645 | 0.645 |
| $A_5$ | 0.645 | 0.645 | 0.645 |
| $A_6$ | 0.645 | 0.645 | 0.645 |
| $A_7$ | 0.645 | 1.290 | 1.936 |
| $A_8$ | 5.806 | 6.452 | 5.161 |
| $z_1$ | 0.812 | 0.737 | 0.627 |
| $x_2$ | 1.251 | 1.454 | 1.386 |
| $z_2$ | 3.137 | 2.957 | 3.237 |
| $x_6$ | 1.041 | 1.1 | 1.076 |
| $z_6$ | 2.994 | 3.195 | 3.421 |
| Weight (kg) | 55.101 | 58.9012 | 57.8614 |
| $\omega_1$ (rad/s) | 193.488 | 250.282 | 250.825 |

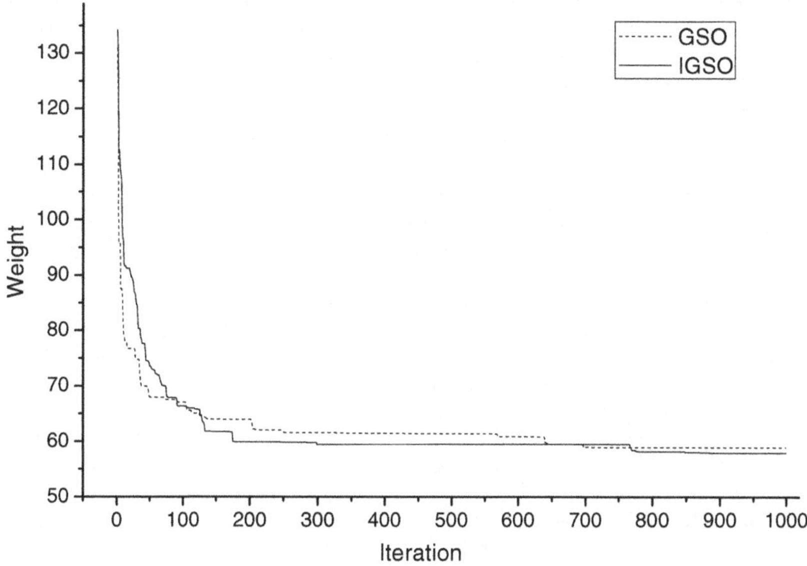


**Fig. 5.35** Convergence curves for the 25-bar spatial truss

Table 5.19 gives the optimal results. It can be seen from Table 5.19 that the optimization result (weight) with 1st frequency constraint is almost the same with the optimal result without the frequency constraint. In other word, the frequency constraint has little influence on the optimal result of this optimization problem. The weight of the optimal structure with shape optimization is 55.101 kg. The weight of the optimal structure of GSO and IGSO with frequency is 57.8614 kg and 58.9012 kg respectively. The 1st frequency of this structure with the frequency constraint is 193.488 rad/s. The 1st frequency of this optimal structure with GSO and IGSO is 250.283 rad/s and 250.82 rad/s respectively. As Fig. 5.35 show, convergence speed of IGSO is much faster than the of GSO, and both two optimizers can approach the optimal solution well with less than 50 iterations, and find the optimal solution with 1000 iterations. The optimal structure with IGSO is shown in Fig. 5.36.

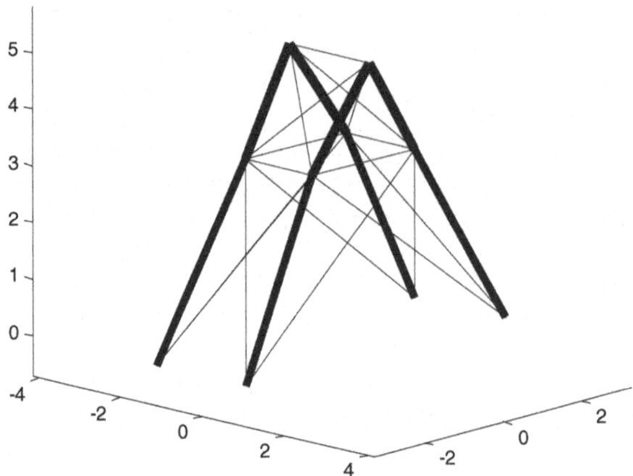


**Fig. 5.36** The optimized 25-bar spatial truss

### *(4) A 10-Bar Planar Truss Structure (Using Model I)*

The 10-bar truss structure (Model I), shown in Fig. 5.37, had been analysed by Luo [20]. The material density is 7680 kg/m.$^3$ and the modulus of elasticity is 210GPa. The members are subjected to the stress limits of ±100 MPa. There are 10 size variables and 5 shape variables. Discrete values considered for this example are taken from the set D=[0.001, 0.01] (m$^2$) with the interval of 0.0005 m$^2$.The shape variable group and the relative boundary are given as: $-2.5 \le y_1 \le 2.5$, $0 \le x_2 \le 2.5$, $-2.5 \le y_2 \le 2.5$, $2.5 \le x_3 \le 5$, $-2.5 \le y_3 \le 2.5$ (m). The vertical downward load of -100 kips on node 5 and 6 are considered. The 1$^{st}$ natural frequency, $\omega_1 \ge 600$ rad/s is taken to as the dynamic constraint of this optimization.

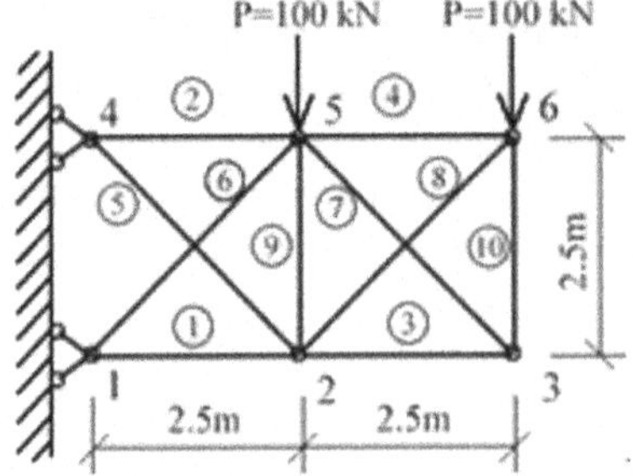


**Fig. 5.37** A 10-bar planar truss

As Table 5.20 listed, the optimization result (weight) with 1st frequency constraint is almost the same as the optimal result without the constraint. The weight of the optimal structure with shape optimization is 235.1748 kg and the corresponding 1st natural frequency is 583.566 rad/s. The 1st frequency of the optimal structure with GSO and IGSO is 636.628 rad/s and 651.630 rad/s respectively, and the weight is 2504276 and 242.1666 kg respectivly. As Fig. 5.38 shown, IGSO's convergence speed is much faster than GSO's, and IGSO can find the optimal solution with about 100 iterations. The optimal structure with IGSO is shown in Fig. 5.39.

**Table 5.20** Comparison of the optimal results for the 10-bar planar truss (Model I)

| Variables | Shape Optimization | GSO | IGSO |
|---|---|---|---|
| $A_1$ | 0.0020 | 0.0020 | 0.0020 |
| $A_2$ | 0.0020 | 0.0015 | 0.0020 |
| $A_3$ | 0.0010 | 0.0010 | 0.0010 |
| $A_4$ | 0.0010 | 0.0010 | 0.0010 |
| $A_5$ | 0.0010 | 0.0015 | 0.0010 |
| $A_6$ | 0.0010 | 0.0010 | 0.0010 |
| $A_7$ | 0.0010 | 0.0010 | 0.0010 |
| $A_8$ | 0.0010 | 0.0010 | 0.0010 |
| $A_9$ | 0.0010 | 0.0010 | 0.0010 |
| $A_{10}$ | 0.0010 | 0.0010 | 0.0010 |
| $y_1$ | -0.438 | -1.057 | -1.077 |
| $x_2$ | 2.476 | 2.353 | 2.229 |
| $y_2$ | 0.332 | 0.099 | 0.322 |
| $x_3$ | 3.663 | 4.633 | 3.828 |
| $y_3$ | 0.979 | 2.094 | 1.067 |
| Weight(Kg) | 235.1748 | 250.4276 | 242.1666 |
| $\omega_1$(Rad/s) | 583.566 | 636.628 | 651.630 |

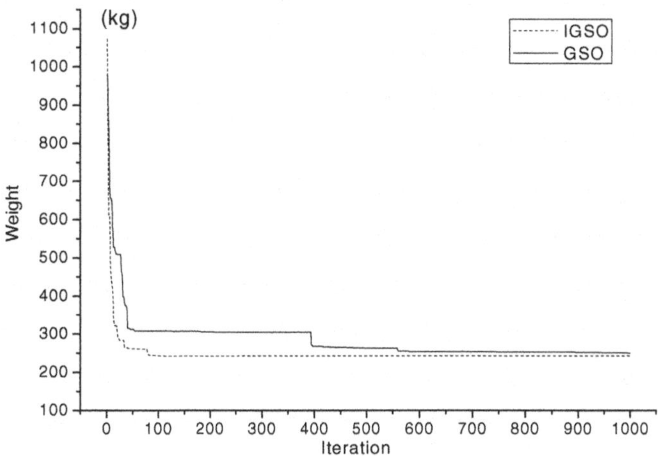


**Fig. 5.38** Convergence curves for the 10-bar planar truss (Model I)

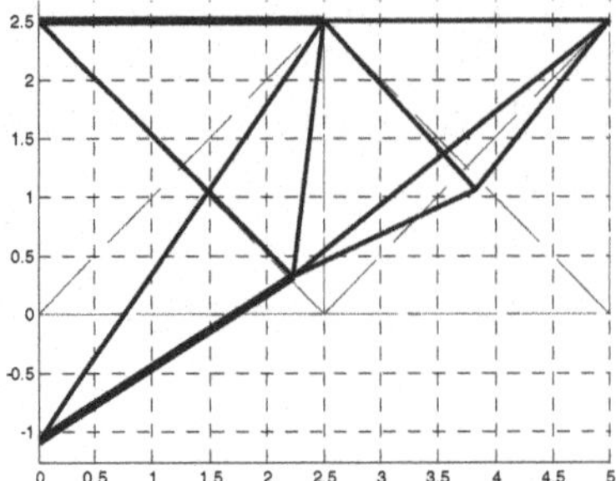

**Fig. 5.39** The optimized 10-bar planar truss (Model I)

### *(5) A 10-Bar Planar Truss Structure (Using Model II)*

The 10-bar truss is the optimization model II, the maximum natural frequency with maximum weight limit. Different from the previous 10-bar truss (model I), the weight of structure is the optimization constraint. The maximum weight limit of original model [20] is 1000 kg. In this section the maximum weight of 600 kg and 800 kg are used as the constraints respectively to test the IGSO and to get the influence of different constraint conditions on the searching ability of IGSO.

**Table 5.21** Comparison of the optimal results for the 10-bar planar truss (Model II)

| Variables | IGSO | | | Luo [20] |
|---|---|---|---|---|
| | < 600kg | < 800kg | < 1000kg | < 1000kg |
| $A_1$ | 0.0065 | 0.0095 | 0.01 | 0.0074 |
| $A_2$ | 0.005 | 0.0065 | 0.0085 | 0.0027 |
| $A_3$ | 0.0015 | 0.0015 | 0.0015 | 0.0059 |
| $A_4$ | 0.001 | 0.001 | 0.001 | 0.0012 |
| $A_5$ | 0.0045 | 0.007 | 0.008 | 0.0033 |
| $A_6$ | 0.0025 | 0.0025 | 0.0055 | 0.0047 |
| $A_7$ | 0.001 | 0.001 | 0.001 | 0.0047 |
| $A_8$ | 0.001 | 0.001 | 0.001 | 0.0011 |
| $A_9$ | 0.001 | 0.002 | 0.0015 | 0.0086 |
| $A_{10}$ | 0.001 | 0.001 | 0.001 | 0.0010 |
| $y_1$ | -2.132 | -2.252 | -2.287 | -2.003 |
| $x_2$ | 2.493 | 2.488 | 2.5 | 2.353 |
| $y_2$ | -0.125 | -0.108 | -0.263 | 0.511 |
| $x_3$ | 4.061 | 4.159 | 4.184 | 3.803 |
| $y_3$ | 1.479 | 1.495 | 1.485 | 1.041 |
| $\omega_1$ (rad/s) | 925.383 | 983.388 | 1012.753 | 893.3 |
| Weight (kg) | 599.999 | 799.998 | 999.998 | 727.1 |

Because the optimal model II is much more complex than the model I, and requires much more computing time, IGSO is only used to optimize the model II. As Table 5.21 shows, the value of weight constraint for trusses is proportional to the 1st frequency of the optimal structure obtained by IGSO. As Fig. 5.40 shows, IGSO can approach the optimal solution with different weight constraints and has good global convergence ability with about 150 iterations. The optimal structures with three weight constraints are shown in Fig. 5.41 by IGSO (1000 iterations).

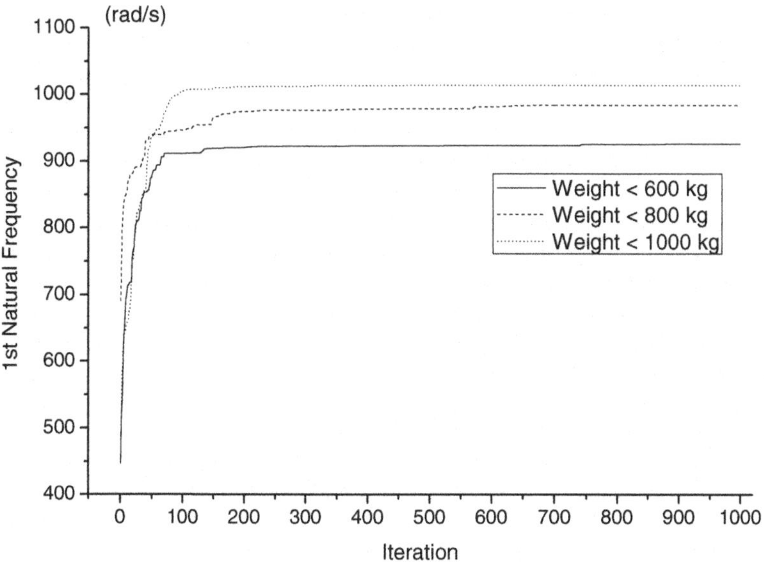


**Fig. 5.40** Convergence curves for the 10-bar planar truss (Model II)

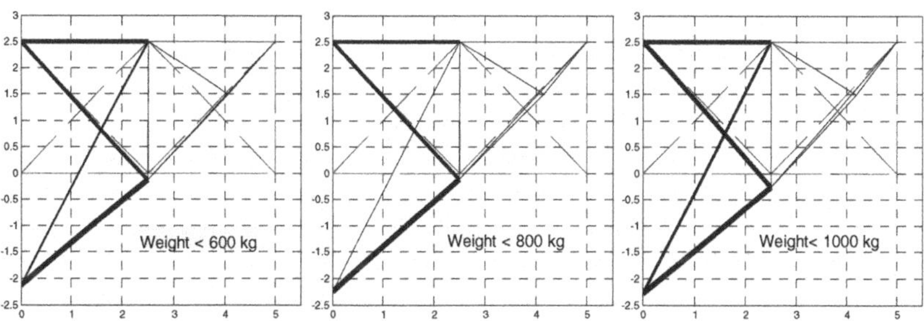


**Fig. 5.41** The optimized 10-bar planar truss (Model II)

## 5.6 The Application of IGSO in Truss Topology Optimization

It is always the main work of the structural engineers to design reliable structures or other optimal objects with a minimum cost. Because of the wide application of truss-structures, the optimal design of truss structures has been an active area of structural optimization. Various techniques based on classical optimization methods have been developed to find optimal truss-structures for a long time. However, these classical optimization methods were unable to solve the nonlinear programming (NLP) problem well, such as sizing, geometry (configuration) optimization, etc. In this chapter, GSO and IGSO are used to the topology optimization of truss structures [35-40] by two topology methods.

### *5.6.1 Topology Optimization Model*

Usually, most of optimization problems can be classified into three main categories: (i) sizing, (ii) geometry (configuration), and (iii) topology optimization. In the sizing optimization of trusses, cross-sectional areas ($A_i$) of members are selected as design variables. The changes in nodal coordinates ($C_j$) are chosen as design variables in the geometry optimization of trusses. However, the connections of members are determined as variables in the strictly topology optimization of a truss. Usually, the generalized topology optimization considers all the above three optimization simultaneously. In this paper, a group of topology variables $\{T\}$ was employed to represent the connectivity of members, the formulation of the truss-structure optimization problem as a nonlinear programming (NLP) problem can be defined as:

$$
\begin{aligned}
&\min .Weight\left(T_i, A_i, C_j\right)=\sum_{i=1}^{N}\rho_i A_i T_i L_i\left(C_j\right) \\
&\text{Subject to } G1 \equiv A_i \in D, \quad i=1,2,\ldots,N; D=\{a_1, a_2, \ldots, a_k\} \\
&\qquad G2 \equiv C_{j\min} \le C_j \le C_{j\max}, \qquad j=1,2,\ldots,M \\
&\qquad G3 \equiv Truss\ is\ acceptable \qquad (about\ T_i) \\
&\qquad G4 \equiv Truss\ is\ kinematically\ stable \quad (about\ T_i) \\
&\qquad G5 \equiv [\sigma]-|\sigma_i| \ge 0, \qquad i=1,2,\ldots,N \\
&\qquad G6 \equiv [\delta]-|\delta_j| \ge 0, \qquad j=1,2,\ldots,M
\end{aligned}
\tag{5.19}
$$

where $T_i$ is the $i$th member connection (1 for presence and 0 for absence). $A_i$ is the $i$th member's cross-sectional areas. $D$ is a set of discrete cross-sectional areas. $Cj$ is the $j$th node's coordinate.

### 5.6.2 Topology Methods

As the topology variable $T_i$ gets rid of the restriction of classical ground structures [35], the random topology of the objective structures can be any possible topology which includes all the workable and failed topology (violating constraints *G3* or *G4*). It means the global optimal topology can be obtained. In comparison, the optimization with ground structure theory always converges to local optima. However, this optimal model, which is built for stochastic optimization methods, has a great deal of opportunity to approach the global optima. For guiding the random topology which may be subject to the constraints *G3* and *G4*, the structural property of the truss topology variables is discussed in this paper. Firstly, we must make sense of several concepts, (i) support node, (ii) load node, (iii) undeleted node, (iv) erasable node. As Fig. 5.42 shows, the 5th point of part (b), the 6th point of part (c) and the 6th point of part (f) are all erasable node, but the 5th of part (e) is an undeleted node.

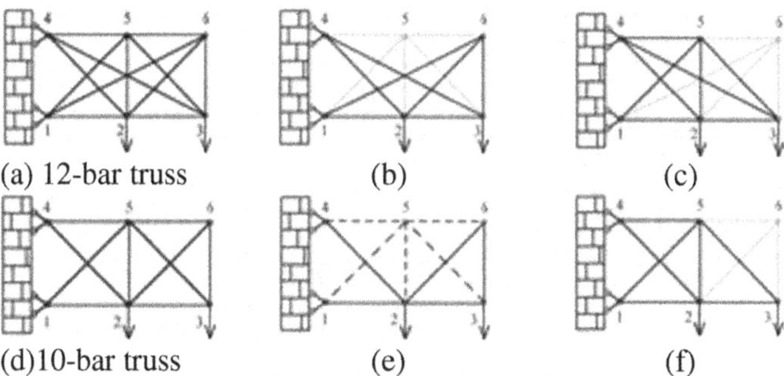


**Fig. 5.42** The erasable node and undeleted node in topology of truss optimization

For a simple truss structure, the phenomenon of failed topology (PFT) may be not so obvious, but for a complex structure, which includes a great number of topology variables, the PFT must be serious and obvious and make the optimization failed. Two truss topology methods, including heuristic topology and discrete topology variables, are proposed here to ease the PFT in this paper. The rules, which can guide the random topology of trusses, can be drawn by analyzing the truss structures, as follows:

*Rule.1:* The erasable nodes, without any loads, can be deleted on the cases that there are less than two bars connecting to it. Then the bars, which connect to the deleted nodes, will be deleted at the same time.

*Rule.2:* The support nodes must be connected with one or more than one bar. For the planar trusses, the total amount of support link-bars must be greater than or equal to 3.

*Rule.3:* The load nodes must be connected with two or more than two bars. (In addition, it can't be connected with two bars which locate on the same straight line.)

*Rule.4:* The undeleted nodes must be connected with three or more than three bars.

### *(1) Heuristic Topology (HT) Method*

Based on the rules about the truss topology property, a presented topology method, heuristic topology (HT) [39], is improved on some details in this paper, and the program of the new heuristic topology is executed as follows:

*Step.1:* checking the erasable nodes, delete the erasable nodes which are connected with less than three bars.

*Step.2:* checking the load nodes, choose the connectable node randomly from the existing erasable nodes and undeleted nodes, then create the bar to connect them, till the load nodes match the rule 3.

*Step.3:* checking the support nodes, choose the connectable node randomly from the existing erasable nodes and undeleted nodes, then create the bar to connect them, till the support nodes match rule 2.

*Step.4:* checking the undeleted nodes and temp undeleted nodes, and stop this step till all the undeleted nodes match rule 4.

In addition, sometimes, there is not any connectable node included in the existing erasable nodes and undeleted nodes on running Step.2 and Step.3 of the above topology program. The connectable node, which belongs to the temp undeleted nodes, will be created to fit the structure on this case.

The topology methods presented in the references, as well as the improved heuristic topology in this paper, can't make sure that the topology structures are all workable, but the unworkable topology can be selected easily to save a lot of unnecessary computational time. The 15-bar truss, which is shown in Fig. 5.43, is a failed topology, whereas it can match the rules of heuristic topology well. Therefore reference [21] and [39] as well as this paper still check the singularity of the structural global stiffness matrix (the constraint *G4*) in order to make sure the topology is workable.

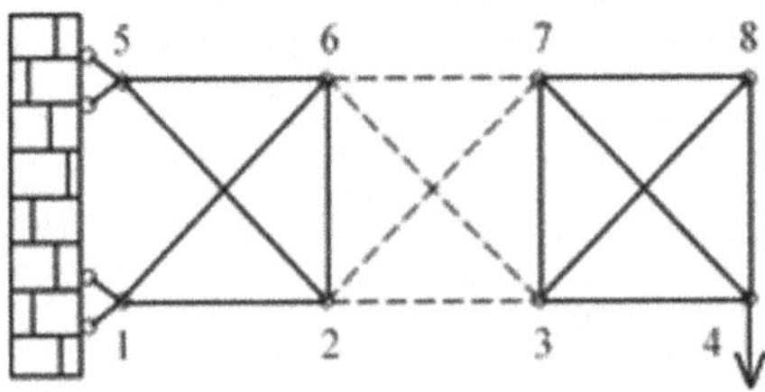


**Fig. 5.43** The failed topology of 15-bar truss

***(2) Discrete Topology (DT) Variable Method***

Now we suppose that there are *n* topology variables (0-1 type) in the structural optimal model shown by Fig. 5.51, as $\{T\}=[T_1\ T_2\ \cdots\ T_n]^T$. So the Part II has four topology variables, as $[T_i\ T_j\ T_k\ T_l]^T$. Considering the property of the truss, we can find there are only five workable bar-combinations, such as $[1\ 1\ 1\ 0]^T$ and other four combinations shown by Fig. 5.44 (a) to (e). Use a discrete topology variable $DT_1 \in [1\ 2\ 3\ 4\ 5]$ to represent all of the five combinations, for example, $DT_1$=1 represents the Fig. 5.51 (a) combinations $[1\ 1\ 1\ 0]^T$, then for other truss structures, we can get $\{DT\}=[DT_1\ DT_2\ \cdots\ DT_n]^T$ with the same rule above. This processing is the discretization of topology variables and is called discrete topology variable method in this paper. By this process, the number of topology variables is reduced greatly, and the efficiency of structural topology optimization is raised obviously, because most of fail topology configurations have been deleted.

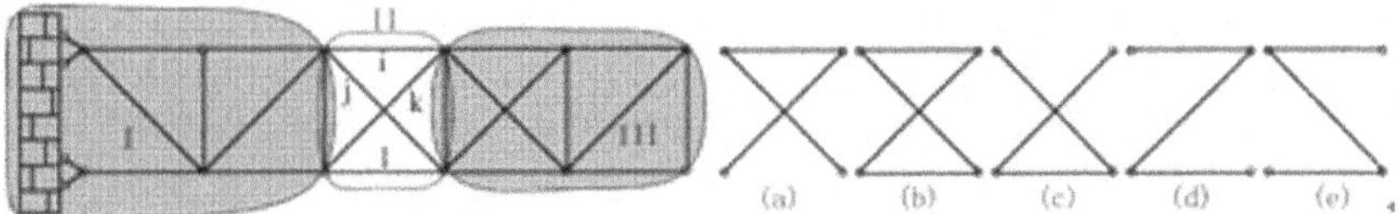


**Fig. 5.44** Discretization of topology variables

Generally, all or part of the topology variables can be transformed to discrete topology variables and the possible fail topologies can be deleted, such as the ten-bar truss example in this paper. When all topology variables are transformed to discrete topology variables, it is the case of Topology Group [41], which is only a special case of this discrete topology variable method. The discrete topology variable method is different from the Heuristic topology method, and the optimal results of IGSO show that it is a simple and effective topology method.

## *5.6.3 Numerical Example*

In this section, five pin-connected structures commonly designed as benchmark problems are used to test the GSO and IGSO. The proposed algorithm is coded in Matlab language and ANSYS APDI language.

The examples given in the simulation analysis include:

- a 5-bar planar truss structure with HT;
- a 12-bar spatial truss structure with HT;
- a 10-bar spatial truss structure with HT;
- a 15-bar planar truss A structure with HT;
- a 25-bar planar truss B structure with DT.

The GSO, IGSO are applied respectively to all these examples and the results are compared in order to evaluate the performance of the new algorithm. For all these two topology algorithms, the maximum number of iterations is limited to 1000 and the population size is set to 50. 20% of the population is selected as rangers; the initial head angle $\varphi_0$ of each individual is set to be $\pi/4$. The constant $a$ is given by round$(\sqrt{n+1})$. The maximum pursuit angle $\theta_{\max}$ is $\pi/a^2$. The maximum turning angle $\alpha$ is set to be $\pi/2a^2$. The maximum pursuit distance $l_{\max}$ is calculated from equation (5.14).

### *(1) A 5-Bar Planar Truss Structure*

The 5-bar truss structure is shown in Fig. 5.45. It was analysed by Sun [22] as a testing structure without practical unit. The material density, the length of bar and the modulus of elasticity all are assumed as 1. The members are subjected to the stress limits of ± 20. There are 5 size variables and 5 topology variables. Discrete values considered for this example are taken from the set D={0.3, 0.5, 1, 1.5, 2.25, 2.5, 2.75, 3, 3.5, 4.6, 5, 5.4, 5.8, 6, 6.5, 7.219, 7.5, 8, 8.9, 10, 11, 12, 14.143, 15, 15, 15.5, 20, 30, 35}. Two working cases are considered: Case 1, $P_{1x}$=5, $P_{1y}$=-50; Case 2, $P_{2x}$=5, $P_{2y}$=-50.

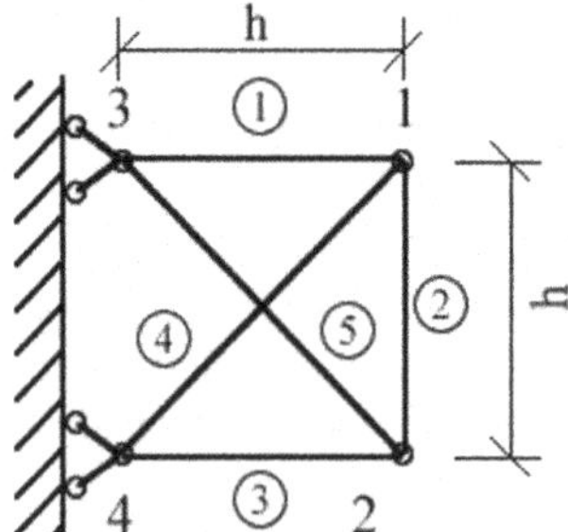


**Fig. 5.45** A 5-bar planar truss

Sun [22] made the topology optimization with different initial values of the particles in traditional optimization algorithm. The GSO and IGSO are used to optimize this 5-bar truss structure with HT. As Table 5.22 lists, the IGSO and GSO both find the same optimal solution as Sun did with less than 20 iterations. Because the 5-bar truss topology optimization is simple and fewer iterations are required, the convergence curves were not given in this example. The optimal truss obtained by IGSO is shown in Fig. 5.46.

**Table 5.22** Comparison of the optimal results for the 5-bar planar truss

| Variables | Sun [22] | GSO | IGSO |
|---|---|---|---|
| $A_1$ | 1 | 1 | 1 |
| $A_2$ | 2.5 | 2.5 | 2.5 |
| $A_3$ | 10 | 10 | 10 |
| $A_4$ | 0 | 0 | 0 |
| $A_5$ | 15 | 15 | 15 |
| Weight | 33.5012 | 33.5012 | 33.5012 |

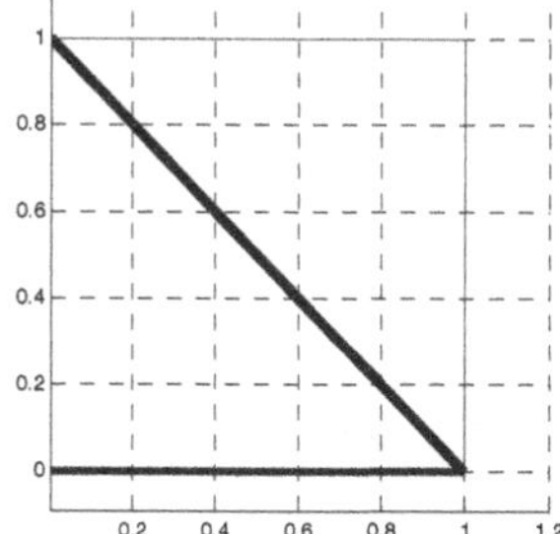


**Fig. 5.46** The optimized 5-bar planar truss

### *(2) A 12-Bar Planar Truss Structure*

The 12-bar truss structure, shown in Fig. 5.47, was analysed by Sun [22]. The material density is 0.02768 N/cm$^3$, and the modulus of elasticity is 6.987×10$^6$ N/cm$^2$. The members are subjected to the stress limits of ±17243.5 N/cm$^2$. And the displacement limit is 5.08 cm in y direction of node 2 and 4. There are 12 size variables and 12 topology variables. As the different section sets used in Sun [22], discrete values considered in this example are taken respectively from the two sets $D_1$={129.03, 167.74, 180.64, 225.81, 264.52, 296.77, 322.58} (cm$^2$). and $D_2$={6.45, 19.35, 32.26, 51.61, 67.74, 77.42, 96.77, 109.68, 141.94, 145.84, 167.74, 180.64, 187.10, 200, 225.81} (cm$^2$). Two working cases are considered: Case 1, $P_{2y}$=-4.45×10$^5$ N, Case 2, $P_{4y}$=-4.45×10$^5$ N.

Sun [22] did not consider the node coordinates variable in this example, only two set of original section variables were used to find the optimal solution. The HT was used in this simple topology optimization example. As the cross-section variable contained in the section set *D1* is relatively small, it is easy for HT to get the optimal topology results which is displayed in Table 5.22. It can be seen from Table 5.29 that the algorithms HT used in this example have achieved better results than the literature [22]. It also shows that for a little complex example, the traditional method has its limitations.

It can be seen from Fig. 5.48 that for the cross-section variables D1, the combination of GSO and IGSO with HT respectively need less than 20 iterations to converge to the optimal solution. Whereas for the variable cross-section D2, 100 times were needed to get the optimal value. IGSO converges faster than GSO does. The optimized truss structures with respect to the two cross-section sets $D_1$ and $D_2$ are shown in Fig. 5.49 respectively.

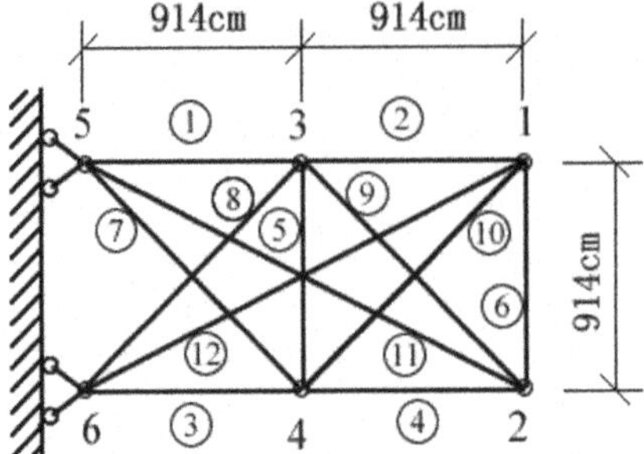


**Fig. 5.47** A 12-bar planar truss

**Table 5.23** Comparison of the optimal results for the 12-bar planar truss

| Variables | Section set $D_1$ | | | Section set $D_2$ | | |
|---|---|---|---|---|---|---|
| | Sun [22] | HT & GSO | HT & IGSO | Sun [22] | HT & GSO | HT & IGSO |
| $A_1$ | 167.74 | 167.74 | 167.74 | 167.74 | 180.64 | 180.64 |
| $A_2$ | - | - | - | - | - | - |
| $A_3$ | 129.03 | 129.03 | 129.03 | 109.68 | 96.77 | 96.77 |
| $A_4$ | 129.03 | 129.03 | 129.03 | 96.77 | 96.77 | 96.77 |
| $A_5$ | 129.03 | 129.03 | 129.03 | 51.61 | 19.35 | 19.35 |
| $A_6$ | - | - | - | - | - | - |
| $A_7$ | - | - | - | 32.26 | 19.35 | 19.35 |
| $A_8$ | 129.03 | 129.03 | 129.03 | 96.77 | 109.68 | 109.68 |
| $A_9$ | 129.03 | 129.03 | 129.03 | 141.94 | 141.94 | 141.94 |
| $A_{10}$ | - | - | - | - | - | - |
| $A_{11}$ | - | - | - | - | - | - |
| $A_{12}$ | - | - | - | - | - | - |
| Weight (N) | 23281 | 23281 | 23281 | 20477 | 19659.8 | 19659.8 |

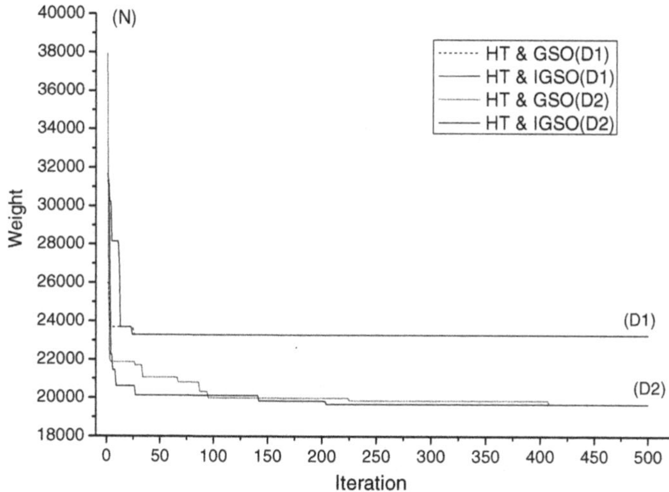


**Fig. 5.48** Convergence curves for the 12-bar planar truss

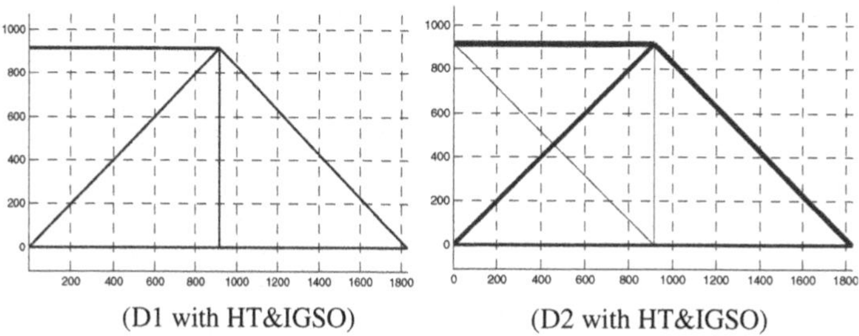


**Fig. 5.49** The optimized 12-bar planar truss

### *(3) A 10-Bar Planar Truss Structure*

In this section, the performance of the topology optimization is studied on a 10-bar truss used by Wu [18] and Tang [21]. The geometry of 10-bar truss is shown in Fig. 5.50. The stress limit is 172.25 MPa in both tension and compression for all members. Young's modulus is specified as $6.89\times10^4$ MPa, and the material density is 2.768 kg/m$^3$. Joints 4, 5 and 6 are allowed to move only in the vertical direction. Discrete values considered for this example are taken from the set D={1.045, 1.161, 1.535, 1.690, 1.858, 1.993, 2.019, 2.181, 2.342, 2.477, 2.497, 2.697, 2.897, 3.097, 3.206, 3.303, 3.703, 4.658, 5.142, 7.419, 8.710, 8.968, 9.161, 10.000, 10.322, 12.129, 12.839, 14.194, 14.774, 17.097, 19.355, 21.613} ($\times10^{-3}$ m$^2$). More details can be found from literature [18].

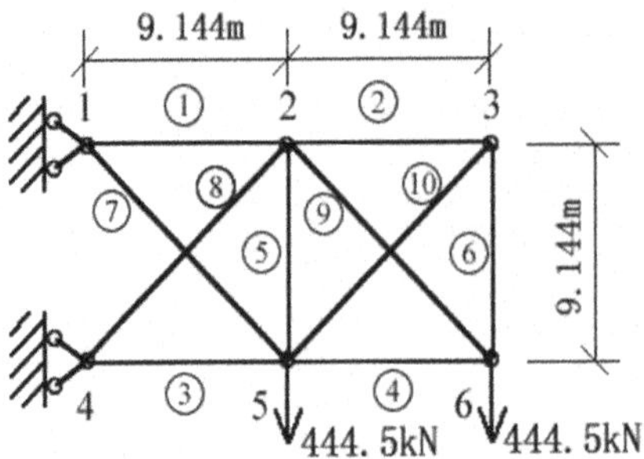


**Fig. 5.50** A 10-bar planar truss

In 10-bar truss topology optimization, two topology methods HT and DT with respective to GSO and IGSO are both used respectively. In the first step, all the topology variables are condensed to one discrete topology variable. In the second step, the heuristic topology method was used with GSO and IGSO respectively. The population of the GSO and IGSO is 50, which contains 2 at the two searching angles (left and right). It can be seen from Table 5.24 that DT&IGSO and HT&IGSO both get the better optimal result (1244.9 kg and 1239.6 kg) compared with references. (DT&IGSO means IGSO with DT, and HT&IGSO means IGSO with HT). The finally optimal trusses are shown respectively in Fig. 5.52.

**Table 5.24** Comparison of the optimal results for the 10-bar planar truss (Model II)

| Variables | Wu [18] | Tang [21] | DT& GSO | DT& IGSO | HT& GSO | HT& IGSO |
|---|---|---|---|---|---|---|
| $A_1$ | 6.387 | 8.710 | 7.419 | 8.710 | 10.000 | 8.710 |
| $A_2$ | 6.064 | 0.000 | 4.658 | 0 | 0 | 0 |
| $A_3$ | 7.419 | 5.142 | 8.710 | 7.419 | 5.142 | 7.419 |
| $A_4$ | 0.968 | 4.658 | 2.342 | 4.658 | 4.658 | 5.142 |
| $A_5$ | 0.000 | 1.045 | 0 | 0 | 3.097 | 0 |
| $A_6$ | 7.742 | 0.000 | 3.703 | 0 | 0 | 0 |
| $A_7$ | 7.419 | 2.897 | 4.658 | 3.703 | 0 | 3.703 |
| $A_8$ | 2.323 | 2.019 | 1.690 | 1.535 | 3.703 | 1.993 |
| $A_9$ | 0.000 | 8.710 | 3.703 | 8.710 | 0.322 | 7.419 |
| $A_{10}$ | 6.710 | 0.000 | 3.303 | 0 | 0 | 0 |
| $Y_1$ | 20.0 | 22.6 | 20.13 | 19.47 | 21.741 | 20.010 |
| $Y_2$ | 14.1 | 13.4 | 13.22 | 11.75 | 12.205 | 12.386 |
| $Y_3$ | 4.7 | - | 46.00 | - | 21.347 | - |
| Weight (kg) | 1476.0 | 1276.3 | 1351.6 | 1244.9 | 1318.9 | 1239.6 |

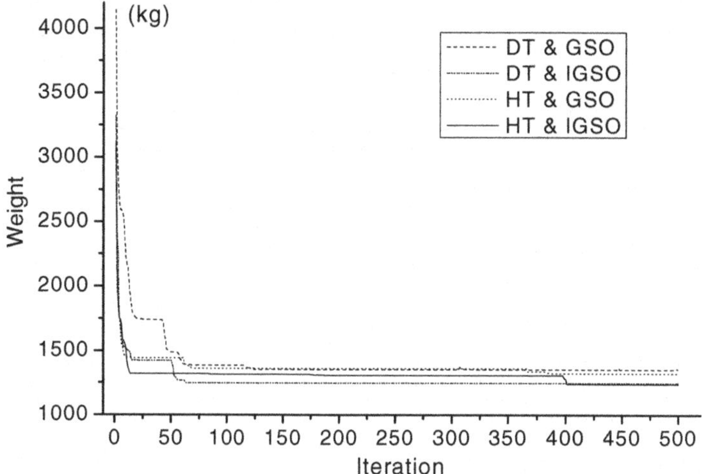


**Fig. 5.51** Convergence curves for the 10-bar planar truss

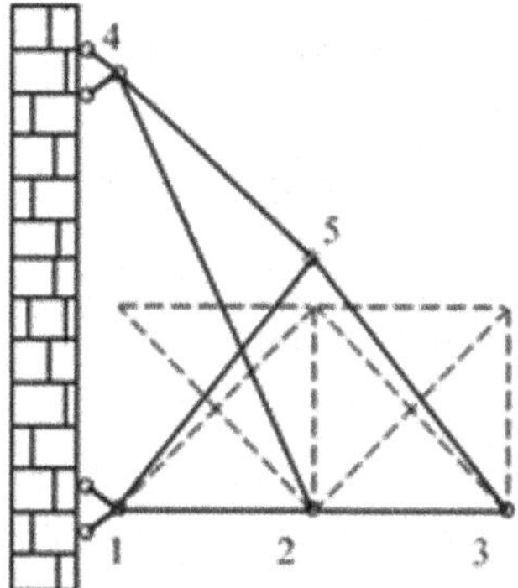


**Fig. 5.52** The optimized 10-bar planar truss

### *(4) A 15-Bar Planar Truss Structure (Model A)*

In this example, the performance of the topology optimization is studied on a 15-bar cantilever truss by Wu [24], Tang [21] and Raj [23]. The geometry of 15-bar truss is shown in Fig. 5.53. A tip load of 44.45 kN is applied to the truss. The stress limit is 172.25 MPa for all members. Young's modulus is specified as $6.89 \times 10^4$ MPa, and the material density as 2.768 kg/m$^3$. The x and y coordinates of joint 2, 3, 6 and 7 are movable. The x coordinates of node 6 and 7 are almost same as that of joint 2 and 3 respectively. Joint 4 and 8 are allowed to move only in the y direction. Hence, this problem has 38 design variables including 15 sizing variables (cross-sectional area of members), 8 configuration variables ($x_2$=$x_6$, $x_3$=$x_7$, $y_2$, $y_3$, $y_4$, $y_6$, $y_7$, $y_8$) and 15 topology variables. Discrete values considered in this example are taken from the set D={0.072, 0.091, 0.112, 0.142, 0.174, 0.185, 0.224, 0.284, 0.384, 0.615, 0.697, 0.757, 0.860, 0.960, 1.138, 1.382, 1.740, 1.806, 2.020, 2.300, 2.460, 3.100, 3.840, 4.240, 4.640, 5.500, 6.000, 7.000, 8.600, 9.219, 11.077, 12.374} ($\times 10^{-3}$ $m^2$). More details can be found from literature [21] and [23].

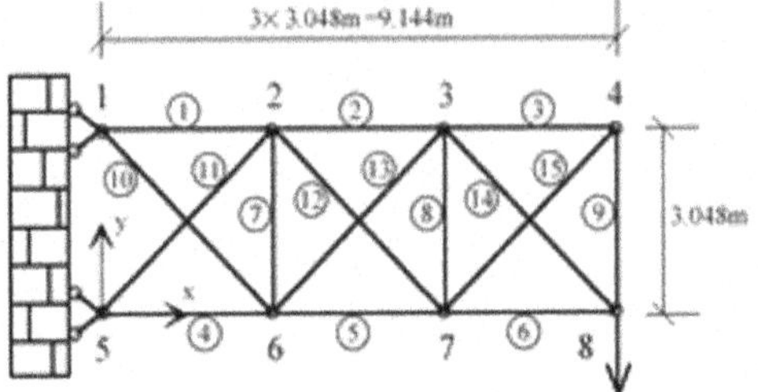


**Fig. 5.53** A 15-bar planar truss

**Table 5.25** Comparison of the optimal results for the 10-bar planar truss (Model A)

| Variables | Wu [18] | Tang [21] | Raj [23] | DT& GSO | DT& IGSO | HT& GSO | HT& IGSO |
|---|---|---|---|---|---|---|---|
| $A_1$ | 0.757 | 0.697 | 0.615 | 8.60 | 6.15 | 8.60 | 6.97 |
| $A_2$ | 0.615 | 0.348 | 0.615 | 6.15 | 6.15 | 2.84 | 6.15 |
| $A_3$ | 0.284 | 0.000 | 0.000 | 0 | 0 | 0 | 0 |
| $A_4$ | 0.86 | 0.697 | 0.697 | 6.15 | 6.97 | 6.15 | 6.97 |
| $A_5$ | 0.615 | 0.615 | 0.348 | 6.15 | 3.48 | 6.15 | 2.84 |
| $A_6$ | 0.112 | 0.284 | 0.348 | 2.84 | 3.48 | 3.48 | 2.84 |
| $A_7$ | 0.284 | 0.000 | 0.000 | 1.74 | 0 | 0.72 | 0 |
| $A_8$ | 0.284 | 0.091 | 0.000 | 1.42 | 1.12 | 1.74 | 0.72 |
| $A_9$ | 0.697 | 0.000 | 0.000 | 0 | 0 | 0 | 0 |
| $A_{10}$ | 0.86 | 0.174 | 0.284 | 0 | 2.84 | 0 | 1.74 |
| $A_{11}$ | 0.112 | 0.174 | 0.142 | 2.24 | 1.42 | 2.84 | 1.74 |
| $A_{12}$ | 0.112 | 0.348 | 0.072 | 3.48 | 0 | 6.15 | 0 |
| $A_{13}$ | 0.224 | 0.091 | 0.224 | 1.74 | 2.84 | 0 | 3.48 |
| $A_{14}$ | 0.224 | 0.284 | 0.348 | 3.48 | 3.48 | 3.48 | 3.48 |
| $A_{15}$ | 0.284 | 0.000 | 0.000 | 0 | 0 | 0 | 0 |
| $X_2$ | 3.129 | 2.841 | 2.728 | 3.008 | 2.54 | 2.602 | 2.54 |
| $X_3$ | 5.883 | 6.158 | 6.209 | 5.589 | 5.829 | 5.872 | 6.142 |
| $Y_2$ | 2.723 | 2.642 | 3.186 | 2.699 | 3.172 | 2.803 | 3.183 |
| $Y_3$ | 3.027 | 2.774 | 2.979 | 3.515 | 2.763 | 3.172 | 2.581 |
| $Y_4$ | 1.536 | - | - | - | - | - | - |
| $Y_6$ | 0.425 | 0.275 | 0.041 | 0.122 | -0.012 | -0.003 | 0.14 |
| $Y_7$ | 0.395 | 0.283 | 0.459 | 0.071 | 0.092 | 0.180 | -0.264 |
| $Y_8$ | 0.931 | 1.241 | 1.275 | 1.385 | 1.356 | 1.524 | 0.193 |
| Weight (kg) | 54.672 | 35.308 | 34.094 | 39.009 | 33.862 | 37.892 | 33.543 |

Both HT and DT are used in this example. Partial discrete topology variables are used in this structure. The best optimal results of twenty random computation about the 15-bar truss by different optimal method is listed in Table 5.24. The finally optimal trusses are shown respectively in Fig. 5.55. The better optimal results 33.862 kg and 33.543 kg of DT&IGSO and HT&IGSO respectively are obtained. Only 80 generations, 48 population and 50 members were needed in computation, which shows the two topology methods incorporated IGSO are powerful techniques for topology optimization. The obvious distinction of Euclidean distance between the results with the two methods, to some extend, shows the complexity of the design space of the topology optimization.

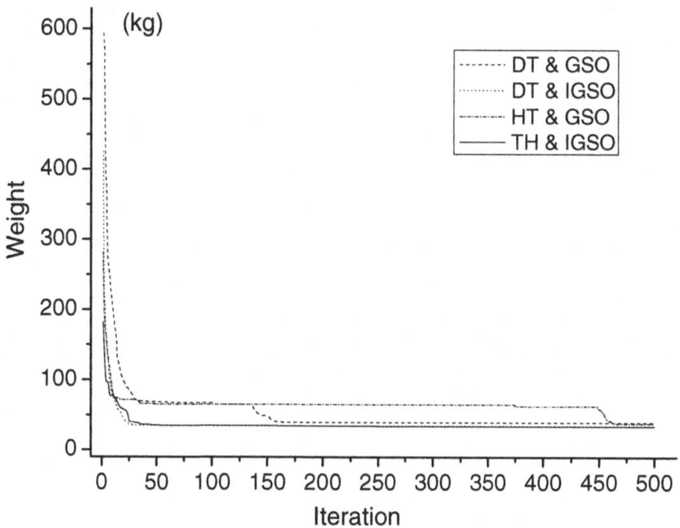


**Fig. 5.54** Convergence curves for the 15-bar planar truss A

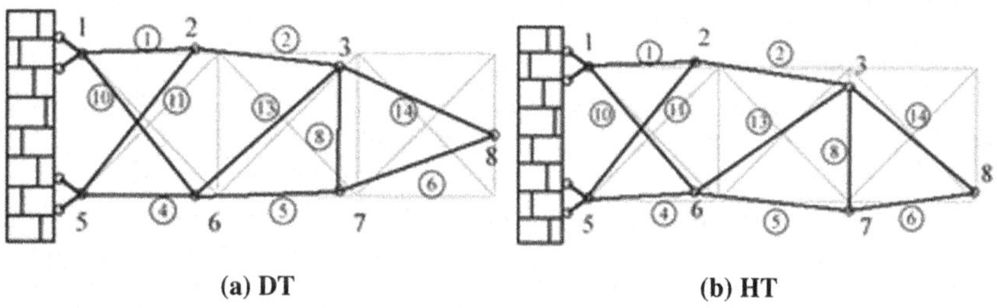


**Fig. 5.55** The optimized 15-bar planar truss A

### *(5) A 15-Bar Planar Truss Structure B*

A different configuration of 15-bar truss topology example is from Sun [22], and the geometry is shown in Fig. 5.56. Young's modulus is specified as $6.897\times10^6$ N/cm$^2$, and the material density as 0.02768 N/cm$^3$. The stress limit is ±17243.5 N/cm$^2$ for all members, and the displacement limit is ±2.032 cm in y direction of node 5. Because of the symmetry, there are 15 design variables including 8 sizing variables ($A_1$=$A_2$, $A_3$=$A_4$, $A_5$=$A_6$, $A_7$=$A_8$, $A_9$=$A_{11}$, $A_{12}$=$A_{13}$, $A_{11}$=$A_{14}$, $A_{15}$), 4 configuration variables ($x_3$=-$x_7$, $x_4$=-$x_8$, $y_4$=$y_8$, $y_6$) and 3 topology variables (2 discrete topology variables). Discrete values considered for this example are taken from the set D={6.452, 9.677, 22.581, 32.258, 45.161, 70.968, 83.871, 103.226, 129.032, 161.29, 193.548} (cm$^2$). This model has two load cases, case 1: node 3, 5, 7 all have a vertical load, $-4.45\times10^5$ N; case 2: node 4, 6, 8 all have a vertical load,$-4.45\times10^5$ N.

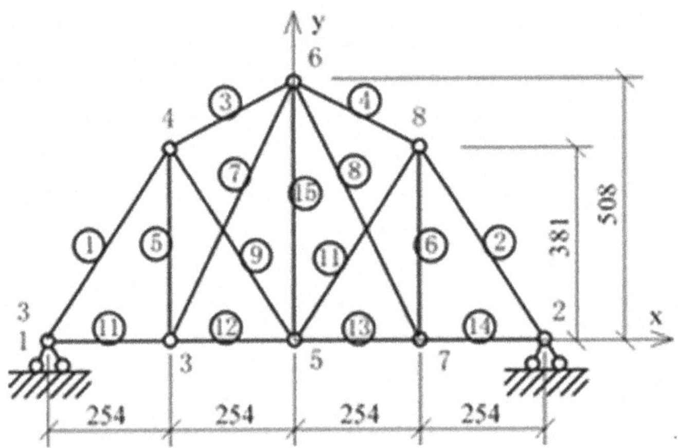

**Fig. 5.56** A 15-bar planar truss B

Table 5.26 is the optimal results of 10-bar planar truss (Model B). Fig. 5.57 is the convergence rate curve.

Because of the symmetry of this truss, [$T_3$, $T_4$, $T_7$, $T_8$, $T_9$, $T_{11}$, $T_{12}$, $T_{13}$], can be discretized to $DT_1$, and [$T_5$, $T_6$] to $DT_2$. Reference [28] did not give the limitation of shaping variables, so the relative boundary are given in this paper as, $-123\le x_3\le123$, $-123\le x_4\le123$, $123\le y_4\le123$ and $123\le y_6\le123$ (cm).

It can be seen from Table 5.32 that the DT&IGSO can get better optimal result (2009.9N) than reference [22] did. As the Fig. 5.57 shows, the optimizer has found a good solution with about 50 iterations. The population of member is 48 and the iteration is 500 times.

The finally optimal truss is shown in Fig. 5.58 and the checking detail of the optimal result with Ansys is listed on Table 5.26.

**Table 26** Comparison of the optimal results for the 10-bar planar truss (Model B)

| Variables | Sun [22] | DT & GSO | DT & IGSO |
|---|---|---|---|
| $A_1$ (cm$^2$) | 70.968 | 70.968 | 70.968 |
| $A_3$ (cm$^2$) | 45.161 | 45.161 | 32.258 |
| $A_5$ (cm$^2$) | 32.258 | 32.258 | 32.258 |
| $A_7$ (cm$^2$) | 0.000 | 0.000 | 6.452 |
| $A_9$ (cm$^2$) | 6.452 | 6.452 | 0.000 |
| $A_{11}$ (cm$^2$) | 6.452 | 6.452 | 6.452 |
| $A_{12}$ (cm$^2$) | 6.452 | 6.452 | 6.452 |
| $A_{15}$ (cm$^2$) | 32.258 | 32.258 | 32.258 |
| $x_3$ (cm) | -189.992 | -417.5 | -445.2 |
| $x_4$ (cm) | -228.224 | -426.2 | -453.7 |
| $y_4$ (cm) | 310.642 | 73.3 | 78.4 |
| $y_6$ (cm) | 401.869 | 222.0 | 273.6 |
| Weight (kg) | 3494 | 2226.1 | 2009.9 |

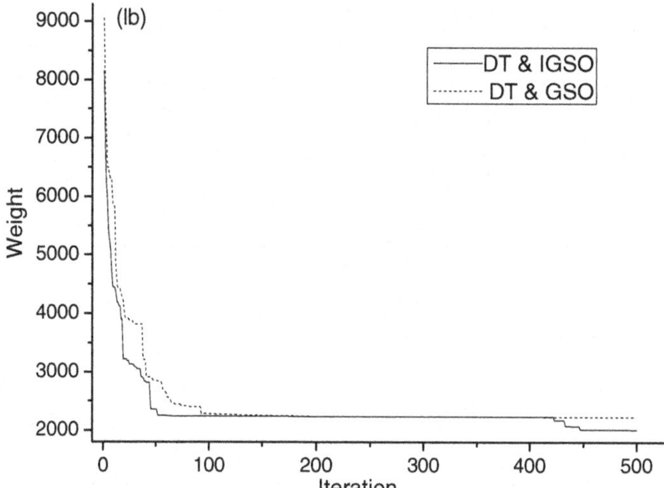


**Fig. 5.57** Convergence curves for the 15-bar planar truss B

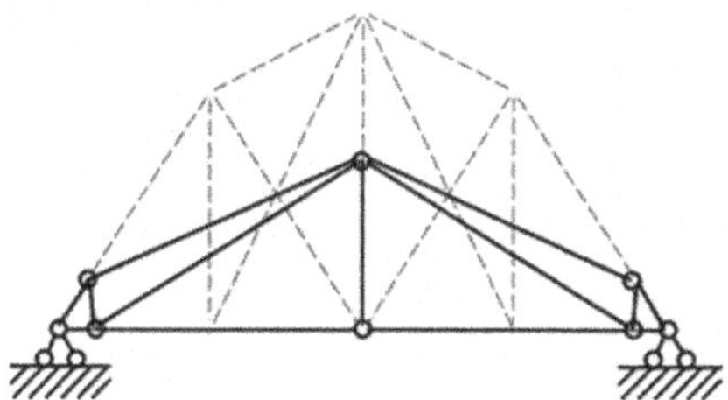

**Fig. 5.58** The optimized 15-bar planar truss B

**Table 5.27** Comparison of the optimal results for the 10-bar planar truss (Model II)

| Load case | Stress /Disp. | Element /Node | Value |
|---|---|---|---|
| | Min stress | 3, 4 | -17239.963 N/cm$^2$ |
| 1 | Max stress | 5, 6 | 13960.399 N/cm$^2$ |
| | Max-displacement | 5 | -2.030 cm |
| | Min stress | 3, 4 | -15684.340 N/cm$^2$ |
| 2 | Max stress | 5, 6 | 702.938 N/cm$^2$ |
| | Max-displacement | 5 | -1.838 cm |

### *(6) A 25-Bar Spatial Truss Structure*

This 25-bar spatial truss example is a sizing and geometry optimization model in [18], and the geometry is shown in Fig. 5.59. However, it becomes a topology optimal problem with adding the topology variables now. Young's modulus is specified as 10000 ksi, and the material density as 0.1lb/in$^3$. The stress limit is ±40 ksi for all members and the displacement limit of node 1~6 is ±0.35 in.. Because of the symmetry of the truss, there are 16 design variables including 8 sizing variables (as table 4 listed) and 5 configuration variables and 3 topology variables. Discrete values considered for this example are taken from the set D={0.1, 0.2, 0.3, 0.4, 0.5, 0.6, 0.7, 0.8, 0.9, 1.0, 1.1, 1.2, 1.3, 1.4, 1.5, 1.6, 1.7, 1.8, 1.9, 2.0, 2.1, 2.2, 2.3, 2.4, 2.5, 2.6, 2.8, 3.0, 3.2, 3.4} (in.$^2$). With the symmetry, the shaping variables grouping and the relative boundary are given in this paper as, $20 \leq x_4 = x_5 = -x_3 = -x_6 \leq 60$, $40 \leq y_3 = y_4 = -y_5 = -y_6 \leq 80$, $90 \leq z_3 = z_4 = z_5 = z_6 \leq 130$, $40 \leq x_8 = x_9 = -x_7 = -x_{10} \leq 80$, $100 \leq y_7 = y_8 = -y_9 = -y_{10} \leq 140$ (in.). The load case is listed on Table 5.5. More details can be found in reference [30].

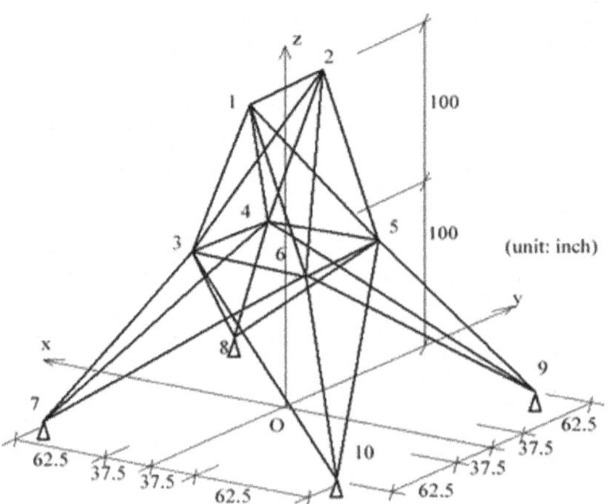


**Fig. 5.59** The optimal topology of 25-bar truss

With the previous analysis, we can know most of bars of the 25-bar spatial truss are indispensable. That means most of topology variables are fixed as $T_k$=1. Therefore, some other topology methods are not fit for this problem, such as DOF [21], Heuristic topology [39] and so on. However, the result shows discretization part of topology variables incorporated with IGSO works well. As Table 5.28 shows, the optimal results of the sizing and geometry optimization are 136.1977 lb [18] and 121.3684 lb with GSO, the results of the DT&GSO and DT&IGSO are120.455 lb and 118.2341 lb respectively. The convergence curve of Fig. 5.67 shows the IGSO is more efficient than GSO. The topology method incorporating GSO and IGSO are both powerful techniques for topology optimization. The finally optimal trusses are shown respectively in Fig. 5.68. As the 15-bar example does, A checking optimal result with Ansys is taken to this example, and the detail is that the min-stress bar is Bar 24 (-15.7831 ksi) and the max-stress bar is Bar 17 (11.2317 ksi). The max-displacement node is node 1 with -0.35 in. value in $y$ direction. The optimized structure is shown in Fig. 5.61.

**Table 5.28** Comparison of the optimal results for the 25-bar planar truss

| Variables | Wu [18] | GSO | DT & GSO | DT & IGSO |
|---|---|---|---|---|
| $A_1$ (in$^2$) | 0.1 | 0.1 | 0.1 | - |
| $A_2$ (in$^2$) | 0.2 | 0.1 | 0.1 | 0.1 |
| $A_3$ (in$^2$) | 1.1 | 1.0 | 1.1 | 1.0 |
| $A_4$ (in$^2$) | 0.2 | 0.1 | - | - |
| $A_5$ (in$^2$) | 0.3 | 0.1 | - | - |
| $A_6$ (in$^2$) | 0.1 | 0.1 | 0.1 | 0.1 |
| $A_7$ (in$^2$) | 0.2 | 0.2 | 0.2 | 0.2 |
| $A_8$ (in$^2$) | 0.9 | 0.9 | 0.9 | 0.9 |
| $x_4$ (in) | 41.07 | 32.149 | 33.743 | 36.026 |
| $y_3$ (in) | 53.47 | 52.742 | 50.597 | 59.044 |
| $z_3$ (in) | 124.60 | 128.23 | 128.847 | 20.085 |
| $x_8$ (in) | 50.80 | 42.401 | 42.500 | 46.717 |
| $y_7$ (in) | 131.48 | 132.603 | 128.956 | 134.817 |
| Weight (lb) | 136.1977 | 121.3684 | 120.4550 | 118.2341 |

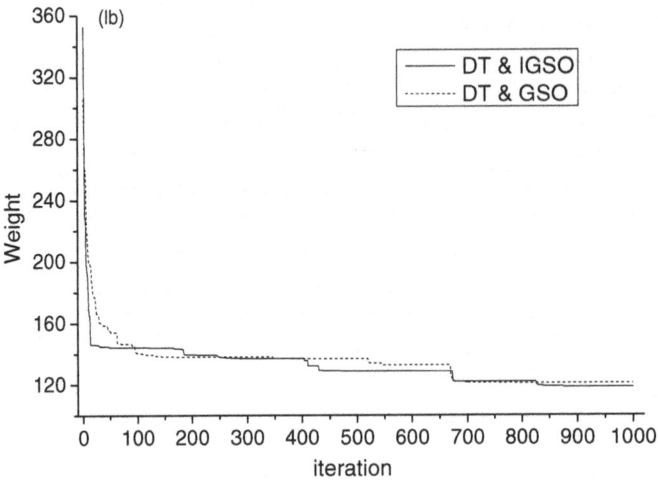


**Fig. 5.60** The optimal topology of 25-bar truss

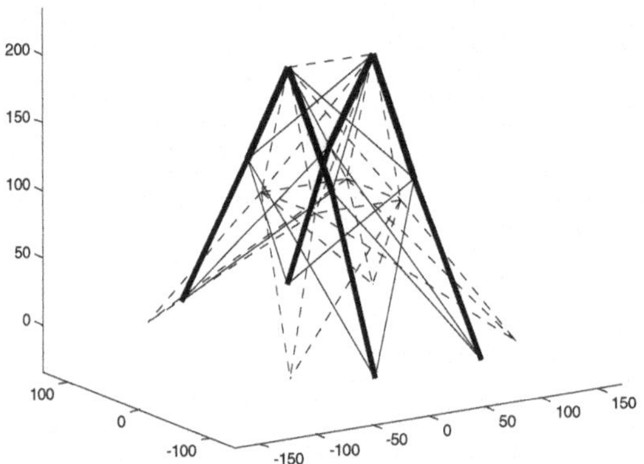


**Fig. 5.61** The optimal topology of 25-bar truss

## 5.7 Conclusions Remarks

In this chapter, an intelligent algorithm, Group Search Optimizer is introduced to deal with civil engineering optimization problem. The GSO is improved (named IGSO) to make sure it is workable and efficient in the complex structural optimal design. According to one of the characteristics of the solution space of structural optimization, which indicates the optimum is usually located near to the boundary, a new searching mechanism, called adhering to the boundary, is presented to improve the searching behaviour of IGSO. The calculation results show IGSO can raise convergence speed greatly. Even if the optimum locates in the boundary, the algorithm will converge to the optimum quickly.

Different optimal problems such as size optimization, geometry optimization, dynamic optimization and topology optimization were solved with the IGSO presented in this chapter. Two topology algorithm, incorporated with IGSO, named HT&IGSO and DT&IGSO were produced and the results proved that the algorithms are workable and robust for the structural topology optimal design.

## References


1. Michell, A.G.M.: The limits of economy of materials in frame structures. Philosophical Magazine 8(47), 589–597 (1904)
2. Prager, W., Rozvany, G.: Optimal layout of grillages. Journal of Structural Mechanics 5(3), 265–294 (1977)
3. Dorn, W., Gomory, R., Greenberg, H.: Automatic design of optimal structures. J. de Mechanique 3(1), 25–52 (1964)
4. Colorni, A., Dorigo, M., Maniezzo, V.: Distributed optimization by ant colonies. In: Proceedings of the First European Conference on Artificial Life, Paris, France, pp. 134–142 (1991)
5. Holland, J.H.: Adaptation in natural and artificial systems. The University of Michigan Press (1975)
6. Schwefel, H.P.: Numerical optimization of computer models. John Wiley, Chichester (1981)
7. Forgel, D.B.: Applying evolutionary programming to selected travelling salesman problems. Cybernetics and Systems 24, 27–36 (1993)
8. Goldberg, D.E.: Genetic algorithms in search. Optimization and Machine Learning. Addison Wesley Publishing Company, MA (1989)
9. Kenndy, J., Eberhart, R.C.: Particle swarm optimization. In: Proceedings of the 1995 IEEE International Conference on Neural Networks, Piscataway, NJ, USA, pp. 1942–1948 (1995)
10. Perez, R.E., Behdinan, K.: Particle swarm approach for structural design optimization. Computers and Structures 85(19-20), 1579–1588 (2007)
11. Li, L.J., Huang, Z.B., Liu, F., Wu, Q.H.: A heuristic particle swarm optimizer for optimization of pin connected structures. Computers and Structures 85(7-8), 340–349 (2007)
12. He, S., Wu, Q.H., Saunder, J.R.: A novel group search optimizer inspired by animal behaviour Ecology. In: 2006 IEEE Congress on Evolutionary Computation, Vancouver, BC, Canada (2006)
13. Li, L.J., Xu, X.T., Liu, F., Wu, Q.H.: The group search optimizer and its application on truss structure design. Advances in Structural Engineering 13(1), 43–51 (2010)
14. Barnard, C.J., Sibly, R.M.: Producers and scroungers: a general model and its application to captive flocks of house sparrows. Animal Behaviour 29(2), 543–550 (1981)
15. Couzin, I.D., Krause, J., Franks, N.R., Levin, S.A.: Effective leadership and decision-making in animal groups on the move. Nature 433(3), 513–516 (2005)
16. O'Brien, W.J., Evans, B.I., Browman, H.I.: Flexible search tactics and efficient foraging in saltatory searching animals. Oecologia 80(1), 100–110 (1989)

17. Geem, Z.W., Kim, J.H., Loganathan, G.V.: A new heuristic optimization algorithm: harmony search. Simulation 76(2), 60–68 (2001)
18. Wu, S.J., Chow, P.T.: Integrated discrete and configuration optimization of trusses using genetic algorithms. Computers & Structures 55(4), 695–702 (1995)
19. Wang, D., Zhang, W.H., Jiang, J.S.: Truss shape optimization with multiple displacement constraints. Computer Methods in Applied Mechanics and Engineering 191(33), 3597–3162 (2002)
20. Luo, Z.F., Rong, J.H., Lu, Y.Z.: Shape optimization for truss dynamics based on genetic algorithm. Machinery design & Manufacture (4), 68–70 (2004) (in Chinese)
21. Tang, W.Y., Yuan, Q.K.: Improved genetic algorithm for topology optimization of truss structures. Chinese Journal of Computational Mechanics 25(1), 79–84 (2008) (in Chinese)
22. Wang, Y.F., Sun, H.C., Huang, L.H.: Studies on optimal topology designs of structures with discrete variables. Acta Mechnica Solida Sinica 19(1), 59–63 (1998) (in Chinese)
23. Rajeev, S., Krishnamoorhty, C.S.: Genetic algorithms based methodologies for design optimization of trusses. Journal of Structural Engineering ASCE 123(3), 350–358 (1997)
24. Hasanqehi, O., Erbatur, F.: Layout optimization of trusses using improved GA methodlogies. ACTA Mechanica 146(1-2), 87–107 (2001)
25. Kaveh, A., Kalatjari, V.: Size/Geometry optimization of trusses by the force method and genetic algorithm. Zeitschrift fur Angewandte Mathematik and Mechanics 84(5), 347–357 (2004)
26. Qian, L.X.: Optimal design of engineering structures. Water and Power Press, Beijing (1983)
27. Thierauf, G., Cai, J.: Parallel evolution strategy for solving structural optimization. Engineering structures 19(4), 318–324 (1997)
28. Cheng, G.D., Gu, Y.X.: Applications of SQP to structural dynamics optimization. Journal of Vibration and Shock 5(1), 12–20 (1986)
29. Xie, Y.M., Steven, G.P.: A simple approaches to structural frequency optimization. Computers & Structures 53(6), 1487–1492 (1994)
30. Zhao, C.B., Steven, G.P., Xie, Y.M.: Simultanceously evolutionary optimization of several natural frequencies of a two dimensional structure. Structure Engineering & Mechanics 7(5), 447–456 (1999)
31. Chen, J.J., Che, J.W., et al.: A review on structural dynamic optimum design. Advances in Mechanics 31(2), 181–192 (2001)
32. Sadek, E.A.: Dynamic optimization of framed structures with variable layout. International Journal for Numerical Methods in Engineering 23(7), 1273–1294 (1986)
33. Lee, K.S., Geem, Z.W.: A new structural optimization method based on the harmony search algorithm. Computers and Structures 82(9-10), 781–798 (2004)
34. Wang, D., Li, J.: Shape optimization of space trusses subject to frequency constraints. Engineering Mechanics 24(4), 129–134 (2007)
35. Kirsch, U.: Optimal topologies of truss structure. Computer Methods in Applied Mechanics and Engineering 72(1), 15–28 (1989)
36. Tan, Z.F., Sun, H.C.: The modified simplex method for topology optimization of space truss structure with multiple loading conditions. Chinese Journal of Theoretical and Applied Mechanics 26(1), 90–97 (1994)
37. Chai, S., Shi, L.S., Sun, H.C.: Topology optimization of truss structures with discrete variables including two kinds of variables. Chinese Journal of Theoretical and Applied Mechanics 31(5), 574–584 (1999) (in Chinese)

38. Cai, W.X., Cheng, G.D.: Simulated annealing algorithm for the topology optimization of truss. Journal of South China University of Technology (Natural Science Edition) 26(9), 78–84 (1998) (in Chinese)
39. Zhu, C.Y., Zhang, X.D., et al.: An improved hybrid genetic algorithm for discrete topology optimization of trusses. Journal of Lanzhou University (Natural Sciences) 41(5), 102–106 (2005) (in Chinese)
40. Jiang, D.J., Zhang, Z.M.: A renew on topology and layout optimization of truss structure. Advances in Science and Technology of Water Resource 26(2), 81–86 (2006)
41. Xu, B.: An investigation on truss structural dynamic topology optimization. Northwestern Polytechnical University (2002)
42. Xie, H.B., Liu, F., Li, L.J.: A topology optimization for truss based on improved group search optimizer. In: 2009 International Conference on Computational Intelligence and Security, Beijing, pp. 244–248 (2009)
43. Xie, H.B., Liu, F., Li, L.J.: Research on topology optimization of truss structure based on the improved group search optimizer. In: 2nd International Symposium on Computational Mechanics. 12th International Conference on Enhancement and Promotion of Computational Methods in Engineering and Science. HK-Macau (2009) Paper number 304

# Chapter 6
# Optimum Design of Structures with Quick Group Search Optimization Algorithm

**Abstract.** Based on the basic principles of an optimization algorithm, group search optimization (GSO) algorithm, two improved GSO, named quick group search optimizer (QGSO) and quick group search optimizer with passive congregation (QGSOPC), are presented in this chapter to deal with structural optimization design tasks. The improvement of QGSO has three main aspects: first, increase the number of 'ranger' when the target stops going forward. Second, use the search strategy of particle swarm optimizer (PSO) by considering the best group member and the best personal member. Employ the step search strategy to replace the visual search strategy. Third, reproduce the 'ranger' with hybrid of the group best member and the personal best member. the QGSOPC is a hybrid QGSO with passive congregation. The QGSO is tested by planar and space truss structures with continuous variables and discrete variables. The QGSOPC is only tested by discrete variables. The calculation results of QGSO and QGSOPC are compared with that of the GSO and HPSO. The results show that the QGSO and QGSOPC algorithms can handle the constraint problems with discrete variables efficiently, and the QGSOPC has more efficient search ability, faster convergent rate and less iterative times to find out the optimum solution.

## 6.1 Introduction

Bionic optimization algorithms, notably Evolutionary Algorithms (EAs) [1] had been widely used to solve various scientific and engineering problems and have been extensively used in structural optimization problems recently. Thereinto, Ant Colony Optimizer (ACO), Particle Swarm Optimizer (PSO) and Group Search Optimization inspired by Dorigo [2], Kenndy & Eberthart [3], and Barnard [4] respectively are three typical representatives. The 'Individual Behavior' of group is mainly considered by ACO and PSO. It is based on the evolutionary theory to consider such evolution behaviour. These two algorithms belong to 'evolutionary strategies' areas in a way. ACO is good at solving complex and combination optimization problems with discrete variables but shows a low evolutionary velocity [5]. PSO suits for continuous and discrete variables optimization problems but is easy to entrap in local minima. Also they are time consuming in optimizing

L. Li & F. Liu: Group Search Optimization for Applications in Structural Design, ALO 9, pp. 161–206.
springerlink.com

complex structures [6-8]. As we know, gregarious is a common phenomenon in the animality, 'information communion' and 'mutual cooperation' is another important aspect of group behavior. Group Search Optimizer (GSO) is such an optimization algorithm which is based upon this group speciality and also has been successfully used in structural optimal design with continuous variables [9-11]. However, as the practical engineering problems use the bars, the areas of cross-sections of which are produced by a certain specifications, structural optimization with discrete variable design works more obvious significance and value of practical application.

To improve the efficiency of GSO for structural optimization, the basic principle of GSO has been ameliorated in this chapter. Then an improved algorithm for structural optimization is proposed, named a quick group search optimizer (QGSO). Compared with the basic GSO algorithm and the improved PSO algorithm named HPSO, this algorithm (QGSO) has preferable convergence rate and accuracy.

## 6.2 The Introduction of GSO

GSO is inspired by the food searching behavior and group living theory of social animals, such as birds, fish and lions. The foraging strategies of these animals mainly include: (1) producing, e.g., searching for food; and (2) joining (scrounging), e.g., joining resources uncovered by others. GSO also employs 'rangers' which perform random walks to avoid entrapment in local minima. Therefore, in GSO, a group consists of three kinds of members: producers, scroungers and rangers. At each iteration, a group member, located in the most promising area, conferring the best fitness value, is chosen as the producer. It locates in the most promising area and stay still. The other group members are selected as scroungers or rangers by random. Then, each scrounger make a random walk towards the producer, and each rangers make a random walk in arbitrary direction. It is also assumed that the producer, scroungers and rangers do not differ in their relevant phenotypic characteristics. Therefore, they can switch among the three roles. The GSO behaves as follows:

In an n-dimensional search space, the $i_{th}$ member at the $k_{th}$ searching bout (iteration) has a current position $X_i^k \in R^n$, a head angle $\varphi_i^k = (\varphi_{i1}^k, \ldots, \varphi_{i(n-1)}^k) \in R^{n-1}$ and a head direction $D_i^k(\varphi_i^k) = (d_{i1}^k, \ldots, d_{in}^k) \in R^n$ which can be calculated from $\varphi_i^k$ via a Polar to Cartesian coordinates transformation:

$$
\begin{aligned}
d_{i1}^k &= \prod_{p=1}^{n-1} \cos(\varphi_{ip}^k) \\
d_{ij}^k &= \sin(\varphi_{i(j-1)}^k) \cdot \prod_{p=i}^{n-1} \cos(\varphi_{ip}^k) \\
d_{in}^k &= \sin(\varphi_{i(n-1)}^k)
\end{aligned}
\tag{6.1}
$$

In GSO, a group consists of three kinds of members: producer scroungers and rangers. In the GSO algorithm, at the $k_{th}$ iteration, the producer $X_p$ behaves as follows:

(1) The producer will scan at zero degree and then scan laterally by randomly sampling three points in the scanning field: one point at zero degree:

$$X_z = X_p^k + r_1 l_{\max} D_p^k(\varphi^k) \tag{6.2}$$

one point in the left hand side hypercube:

$$X_l = X_p^k + r_1 l_{\max} D_p^k(\varphi^k - r_2\theta_{\max}/2) \tag{6.3}$$

and one point in the right hand side hypercube:

$$X_r = X_p^k + r_1 l_{\max} D_p^k(\varphi^k + r_2\theta_{\max}/2) \tag{6.4}$$

where $r_1 \in R^1$ is a normally distributed random number with mean 0 and standard deviation 1 and $r_2 \in R^{n-1}$ is a random sequence in the range (0, 1). The maximum pursuit distance $l_{\max}$ is calculated from:

$$l_{\max} = |U_i - L_i| = \sqrt{\sum_{i=1}^{n}(U_i - L_i)^2} \tag{6.5}$$

where $L_i$ and $U_i$ are the lower and upper bounds for the $i_{th}$ dimension.

(2) The producer will then find the best point with the best resource (fitness value). If the best point has a better resource than its current position, then it will fly to this point. Or it will stay in its current position and turn its head to a new angle:

$$\varphi^{k+1} = \varphi^k + r_2\alpha_{\max} \tag{6.6}$$

where $\alpha_{\max}$ is the maximum turning angle.

(3) If the producer cannot find a better area after *a* iterations, it will turn its head back to zero degree:

$$\varphi^{k+a} = \varphi^k \tag{6.7}$$

where *a* is a constant.

At the $k_{th}$ iteration, the area copying behavior of the $i_{th}$ scrounger can be modeled as a random walk towards the producer:

$$X_i^{k+1} = X_i^k + r_3(X_p^k - X_i^k) \tag{6.8}$$

where $r_3 \in R^n$ is a uniform random sequence in the range (0, 1).

Besides the producer and the scroungers, a small number of rangers have been also introduced into GSO algorithm. Random walks, which are thought to be the most efficient searching method for randomly distributed resources, are employed by rangers. If the $i_{th}$ group member is selected as a ranger, at the $k_{th}$ iteration, firstly, it generates a random head angle $\varphi_i$ :

$$\varphi^{k+1} = \varphi^k + r_2\alpha_{\max} \tag{6.9}$$

where $\alpha_{\max}$ is the maximum turning angle; and secondly, it chooses a random distance:

$$l_i = a \cdot r_1 l_{\max} \tag{6.10}$$

and move to the new point:

$$X_i^{k+1} = X_i^k + l_i D_i^k(\varphi^{k+1}) \tag{6.11}$$

## 6.3 Constraint Handling Method

In various fields of science and engineering, extremal problem solving is difficult because of the particle constraint. Solving these extremal optimal problems with constraint is called constrained optimization (CO).

A constraint condition for the minimization problem is concerned, not only to make the objective function value continuous reduction in the iterative process, but also to take note of the feasibility of the solution. In general, the conditions for solving constrained extremum problems are usually changed into unconstrained optimization problem, nonlinear programming problem into a linear programming problem, complex problems into simple problems.

There are many traditional methods such as Feasible Direction, Gradient Projection Method, and Active Set Method etc, to solve constrained optimization problems. These methods have different scope and limitations, most of them requires gradient information, the objective function or constraints continuously differentiable. However, in practical engineering, it's incapable of knowing the objective function or it can't be expressed in an explicit function.

Penalty functions have been commonly used to deal with constraints. However, the major disadvantage of using the penalty functions is that some tuning parameters are added in the algorithm and the penalty coefficients have to be tuned in order to balance the objective and penalty functions [12].

Another way to handle constraints is 'fly-back mechanism' [13]. For most of the optimization problems containing constraints, the global minimum locates on or close to the boundary of a feasible design space. The particles are initialized in the feasible region. When the optimization process starts, the particles fly in the feasible space to search the solution. If any one of the particles flies into the infeasible region, it will be forced to fly back to the previous position to guarantee a feasible solution. The particle which flies back to the previous position may be closer to the boundary at the next iteration. This makes the particles to fly to the global minimum in a great probability. Therefore, such a 'fly-back mechanism' technique is suitable for handling the optimization problem containing the constraints. Compared with the other constraint handling techniques, this method is relatively simple and easy to implement. Some experimental results show that it can find a better solution with a fewer iterations than the other techniques.

The constraint handling methods used in the QGSO will be introduced in this chapter.

## 6.4 The Quick Group Search Optimizer (QGSO)

The QGSO inherits the Producer-Scrounger model from the GSO, and employs the random search strategy. The improvement has three main aspects: first, increase the number of 'ranger' when the target stops going forward. Second, use the search strategy of PSO by considering the best group member and the best personal member. Employ the step search strategy to replace the visual search strategy. Third, reproduce the 'ranger' with hybrid of the group best member and the personal best member. The QGSO behaves as follows [14]:

In a $n$-dimensional search space, the $i_{th}$ member at the $k_{th}$ searching bout (iteration) has a current position $X_i^k \in R^n$, which is randomly initialized before the iterative process begins. A group member, conferred the best fitness value, is chosen as the producer. Its position is made of $X_{Gbest}^k$. Randomly, the rest members are selected by a certain probability ( $w_3$ ) as scroungers. Every scrounger goes forward by a random walk:

$$X_i^{k+1} = X_i^k + w_1 r(X_{Gbest}^k - X_i^k) + w_2 r(X_{i,Pbest}^k - X_i^k) \tag{6.12}$$

where $r_1 \in R^n$ is a random sequence in the range (0, 1), $w_1$ and $w_2$ are the information transfer factor like the acceleration constants in HPSO. While taking a walk, the scroungers are not along to accept the information of the producer, but also consider the best position they have ever been ( $X_{i,Pbest}^k$ ). $X_{i,Pbest}^k$ is the best position the $i_{th}$ member at the $k_{th}$ searching bout has ever been.

Purposefully looking for the next producer, the other members selected as rangers will then perform ranging:

$$X_i^{k+1} = 0.9 \times leftflag \times X_{Pbest}^k + changeflag \times X_{Gbest}^k \tag{6.13}$$

where 0.9 is a mutation of dimension, considering that ranger is hybridized by the best group member and the best personal member. *changeflag* is the flag to permit the dimension to mutate, *leftflag* is the flag without mutation. They are calculated from:

$$changeflag = rand(n,1) < w4 \tag{6.14}$$

$$leftflag = ones(n,1) - changeflag \tag{6.15}$$

where rand(n, 1) is a function to provide a random sequence of n dimensions; and ones(n,1) is the function to provide a n dimensions with 1. $w_4$ is the component mutation probability. The equation (6.14) returns a vector of Boolean value by comparing the random numbers and the component mutation probability ( $w_4$ ) of operation '<'. The equation (6.13) shows that based on the best place they have ever been, the rangers exchange information with producer by a certain probability. Therefore, there is enough information to make the rangers find the next producer quickly.

**Table 6.1** The pseudo-code for the QGSO

```
Set k=1;
Randomly initialize positions and velocities of all member;
   FOR (each particle i in the initial group)
         WHILE (the constraints are violated)
               Randomly re-generate the current particle Xi
         END WHILE
   END FOR
WHILE (the termination conditions are not met)
   FOR (each member i in the group)
         Calculate the fitness value of current member: f(X_i), find the producer of the group,
      and make its position X^k_best . If a random number < w_3 , then X_i is selected as scrounger
      to execute equation (6.12). Otherwise, X_i is selected as ranger to execute equation
      (6.13).
         Check feasibility stage I: Check whether each component of the current vector
      violates its corresponding boundary or not. If it does, select the corresponding
      component of the vector from pbest swarm randomly.
         Check feasibility stage II: Check whether the current particle violates the problem
      specified constraints or not. If it does, reset it to the previous position X_ik-1.
         Calculate the fitness value f(X_ik) of the current particle.
         Update pbest: Compare the fitness value of pbest with f(X_ik). If the f(X_ik) is better
than the fitness value of pbest, set pbest to the current position X_ik.
         Update gbest: Find the global best position in the swarm. If the f(X_ik) is better than the
      fitness value of gbest, then gbest is set to the position of the current particle X_ik.
   END FOR
   Set k=k+1
END WHILE
```

To handle the constraint problems, two methods will be used:

Firstly, for the members out of the variables' boundary, a harmony search (HS) scheme will be employed. There is a matrix named "*Pbest*" stores the best fitness value of each member. When one of the components of the vector (member) violates its variables' boundary, it will be replaced by corresponding component of the vector from "*Pbest*" matrix randomly.

Secondly, for the members out of the stress or the displacement boundary, they will be punished by given a biggish value.

### 6.4.1 *The QGSO Algorithm for the Discrete Variables*

A structural optimization design problem with discrete variables can be formulated as a nonlinear programming problem. In the size optimization for a truss structure, the cross-section areas of the truss members are selected as the design variables. Each of the design variables is chosen from a list of discrete cross-sections based on production standard. The objective function is the structure weight. The design cross-sections must also satisfy some inequality constraints equations, which restrict the discrete variables. The optimization design problem for discrete variables can be expressed as follows [15]:

$$\min f\left(x^1, x^2, ..., x^d\right),\quad d = 1, 2, \cdots, D$$

$$\text{subjected to: } g_q\left(x^1, x^2, ..., x^d\right) \le 0,\quad d = 1, 2, \cdots, D,\quad q = 1, 2, \cdots, M$$

$$x^d \in S_d = \left\{X_1, X_2, \cdots, X_p\right\}$$

where $f\left(x^1, x^2, ..., x^d\right)$ is the truss's weight function, which is a scalar function. And $x^1, x^2, ..., x^d$ represent a set of design variables. The design variable $x^d$ belongs to a scalar $S_d$, which includes all permissive discrete variables $\left\{X_1, X_2, ... X_p\right\}$. The inequality $g_q\left(x^1, x^2, ..., x^d\right) \le 0$ represents the constraint functions. The letter *D* and *M* are the number of the design variables and inequality functions respectively. The letter *p* is the number of available variables.

Considering the areas of cross-sections aren't continuum, when the QGSO algorithm is used to optimize problems with discrete variables, a mapping function is usually created to make the discrete section areas correspond to the continuum integers from small to large. Suppose a discrete set $A_n$ with n discrete variables, by arranging from small to large:

$$A_n = \{X_1, X_2, \cdots, X_j, \cdots X_n\},\quad 1 \le j \le n$$

Employ a mapping function to replace the discrete values of $A_n$ with its serial numbers like this:

$$h(j) = X_j$$

The discrete values were replaced by the serial numbers to keep the searching with continuum values and avoid declining of search efficiency. Suppose that there are p members in the search space with D dimension. And the position of the $i_{th}$ member is denoted with vector $x_i$ as:

$$x_i = (x_i^1, x_i^2, \cdots, x_i^d, \cdots, x_i^D),\quad 1 \le d \le D,\quad i = 1, \cdots, p$$

in which, $x_i^d \in \{1, 2, \cdots, j, \cdots, n\}$ corresponds to the discrete variables $\{X_1, X_2, \cdots, X_j, \cdots X_n\}$ by mapping function $h(j)$. After that, all of the members will search in the continuum space which is the integer space. Each component of vector $x_i$ is integer. Accordingly, expressions (12) and (13) become:

$$X_i^{k+1} = Floor(X_i^k + w_1 r(P_g^k - X_i^k) + w_2 r(P_i^k - X_i^k)) \tag{6.16}$$

$$X_i^{k+1} = Floor(0.9 \times \text{leftflag} \times P_i^k + \text{changeflag} \times P_g^k) \tag{6.17}$$

in which *Floor* is a function rounding to negative infinity. For the discrete variable cases, there is no change for the objective function and constraints before being substitution into equation. The iterated integers are turn into areas of cross-sections correspondingly by the mapping function.

## 6.5 The Application of the QGSO on Truss Structures with Continuous Variables

In this section, five pin-connected structures commonly used in literature are selected as benchmark problems to test the QGSO.

The examples given in the simulation studies include

- a 10-bar planar truss structure subjected to four concentrated loads;
- a 17-bar planar truss structure subjected to a single concentrated load at its free end;
- a 22-bar spatial truss structure subjected to three load cases;
- a 25-bar spatial truss structure subjected to two load cases;
- a 72-bar spatial truss structure subjected to two load cases.

All these truss structures are analysed by the finite element method (FEM). For each structural optimization problems, the objective function is the weight of a truss. The areas of cross-sections of bar members are normally selected as the design variables.

The QGSO, GSO, HPSO schemes are applied respectively to the examples and the results are compared in order to evaluate the performance of the modified algorithm. The population size is set to at 50. For the GSO algorithm, 20% of the population is selected as rangers; the initial head angle $\varphi_0$ of each individual is set to be $\pi/4$. The constant a is given by round$(\sqrt{n+1})$. The maximum pursuit angle $\theta_{max}$ is $\pi/a^2$. The maximum turning angle $\alpha$ is set to be $\pi/2a^2$. For the HPSO algorithm, the inertia weight ($w$) is starting at 0.9 and ending at 0.4 by linearity descending. The acceleration constants $c_1$ and $c_2$ is set to be 0.8. The passive congregation coefficient $c_3$ is 0.6. For the QGSO algorithm, when target goes forward, the parameters are set as: information transfer factor $w_1 = w_2$ =4, selected probability $w_3$ =0.2, component mutation probability $w_4$ =0.65. Otherwise the parameters are set as: information transfer factor $w_1$ =0.8, $w_2$ =1.5, selected probability $w_3$ =0.35, component mutation probability $w_4$ =0.85.

### 6.5.1 Numerical Examples

#### (1) The 10-Bar Planar Truss Structure

The 10-bar truss structure, shown in Fig. 6.1, has previously been analyzed by many researchers, such as Lee [16], Schmit [17], Rizzi [18], and Li [19]. The material density is 0.1 lb/in$^3$ and the modulus of elasticity is 10,000 ksi. The members are subjected to the stress limits of ±25 ksi. All nodes in both vertical and horizontal directions are subjected to the displacement limits of ±2.0 in.. There are 10 design variables in this example and the minimum permitted cross-sectional area of each member is 0.1 in.$^2$. $P_1$=100 kips and $P_2$=0.

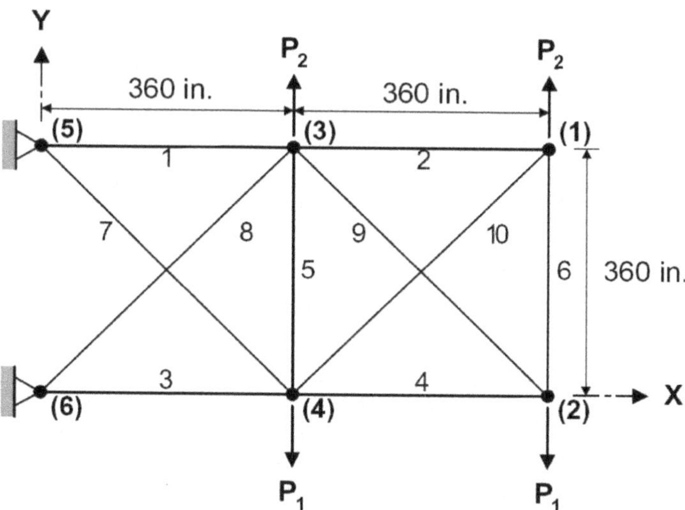


**Fig. 6.1** A 10-bar planar truss structure

For these case, the GSO and the QGSO algorithms achieve the best solutions after 3,000 iterations. However, the latter is closer to the best solution than the former after about 500 iterations. The QGSO algorithm displays a faster convergence rate than the GSO and HPSO algorithm in this example. The performance of the GSO algorithm is the worst among the three. Tables 6.2 shows the solutions. Fig. 6.2 provides a comparison of the convergence rates of the three algorithms.

**Table 6.2** Comparison of the designs for the 10-bar planar truss (Case 1)

| | | Optimal cross-sectional areas (in.$^2$) | | | | | |
|---|---|---|---|---|---|---|---|
| Variables | | Schmit [17] | Rizzi [18] | Lee [16] | Li [19] HPSO | GSO | QGSO |
| 1 | $A_1$ | 33.43 | 30.73 | 30.15 | 30.704 | 30.569 | 30.949 |
| 2 | $A_2$ | 0.100 | 0.100 | 0.102 | 0.100 | 0.100 | 0.100 |
| 3 | $A_3$ | 24.26 | 23.93 | 22.71 | 23.167 | 22.974 | 23.010 |
| 4 | $A_4$ | 14.26 | 14.73 | 15.27 | 15.183 | 15.148 | 15.049 |
| 5 | $A_5$ | 0.100 | 0.100 | 0.102 | 0.100 | 0.100 | 0.100 |
| 6 | $A_6$ | 0.100 | 0.100 | 0.544 | 0.551 | 0.547 | 0.533 |
| 7 | $A_7$ | 8.388 | 8.542 | 7.541 | 7.460 | 7.493 | 7.540 |
| 8 | $A_8$ | 20.74 | 20.95 | 21.56 | 20.978 | 21.159 | 21.388 |
| 9 | $A_9$ | 19.69 | 21.84 | 21.45 | 21.508 | 21.556 | 21.089 |
| 10 | $A_{10}$ | 0.100 | 0.100 | 0.100 | 0.100 | 0.100 | 0.101 |
| Weight (lb) | | 5089.0 | 5076.66 | 5057.88 | 5060.92 | 5128.94 | 5062.29 |

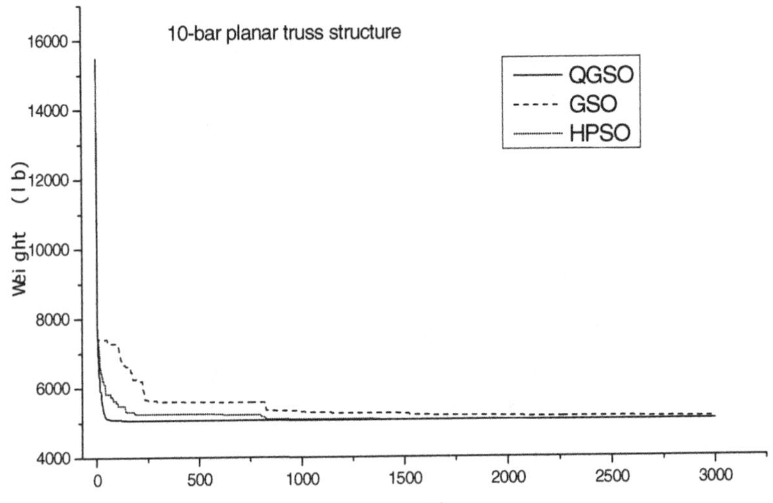


**Fig. 6.2** Comparison of the convergence rates of the three algorithms for the 10-bar truss structure

### *(2) The 17-Bar Planar Truss Structure*

The 17-bar truss structure, shown in Fig. 6.3, had been analyzed by Khot [20], Adeli [21], Lee [16] and Li [19]. The material density is 0.268 lb/in$^3$ and the modulus of elasticity is 30,000 ksi. The members are subjected to the stress limits of ±50 ksi. All nodes in both directions are subjected to the displacement limits of ±2.0 in.. There are 17 design variables in this example and the minimum permitted cross-sectional area of each member is 0.1 in.$^2$. A single vertical downward load of

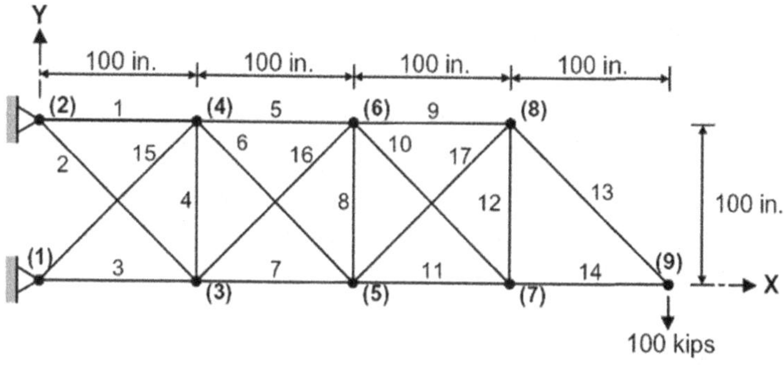


Fig. 6.3 A 17-bar planar truss structure

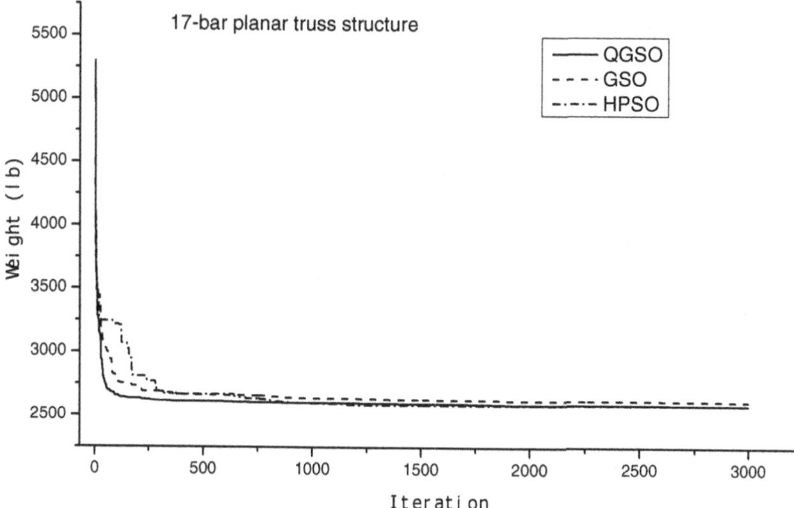


Fig. 6.4 Comparison of the convergence rates of the three algorithms for the 17-bar truss structure

100 kips at node 9 is considered. Table 6.3 shows the solutions and Fig. 6.4 compares the convergence rates of the three algorithms.

Both the GSO and QGSO algorithms achieve a good solution after 3,000 iterations and the latter shows a better convergence rate than the former, especially at the early stage of iterations.

### *(3) The 22-Bar Spatial Truss Structure*

The 22-bar spatial truss structure, shown in Fig. 6.5, had been studied by Lee [16] and Li [19, 9]. The material density is 0.1 lb/in.$^3$ and the modulus of elasticity is 10,000 ksi. The stress limits of the members are listed in Table 6.4. All nodes in all three directions are subjected to the displacement limits of ±2.0 in. Three load cases are

**Table 6.3** Comparison of the designs for the 17-bar planar truss

| Variables | | Optimal cross-sectional areas (in.$^2$) | | | | | |
|---|---|---|---|---|---|---|---|
| | | Khot [20] | Adeli [21] | Lee [16] | Li [19] HPSO | GSO | QGSO |
| 1 | $A_1$ | 15.930 | 16.029 | 15.821 | 15.896 | 15.940 | 16.133 |
| 2 | $A_2$ | 0.100 | 0.107 | 0.108 | 0.103 | 0.646 | 0.150 |
| 3 | $A_3$ | 12.070 | 12.183 | 11.996 | 12.092 | 12.541 | 12.151 |
| 4 | $A_4$ | 0.100 | 0.110 | 0.100 | 0.100 | 0.331 | 0.100 |
| 5 | $A_5$ | 8.067 | 8.417 | 8.150 | 8.063 | 7.361 | 8.263 |
| 6 | $A_6$ | 5.562 | 5.715 | 5.507 | 5.591 | 4.920 | 5.485 |
| 7 | $A_7$ | 11.933 | 11.331 | 11.829 | 11.915 | 11.072 | 11.822 |
| 8 | $A_8$ | 0.100 | 0.105 | 0.100 | 0.100 | 0.335 | 0.103 |
| 9 | $A_9$ | 7.945 | 7.301 | 7.934 | 7.965 | 8.535 | 7.686 |
| 10 | $A_{10}$ | 0.100 | 0.115 | 0.100 | 0.100 | 0.385 | 0.101 |
| 11 | $A_{11}$ | 4.055 | 4.046 | 4.093 | 4.076 | 4.525 | 4.087 |
| 12 | $A_{12}$ | 0.100 | 0.101 | 0.100 | 0.100 | 0.237 | 0.101 |
| 13 | $A_{13}$ | 5.657 | 5.611 | 5.660 | 5.670 | 6.034 | 5.760 |
| 14 | $A_{14}$ | 4.000 | 4.046 | 4.061 | 3.998 | 3.916 | 4.000 |
| 15 | $A_{15}$ | 5.558 | 5.152 | 5.656 | 5.548 | 5.149 | 5.427 |
| 16 | $A_{16}$ | 0.100 | 0.107 | 0.100 | 0.103 | 0.605 | 0.142 |
| 17 | $A_{17}$ | 5.579 | 5.286 | 5.582 | 5.537 | 5.416 | 5.515 |
| Weight (lb) | | 2581.89 | 2594.42 | 2580.81 | 2581.94 | 2582.85 | 2582.93 |

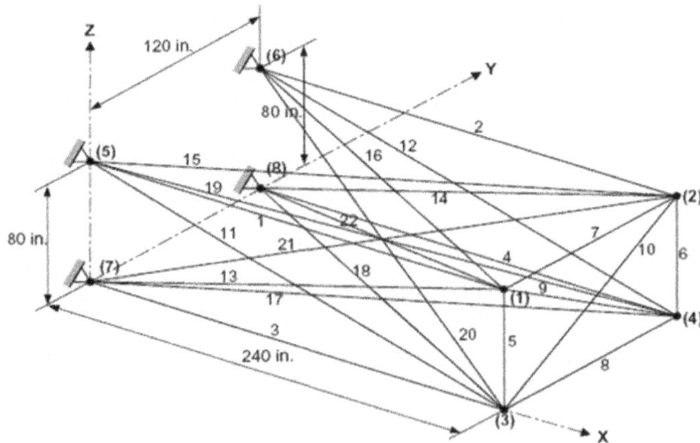


**Fig. 6.5** A 22-bar spatial truss structure

listed in Table 6.5. There are 22 members, which fall into 7 groups, as follows: (1) $A_1$~$A_4$, (2) $A_5$~$A_6$, (3) $A_7$~$A_8$, (4) $A_9$~$A_{10}$, (5) $A_{11}$~$A_{14}$, (6) $A_{15}$~$A_{18}$, and (7) $A_{19}$~$A_{22}$. The minimum permitted cross-sectional area of each member is 0.1 in.$^2$.

**Table 6.4** Member stress limits for the 22-bar spatial truss structure

| Variables | | Compressive stress limitations (ksi) | Tensile stress Limitation (ksi) |
|---|---|---|---|
| 1 | $A_1$ | 24.0 | 36.0 |
| 2 | $A_2$ | 30.0 | 36.0 |
| 3 | $A_3$ | 28.0 | 36.0 |
| 4 | $A_4$ | 26.0 | 36.0 |
| 5 | $A_5$ | 22.0 | 36.0 |
| 6 | $A_6$ | 20.0 | 36.0 |
| 7 | $A_7$ | 18.0 | 36.0 |

**Table 6.5** Load cases for the 22-bar spatial truss structure

| Node | Case 1 (kips) | | | Case 2 (kips) | | | Case 3 (kips) | | |
|---|---|---|---|---|---|---|---|---|---|
| | $P_X$ | $P_Y$ | $P_Z$ | $P_X$ | $P_Y$ | $P_Z$ | $P_X$ | $P_Y$ | $P_Z$ |
| 1 | -20.0 | 0.0 | -5.0 | -20.0 | -5.0 | 0.0 | -20.0 | 0.0 | 35.0 |
| 2 | -20.0 | 0.0 | -5.0 | -20.0 | -50.0 | 0.0 | -20.0 | 0.0 | 0.0 |
| 3 | -20.0 | 0.0 | -30.0 | -20.0 | -5.0 | 0.0 | -20.0 | 0.0 | 0.0 |
| 4 | -20.0 | 0.0 | -30.0 | -20.0 | -50.0 | 0.0 | -20.0 | 0.0 | -35.0 |

**Table 6.6** Comparison of the designs for the 22-bar spatial truss structure

| Variables | | Optimal cross-sectional areas (in.$^2$) | | | |
|---|---|---|---|---|---|
| | | Lee [16] | Li [19] HPSO | Li [9] GSO | QGSO |
| 1 | $A_1$ | 2.588 | 2.613 | 2.803 | 2.772 |
| 2 | $A_2$ | 1.083 | 1.151 | 1.197 | 1.2243 |
| 3 | $A_3$ | 0.363 | 0.346 | 0.332 | 0.3694 |
| 4 | $A_4$ | 0.422 | 0.419 | 0.458 | 0.4042 |
| 5 | $A_5$ | 2.827 | 2.797 | 2.634 | 2.6182 |
| 6 | $A_6$ | 2.055 | 2.093 | 2.104 | 1.8667 |
| 7 | $A_7$ | 2.044 | 2.022 | 2.003 | 2.2652 |
| Weight (lb) | | 1022.23 | 1023.90 | 1026.02 | 1025.22 |

In this example, all the algorithms have converged after 50 iterations. The optimum speed obtained by using the QGSO algorithm is better than that obtained by the HPSO and GSO algorithms. Table 6.6 shows the optimal solutions of the four algorithms and Fig. 6.6 and 6.7 provide the convergence rates of three of the four algorithms.

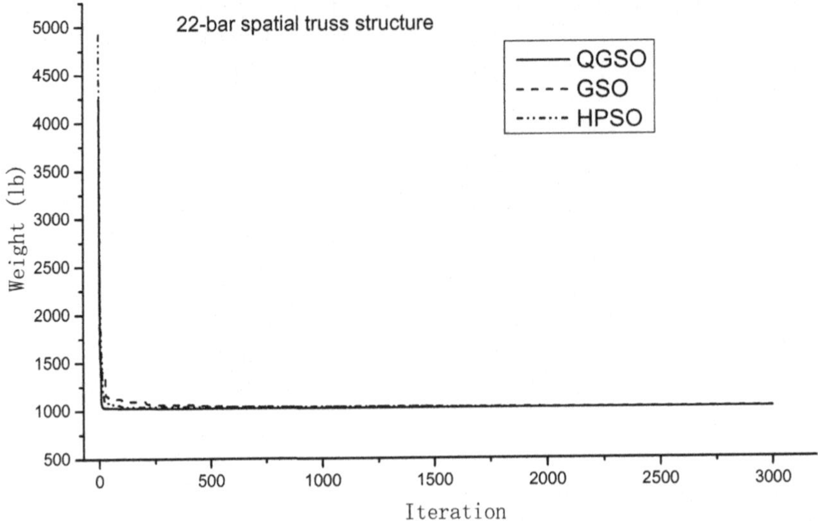


**Fig. 6.6** Comparison of the convergence rates of the three algorithms for the 22-bar truss structure

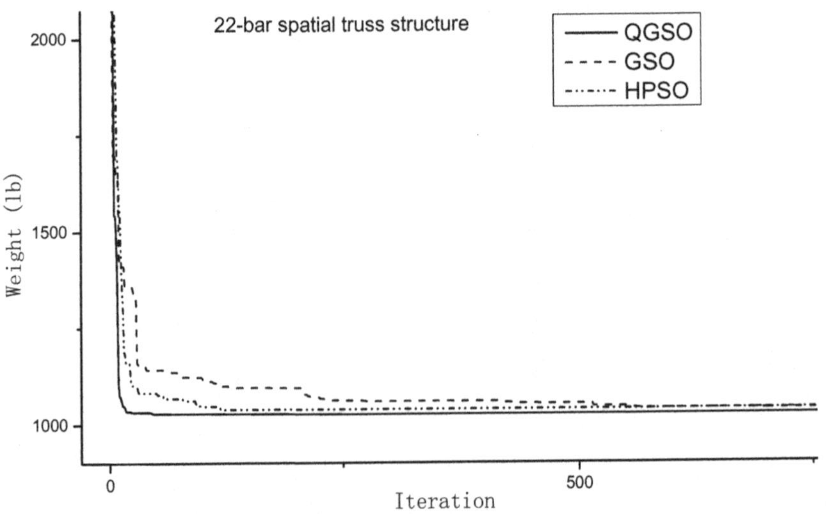


**Fig. 6.7** Comparison of convergence rates of three algorithms for the 22-bar truss structure (Magnified partly)

### *(4) The 25-Bar Spatial Truss Structure*

The 25-bar spatial truss structure shown in Fig. 6.8 had been studied by several researchers, such as Schmit [17], Rizzi [18], Lee [16] and Li [19, 9]. The material density is 0.1 lb/in.$^3$ and the modulus of elasticity is 10,000 ksi. The stress limits of the members are listed in Table 6.8. All nodes in all directions are subjected to the displacement limits of ±0.35 in.. Two load cases listed in Table 6.7 are considered. There are 25 members, which are divided into 8 groups, as follows: (1) $A_1$, (2) $A_2$~$A_5$, (3) $A_6$~$A_9$, (4) $A_{10}$~$A_{11}$, (5) $A_{12}$~$A_{13}$, (6) $A_{14}$~$A_{17}$, (7) $A_{18}$~$A_{21}$ and (8) $A_{22}$~$A_{25}$. The minimum permitted cross-sectional area of each member is 0.01 in.$^2$.

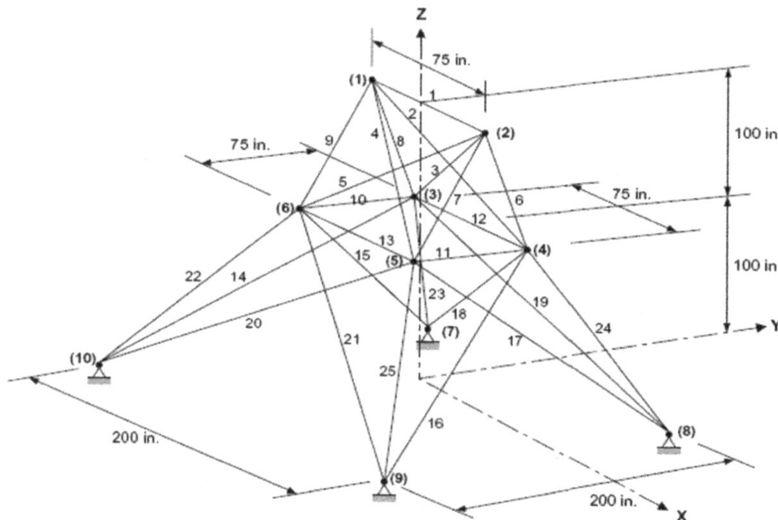


**Fig. 6.8** A 25-bar spatial truss structure

**Table 6.7** Member stress limits for the 25-bar spatial truss structure

| Variables | | Compressive stress limitations (ksi) | Tensile stress limitation (ksi) |
|---|---|---|---|
| 1 | $A_1$ | 35.092 | 40.0 |
| 2 | $A_2$ | 11.590 | 40.0 |
| 3 | $A_3$ | 17.305 | 40.0 |
| 4 | $A_4$ | 35.092 | 40.0 |
| 5 | $A_5$ | 35.902 | 40.0 |
| 6 | $A_6$ | 6.759 | 40.0 |
| 7 | $A_7$ | 6.959 | 40.0 |
| 8 | $A_8$ | 11.802 | 40.0 |

**Table 6.8** Load cases for the 25-bar spatial truss structure

| Node | Case 1 $P_X$ (kips) | Case 1 $P_Y$ (kips) | Case 1 $P_Z$ (kips) | Case 2 $P_X$ (kips) | Case 2 $P_Y$ (kips) | Case 2 $P_Z$ (kips) |
|---|---|---|---|---|---|---|
| 1 | 0.0 | 20.0 | -5.0 | 1.0 | 10.0 | -5.0 |
| 2 | 0.0 | -20.0 | -5.0 | 0.0 | 10.0 | -5.0 |
| 3 | 0.0 | 0.0 | 0.0 | 0.5 | 0.0 | 0.0 |
| 6 | 0.0 | 0.0 | 0.0 | 0.5 | 0.0 | 0.0 |

For this spatial truss structure, it takes about 50 iterations for the QGSO algorithm to converge. Table 6.9 shows the solutions and Fig. 6.9 and 6.10 compares the convergence rate of the three algorithms.

**Table 6.9** Comparison of the designs for the 25-bar spatial truss structure

| | Variables | Optimal cross-sectional areas (in.$^2$) Schmit [17] | Rizzi [18] | Lee [16] | Li [19] HPSO | Li [9] GSO | QGSO |
|---|---|---|---|---|---|---|---|
| 1 | $A_1$ | 0.010 | 0.010 | 0.047 | 0.010 | 0.010 | 0.011 |
| 2 | $A_2$~$A_5$ | 1.964 | 1.988 | 2.022 | 1.970 | 1.948 | 1.783 |
| 3 | $A_6$~$A_9$ | 3.033 | 2.991 | 2.950 | 3.016 | 3.054 | 3.289 |
| 4 | $A_{10}$~$A_{11}$ | 0.010 | 0.010 | 0.010 | 0.010 | 0.010 | 0.010 |
| 5 | $A_{12}$~$A_{13}$ | 0.010 | 0.010 | 0.014 | 0.010 | 0.010 | 0.011 |
| 6 | $A_{14}$~$A_{17}$ | 0.670 | 0.684 | 0.688 | 0.694 | 0.684 | 0.716 |
| 7 | $A_{18}$~$A_{21}$ | 1.680 | 1.677 | 1.657 | 1.681 | 1.683 | 1.733 |
| 8 | $A_{22}$~$A_{25}$ | 2.670 | 2.663 | 2.663 | 2.643 | 2.644 | 2.549 |
| Weight (lb) | | 545.22 | 545.36 | 544.38 | 545.19 | 552.20 | 545.70 |

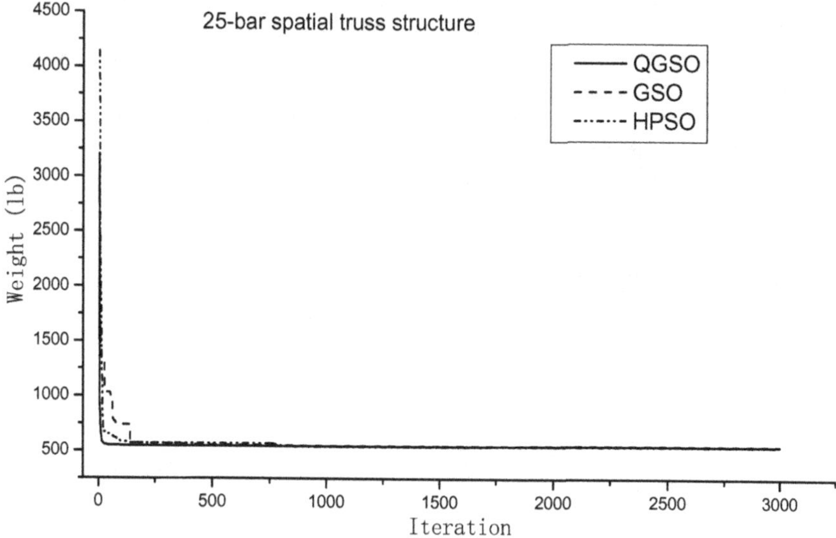


**Fig. 6.9** Convergence rate of the three algorithms for the 25-bar truss structure

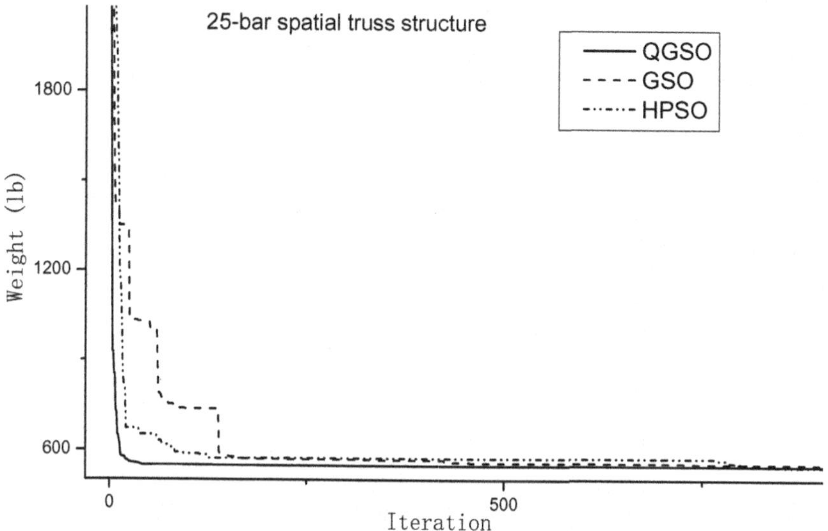


**Fig. 6.10** Convergence rates of three algorithms for the 25-bar truss structure (Magnified in part)

### *(5) The 72-Bar Spatial Truss Structure*

The 72-bar spatial truss structure shown in Fig. 6.11 had also been studied by many researchers, such as Schmit [17], Khot [20], Adeli [21], Lee [16], Sarma [22] and Li [19, 9]. The material density is 0.1 lb/in.$^3$ and the modulus of elasticity is 10,000 ksi. The members are subjected to the stress limits of ±25 ksi. The uppermost nodes are subjected to the displacement limits of ±0.25 in. in both the x and y directions. Two load cases are listed in Table 6.10. There are 72 members classified into 16 groups: (1) $A_1$~$A_4$, (2) $A_5$~$A_{12}$, (3) $A_{13}$~$A_{16}$, (4) $A_{17}$~$A_{18}$, (5) $A_{19}$~$A_{22}$, (6) $A_{23}$~$A_{30}$ (7) $A_{31}$~$A_{34}$, (8) $A_{35}$~$A_{36}$, (9) $A_{37}$~$A_{40}$, (10) $A_{41}$~$A_{48}$, (11) $A_{49}$~$A_{52}$, (12) $A_{53}$~$A_{54}$, (13) $A_{55}$~$A_{58}$, (14) $A_{59}$~$A_{66}$ (15) $A_{67}$~$A_{70}$, (16) $A_{71}$~$A_{72}$. For case 1, the minimum permitted cross-sectional area of each member is 0.1 in.$^2$. For case 2, the value is 0.01 in.$^2$

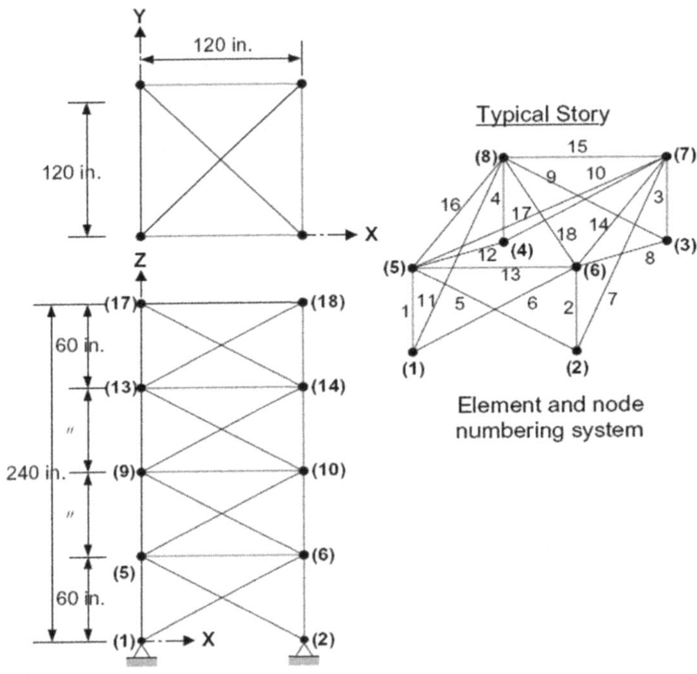


**Fig. 6.11** A 72-bar spatial truss structure

**Table 6.10** Load cases for the 72-bar spatial truss structure

| Node | Case 1 | | | Case 2 | | |
|---|---|---|---|---|---|---|
| | $P_X$ (kips) | $P_Y$ (kips) | $P_Z$ (kips) | $P_X$ (kips) | $P_Y$ (kips) | $P_Z$ (kips) |
| 17 | 5.0 | 5.0 | -5.0 | 0.0 | 0.0 | -5.0 |
| 18 | 0.0 | 0.0 | 0.0 | 0.0 | 0.0 | -5.0 |
| 19 | 0.0 | 0.0 | 0.0 | 0.0 | 0.0 | -5.0 |
| 20 | 0.0 | 0.0 | 0.0 | 0.0 | 0.0 | -5.0 |

For the both loading cases, the QGSO algorithm can achieve the optimal solution after 50 iterations. it shows the fastest convergence rate in the three algorithms, especially at the early stage of iterations. The solutions of the two loading cases are given in Tables 6.11 and 6.12 respectively. Figs. 6.12, 6.13 and 6.14 compare the convergence rate of the three algorithms under the two loading cases.

**Table 6.11** Comparison of the designs for the 72-bar spatial truss structure (Case 1)

| Variables | | Optimal cross-sectional areas (in.$^2$) | | | | | | |
|---|---|---|---|---|---|---|---|---|
| | | Schmit [17] | Adeli [21] | Khot [20] | Lee [16] | Li [19] HPSO | Li [9] GSO | QGSO |
| 1 | $A_1$~$A_4$ | 2.078 | 2.026 | 1.893 | 1.7901 | 1.889 | 3.129 | 1.880 |
| 2 | $A_5$~$A_{12}$ | 0.503 | 0.533 | 0.517 | 0.521 | 0.510 | 0.539 | 0.513 |
| 3 | $A_{13}$~$A_{16}$ | 0.100 | 0.100 | 0.100 | 0.100 | 0.100 | 0.133 | 0.100 |
| 4 | $A_{17}$~$A_{18}$ | 0.100 | 0.100 | 0.100 | 0.100 | 0.100 | 0.152 | 0.102 |
| 5 | $A_{19}$~$A_{22}$ | 1.107 | 1.157 | 1.279 | 1.229 | 1.265 | 1.161 | 1.244 |
| 6 | $A_{23}$~$A_{30}$ | 0.579 | 0.569 | 0.515 | 0.522 | 0.510 | 0.410 | 0.511 |
| 7 | $A_{31}$~$A_{34}$ | 0.100 | 0.100 | 0.100 | 0.100 | 0.100 | 0.102 | 0.100 |
| 8 | $A_{35}$~$A_{36}$ | 0.100 | 0.100 | 0.100 | 0.100 | 0.100 | 0.101 | 0.106 |
| 9 | $A_{37}$~$A_{40}$ | 0.264 | 0.514 | 0.508 | 0.517 | 0.523 | 0.489 | 0.523 |
| 10 | $A_{41}$~$A_{48}$ | 0.548 | 0.479 | 0.520 | 0.504 | 0.519 | 0.366 | 0.517 |
| 11 | $A_{49}$~$A_{52}$ | 0.100 | 0.100 | 0.100 | 0.100 | 0.100 | 0.101 | 0.100 |
| 12 | $A_{53}$~$A_{54}$ | 0.151 | 0.100 | 0.100 | 0.101 | 0.100 | 0.100 | 0.111 |
| 13 | $A_{55}$~$A_{58}$ | 0.158 | 0.158 | 0.157 | 0.156 | 0.156 | 0.152 | 0.156 |
| 14 | $A_{59}$~$A_{66}$ | 0.594 | 0.550 | 0.539 | 0.547 | 0.548 | 0.676 | 0.553 |
| 15 | $A_{67}$~$A_{70}$ | 0.341 | 0.345 | 0.416 | 0.442 | 0.411 | 0.590 | 0.407 |
| 16 | $A_{71}$~$A_{72}$ | 0.608 | 0.498 | 0.551 | 0.590 | 0.568 | 0.633 | 0.569 |
| Weight (lb) | | 388.63 | 379.31 | 379.67 | 379.27 | 379.63 | 409.86 | 379.99 |

**Table 6.12** Comparison of the designs for the 72-bar spatial truss structure (Case 2)

| Variables | | Optimal cross-sectional areas (in.$^2$) | | | | | | |
|---|---|---|---|---|---|---|---|---|
| | | Adeli [21] | Sarma [22] Simple GA | Sarma [22] Fuzzy GA | Lee [16] | Li [19] HPSO | Li [9] GSO | QGSO |
| 1 | $A_1$~$A_4$ | 2.755 | 2.141 | 1.732 | 1.963 | 1.907 | 3.056 | 1.8704 |
| 2 | $A_5$~$A_{12}$ | 0.510 | 0.510 | 0.522 | 0.481 | 0.524 | 0.356 | 0.5151 |
| 3 | $A_{13}$~$A_{16}$ | 0.010 | 0.054 | 0.010 | 0.010 | 0.010 | 0.014 | 0.0104 |
| 4 | $A_{17}$~$A_{18}$ | 0.010 | 0.010 | 0.013 | 0.011 | 0.010 | 0.083 | 0.01 |
| 5 | $A_{19}$~$A_{22}$ | 1.370 | 1.489 | 1.345 | 1.233 | 1.288 | 1.347 | 1.281 |
| 6 | $A_{23}$~$A_{30}$ | 0.507 | 0.551 | 0.551 | 0.506 | 0.523 | 0.432 | 0.5182 |
| 7 | $A_{31}$~$A_{34}$ | 0.010 | 0.057 | 0.010 | 0.011 | 0.010 | 0.072 | 0.01 |
| 8 | $A_{35}$~$A_{36}$ | 0.010 | 0.013 | 0.013 | 0.012 | 0.010 | 0.040 | 0.0102 |
| 9 | $A_{37}$~$A_{40}$ | 0.481 | 0.565 | 0.492 | 0.538 | 0.544 | 0.431 | 0.5165 |
| 10 | $A_{41}$~$A_{48}$ | 0.508 | 0.527 | 0.545 | 0.533 | 0.528 | 0.488 | 0.515 |
| 11 | $A_{49}$~$A_{52}$ | 0.010 | 0.010 | 0.066 | 0.010 | 0.019 | 0.056 | 0.011 |
| 12 | $A_{53}$~$A_{54}$ | 0.643 | 0.066 | 0.013 | 0.167 | 0.020 | 0.096 | 0.1188 |
| 13 | $A_{55}$~$A_{58}$ | 0.215 | 0.174 | 0.178 | 0.161 | 0.176 | 0.177 | 0.1657 |
| 14 | $A_{59}$~$A_{66}$ | 0.518 | 0.425 | 0.524 | 0.542 | 0.535 | 0.726 | 0.5423 |
| 15 | $A_{67}$~$A_{70}$ | 0.419 | 0.437 | 0.396 | 0.478 | 0.426 | 0.430 | 0.4381 |
| 16 | $A_{71}$~$A_{72}$ | 0.504 | 0.641 | 0.595 | 0.551 | 0.612 | 1.015 | 0.5972 |
| Weight (lb) | | 376.50 | 372.40 | 364.40 | 364.33 | 364.86 | 404.45 | 363.91 |

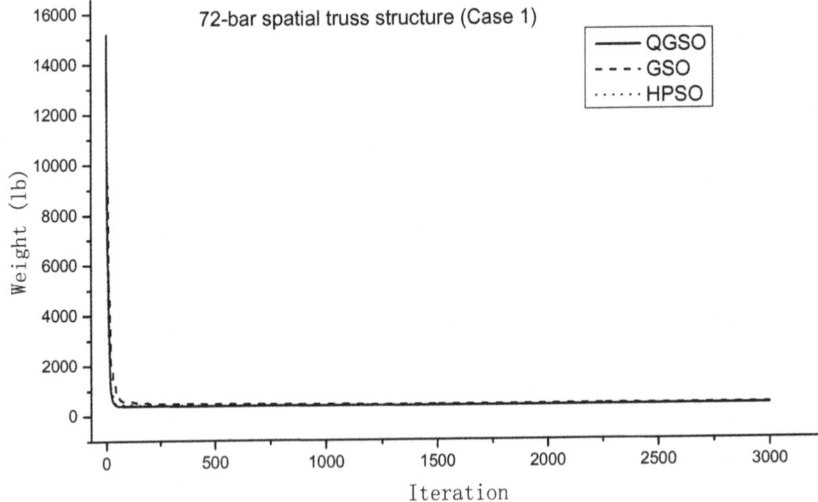


**Fig. 6.12** Convergence rates of the three algorithms for the 72-bar truss structure (Case 1)

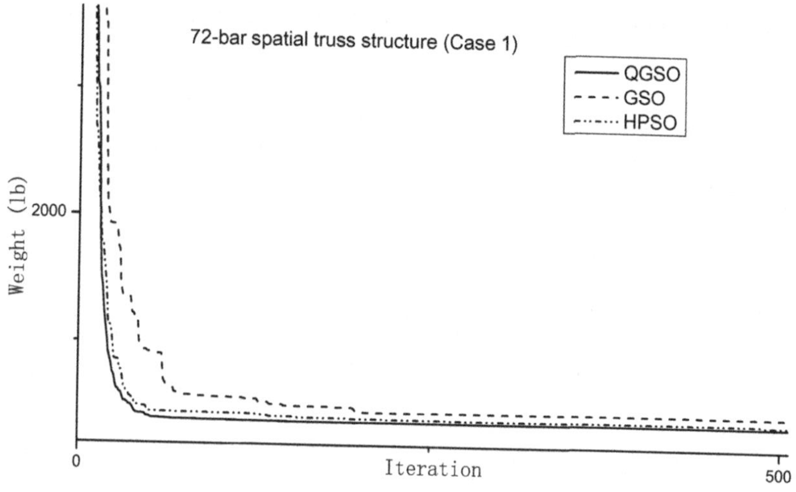


**Fig. 6.13** Convergence rates of three algorithms for the 72-bar truss structure (Case 1, Magnified in part)

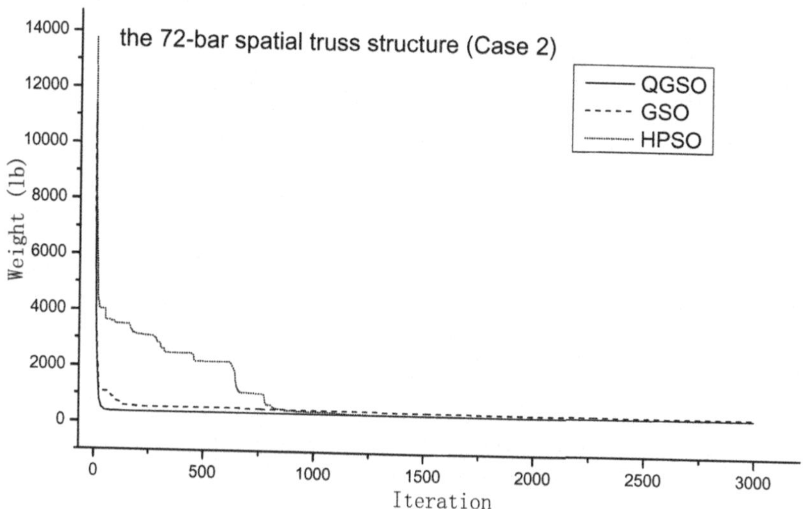


**Fig. 6.14** Convergence rates between the three algorithms for the 72-bar spatial truss structure (Case 2)

## 6.6 The Application of the QGSO on Truss Structures with Discrete Variables

In the past thirty years, many algorithms have been developed to solve structural engineering optimization problems. Most of these algorithms are based on the assumption that the design variables are continuously valued and the gradients of functions and the convexity of the design problem satisfied. However, in reality, the design variables of optimization problems such as the cross-section areas are discretely valued. They are often chosen from a list of discrete variables. Furthermore, the function of the problems is hard to express in an explicit form. Traditionally, the discrete optimization problems are solved by mathematical methods by employing round-off techniques based on the continuous solutions. However, the solutions obtained by this method may be infeasible or far from the optimum solutions.

So there is more engineering significance in structural optimization with discrete variables [23].

The QGSO, GSO, HPSO schemes are applied respectively to the examples and the results are compared in order to evaluate the performance of the modified algorithm. For both algorithms, the population size is set to at 50. For the GSO algorithm, 20% of the population is selected as rangers; the initial head angle $\varphi_0$ of each individual is set to be $\pi/4$. The constant a is given by $\text{round}(\sqrt{n+1})$. The maximum pursuit angle $\theta_{max}$ is $\pi/a^2$. The maximum turning angle $\alpha$ is set to be $\pi/2a^2$. For the HPSO algorithm, the inertia weight (w) is starting at 0.9 and ending at 0.4 by linearity descending. The acceleration constants $c_1$ and $c_2$ is set to be 0.5. The passive congregation coefficient $c_3$ is 0.6. For the QGSO algorithm, when target goes forward, the parameters are set in this: information transfer factor $w_1 = w_2 = 4$, selected probability $w_3 = 0.2$, component mutation probability $w_4 = 0.65$.Otherwise the parameters are set in this: information transfer factor $w_1 = 0.8$, $w_2 = 1.5$, selected probability $w_3 = 0.35$, component mutation probability $w_4 = 0.85$.

In this section, five pin connected structures commonly used in literature are selected as benchmark problems to test the QGSO. The examples given in the simulation studies include:

(1) A 10-bar planar truss structure (two cases)；
(2) A 15-bar planar truss structure;
(3) A 25-bar spatial truss structure;
(4) A 52-bar spatial truss structure;
(5) A 72-bar spatial truss structure;
(6) A double-layer grid steel shell structure;

### 6.6.1 Numerical Examples

#### (1) A 10-Bar Planar Truss Structure

A 10-bar truss structure, shown in Fig. 6.15, has previously been analyzed by many researchers, such as Wu [24], Rajeev [25], Ringertz [26] and Li [14]. The material density is 0.1 lb/in$^3$ and the modulus of elasticity is 10,000 ksi. The members are subjected to stress limitations of ±25 ksi. All nodes in both directions are subjected to displacement limitations of ±2.0 in.. $P_1$=105 lb, $P_2$=0. There are 10 design variables and two load cases in this example to be optimized. For case 1: the discrete variables are selected from the set D={1.62, 1.80, 1.99, 2.13, 2.38, 2.62, 2.63, 2.88, 2.93, 3.09, 3.13, 3.38, 3.47, 3.55, 3.63, 3.84, 3.87, 3.88, 4.18, 4.22, 4.49, 4.59, 4.80, 4.97, 5.12, 5.74, 7.22, 7.97, 11.50, 13.50, 13.90, 14.20, 15.50, 16.00, 16.90, 18.80, 19.90, 22.00, 22.90, 26.50, 30.00, 33.50} (in.$^2$); For case 2: the discrete variables are selected from the set D={0.1, 0.5, 1.0, 1.5, 2.0, 2.5, 3.0, 3.5, 4.0, 4.5, 5.0, 5.5, 6.0, 6.5, 7.0, 7.5, 8.0, 8.5, 9.0, 9.5, 10.0, 10.5, 11.0, 11.5, 12.0, 12.5, 13.0, 13.5, 14.0, 14.5, 15.0, 15.5, 16.0, 16.5, 17.0, 17.5, 18.0, 18.5, 19.0, 19.5, 20.0, 20.5, 21.0, 21.5, 22.0, 22.5, 23.0, 23.5, 24.0, 24.5, 25.0, 25.5, 26.0, 26.5, 27.0, 27.5, 28.0, 28.5, 29.0, 29.5, 30.0, 30.5, 31.0, 31.5} (in.$^2$). A maximum number of 1000 iterations is imposed.

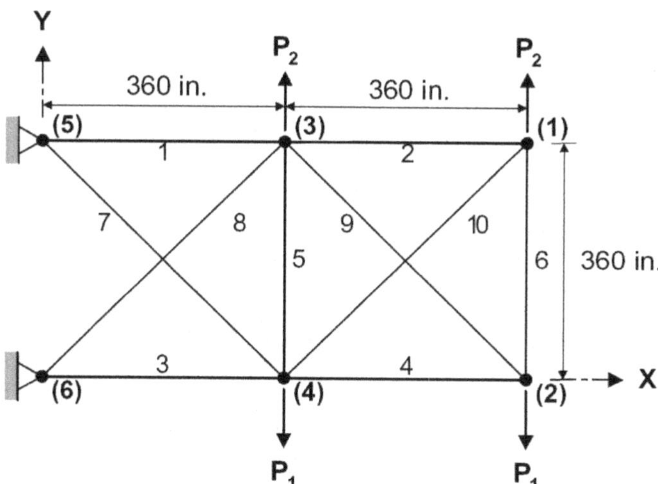


**Fig. 6.15** A 10-bar planar truss structure

Table 6.13 and Table 6.14 give the comparison of optimal design results for the 10-bar planar truss structure under two load cases respectively. Fig.6.16 and Fig.6.17 show the comparison of convergence rates for the 10-bar truss structure. From the Table 6.13 and Table 6.14, It can be find that the results obtained by the three algorithms are larger than those of Wu's. However, it is found that Wu's results do not satisfy the constraints of this problem. It is believed that Wu's results need to be

**Table 6.13** Comparison of optimal designs for the 10-bar planar truss structure (case 1)

| Variables ($in^2$) | Wu [24] | Rajeev [25] | Li [14] HPSO | GSO | QGSO |
|---|---|---|---|---|---|
| $A_1$ | 26.50 | 33.50 | 30.00 | 26.500 | 33.500 |
| $A_2$ | 1.62 | 1.62 | 1.62 | 1.620 | 1.620 |
| $A_3$ | 16.00 | 22.00 | 22.90 | 26.500 | 22.900 |
| $A_4$ | 14.20 | 15.50 | 13.50 | 15.500 | 14.200 |
| $A_5$ | 1.80 | 1.62 | 1.62 | 1.620 | 1.620 |
| $A_6$ | 1.62 | 1.62 | 1.62 | 1.620 | 1.620 |
| $A_7$ | 5.12 | 14.20 | 7.97 | 11.500 | 7.970 |
| $A_8$ | 16.00 | 19.90 | 26.50 | 22.000 | 22.900 |
| $A_9$ | 18.80 | 19.90 | 22.00 | 22.000 | 22.000 |
| $A_{10}$ | 2.38 | 2.62 | 1.80 | 1.800 | 1.620 |
| Weight (lb) | 4376.20 | 5613.84 | 5531.98 | 5558.02 | 5490.74 |

**Table 6.14** Comparison of optimal designs for the 10-bar planar truss structure (case 2)

| Variables ($in^2$) | Wu [24] | Ringertz [26] | Li [14] HPSO | GSO | QGSO |
|---|---|---|---|---|---|
| A1 | 30.50 | 30.50 | 31.50 | 28.500 | 29.500 |
| A2 | 0.50 | 0.10 | 0.10 | 0.100 | 0.100 |
| A3 | 16.50 | 23.00 | 24.50 | 23.000 | 23.500 |
| A4 | 15.00 | 15.50 | 15.50 | 16.500 | 15.500 |
| A5 | 0.10 | 0.10 | 0.10 | 0.100 | 0.100 |
| A6 | 0.10 | 0.50 | 0.50 | 0.500 | 0.500 |
| A7 | 0.50 | 7.50 | 7.50 | 7.500 | 7.500 |
| A8 | 18.00 | 21.0 | 20.50 | 22.000 | 21.500 |
| A9 | 19.50 | 21.5 | 20.50 | 21.500 | 21.500 |
| A10 | 0.50 | 0.10 | 0.10 | 0.100 | 0.100 |
| Weight (lb) | 4217.30 | 5059.9 | 5073.51 | 5074.787 | 5067.33 |

further valuated. For both cases of this structure, the HPSO, GSO and QGSO algorithms have achieved the optimal solutions after 1,000 iterations. But the QGSO has the best solution than the GSO in the early iterations and has the fastest convergence rate.

***(2) A 15-Bar Planar Truss Structure***

A 15-bar planar truss structure, shown in Fig. 6.18, has previously been analyzed by Zhang [27] and Li [14]. The material density is 7800 kg/m$^3$ and the modulus of elasticity is 200 GPa. The members are subjected to stress limitations of ±120 MPa. All nodes in both directions are subjected to displacement limitations of ±10 mm.

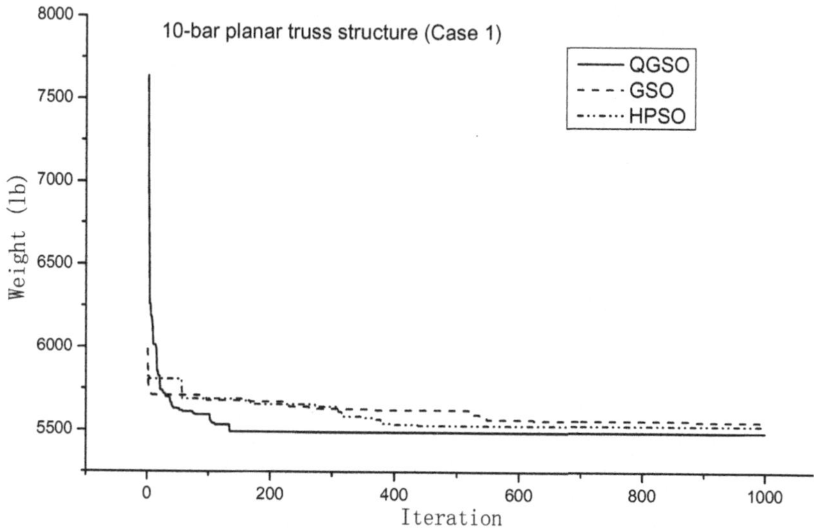


**Fig. 6.16** Comparison of convergence rates for the 10-bar planar truss structure (Case 1)

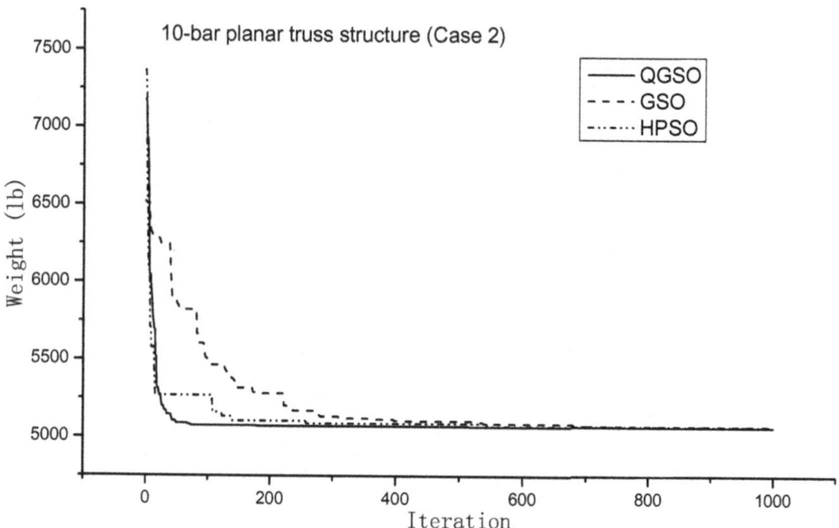


**Fig. 6.17** Comparison of convergence rates for the 10-bar planar truss structure (Case 2)

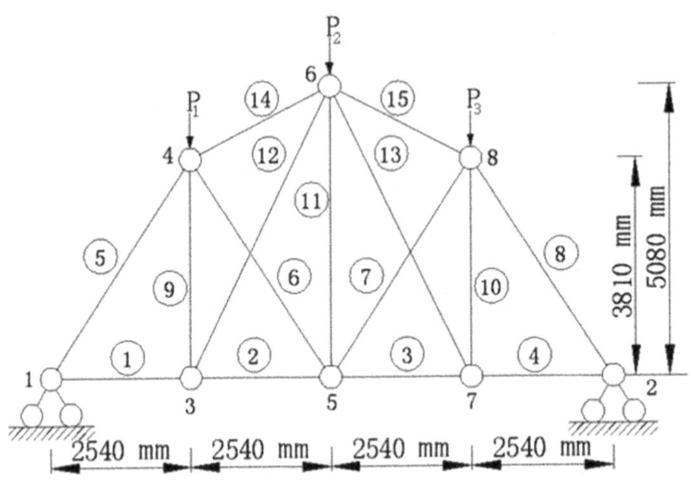


**Fig. 6.18** A 15-bar planar truss structure

**Table 6.15** Comparison of optimal designs for the 15-bar planar truss structure

| Variables ($mm^2$) | Zhang [27] | Li [14] HPSO | GSO | QGSO |
|---|---|---|---|---|
| $A_1$ | 308.6 | 113.2 | 113.200 | 113.200 |
| $A_2$ | 174.9 | 113.2 | 113.200 | 113.200 |
| $A_3$ | 338.2 | 113.2 | 113.200 | 113.200 |
| $A_4$ | 143.2 | 113.2 | 113.200 | 113.200 |
| $A_5$ | 736.7 | 736.7 | 736.700 | 736.700 |
| $A_6$ | 185.9 | 113.2 | 113.200 | 113.200 |
| $A_7$ | 265.9 | 113.2 | 113.200 | 113.200 |
| $A_8$ | 507.6 | 736.7 | 736.700 | 736.700 |
| $A_9$ | 143.2 | 113.2 | 113.200 | 113.200 |
| $A_{10}$ | 507.6 | 113.2 | 113.200 | 113.200 |
| $A_{11}$ | 279.1 | 113.2 | 113.200 | 113.200 |
| $A_{12}$ | 174.9 | 113.2 | 113.200 | 113.200 |
| $A_{13}$ | 297.1 | 113.2 | 113.200 | 113.200 |
| $A_{14}$ | 235.9 | 334.3 | 334.300 | 334.300 |
| $A_{15}$ | 265.9 | 334.3 | 334.300 | 334.300 |
| Weight (kg) | 142.117 | 105.735 | 105.735 | 105.735 |

There are 15 design variables in this example. The discrete variables are selected from the set D= {113.2, 143.2, 145.9, 174.9, 185.9, 235.9, 265.9, 297.1, 308.6, 334.3, 338.2, 497.8, 507.6, 736.7, 791.2, 1063.7} ($mm^2$). Three load cases are considered: Case 1: $P_1$=35 kN, $P_2$=35 kN, $P_3$=35 kN; Case 2: $P_1$=35 kN, $P_2$=0 kN, $P_3$=35 kN; Case 3: $P_1$=35 kN, $P_2$=35 kN, $P_3$=0 kN. A maximum number of 500 iterations is imposed.

Table 6.15 and Fig. 6.19 give the comparison of optimal design results and convergence rates of 15-bar planar truss structure respectively. It can be seen that, after 500 iterations, three algorithms have obtained good results, which are better than the Zhang's. The Fig. 6.19 shows that the QGSO algorithm has the fastest convergence rate, especially in the early iterations.

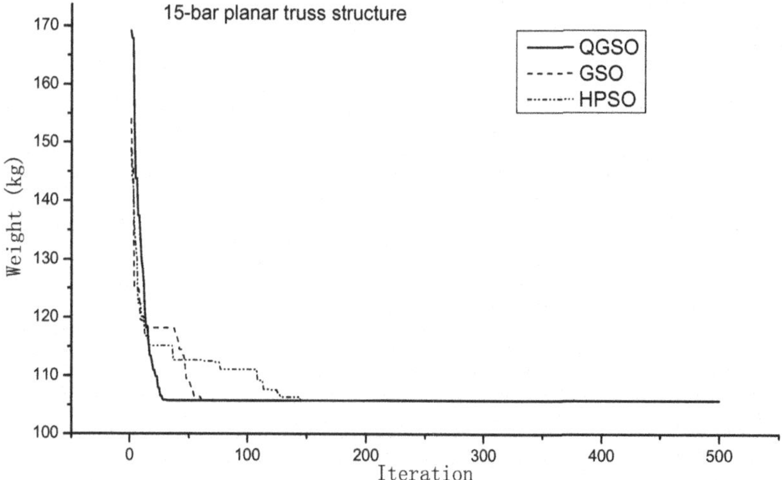


**Fig. 6.19** Comparison of convergence rates for the 15-bar planar truss structure

### *(3) A 25-Bar Spatial Truss Structure*

A 25-bar spatial truss structure, shown in Fig. 6.20, has been studied by Wu [24] and Li [14, 9]. The material density is 0.1 lb/in.$^3$ and the modulus of elasticity is 10,000 ksi. The stress limitations of the members are ±40000 psi. All nodes in three directions are subjected to displacement limitations of ±0.35 in. The structure includes 25 members, which are divided into 8 groups, as follows: (1) $A_1$, (2) $A_2$~$A_5$, (3) $A_6$~$A_9$, (4) $A_{10}$~$A_{11}$, (5) $A_{12}$~$A_{13}$, (6) $A_{14}$~$A_{17}$, (7) $A_{18}$~$A_{21}$ and (8) $A_{22}$~$A_{25}$. The discrete variables are selected from the American Institute of Steel Construction (AISC) Code, which is shown in Table 6.16. The loads are shown in Table 6.17. A maximum number of 500 iterations is imposed for three cases.

Table 6.18 shows the comparison of optimal design results for the 25-bar spatial truss structure under three load cases. While Fig. 6.21 shows comparison of convergence rates for the 25-bar spatial truss structure under three load cases. For this structure, three algorithms can achieve the optimal solution after 500 iterations. But Fig. 6.21 shows that the QGSO algorithm has the fastest convergence rate.

**Table 6.16** The available cross-section areas of the ASIC code

| No. | $in^2$ | $mm^2$ | No. | $in^2$ | $mm^2$ |
|---|---|---|---|---|---|
| 1 | 0.111 | 71.613 | 33 | 3.840 | 2477.414 |
| 2 | 0.141 | 90.968 | 34 | 3.870 | 2496.769 |
| 3 | 0.196 | 126.451 | 35 | 3.880 | 2503.221 |
| 4 | 0.250 | 161.290 | 36 | 4.180 | 2696.769 |
| 5 | 0.307 | 198.064 | 37 | 4.220 | 2722.575 |
| 6 | 0.391 | 252.258 | 38 | 4.490 | 2896.768 |
| 7 | 0.442 | 285.161 | 39 | 4.590 | 2961.284 |
| 8 | 0.563 | 363.225 | 40 | 4.800 | 3096.768 |
| 9 | 0.602 | 388.386 | 41 | 4.970 | 3206.445 |
| 10 | 0.766 | 494.193 | 42 | 5.120 | 3303.219 |
| 11 | 0.785 | 506.451 | 43 | 5.740 | 3703.218 |
| 12 | 0.994 | 641.289 | 44 | 7.220 | 4658.055 |
| 13 | 1.000 | 645.160 | 45 | 7.970 | 5141.925 |
| 14 | 1.228 | 792.256 | 46 | 8.530 | 5503.215 |
| 15 | 1.266 | 816.773 | 47 | 9.300 | 5999.988 |
| 16 | 1.457 | 939.998 | 48 | 10.850 | 6999.986 |
| 17 | 1.563 | 1008.385 | 49 | 11.500 | 7419.340 |
| 18 | 1.620 | 1045.159 | 50 | 13.500 | 8709.660 |
| 19 | 1.800 | 1161.288 | 51 | 13.900 | 8967.724 |
| 20 | 1.990 | 1283.868 | 52 | 14.200 | 9161.272 |
| 21 | 2.130 | 1374.191 | 53 | 15.500 | 9999.980 |
| 22 | 2.380 | 1535.481 | 54 | 16.000 | 10322.560 |
| 23 | 2.620 | 1690.319 | 55 | 16.900 | 10903.204 |
| 24 | 2.630 | 1696.771 | 56 | 18.800 | 12129.008 |
| 25 | 2.880 | 1858.061 | 57 | 19.900 | 12838.684 |
| 26 | 2.930 | 1890.319 | 58 | 22.000 | 14193.520 |
| 27 | 3.090 | 1993.544 | 59 | 22.900 | 14774.164 |
| 28 | 1.130 | 729.031 | 60 | 24.500 | 15806.420 |
| 29 | 3.380 | 2180.641 | 61 | 26.500 | 17096.740 |
| 30 | 3.470 | 2238.705 | 62 | 28.000 | 18064.480 |
| 31 | 3.550 | 2290.318 | 63 | 30.000 | 19354.800 |
| 32 | 3.630 | 2341.931 | 64 | 33.500 | 21612.860 |

**Table 6.17** The load case 2 and case 3 for the 25-bar spatial truss structure

| Load Cases | Nodes | Loads | | |
|---|---|---|---|---|
| | | $P_x$ (kips) | $P_y$ (kips) | $P_z$ (kips) |
| 1 | 1 | 0.0 | 20.0 | -5.0 |
| | 2 | 0.0 | -20.0 | -5.0 |
| 2 | 1 | 1.0 | 10.0 | -5.0 |
| | 2 | 0.0 | 10.0 | -5.0 |
| | 3 | 0.5 | 0.0 | 0.0 |
| | 6 | 0.5 | 0.0 | 0.0 |

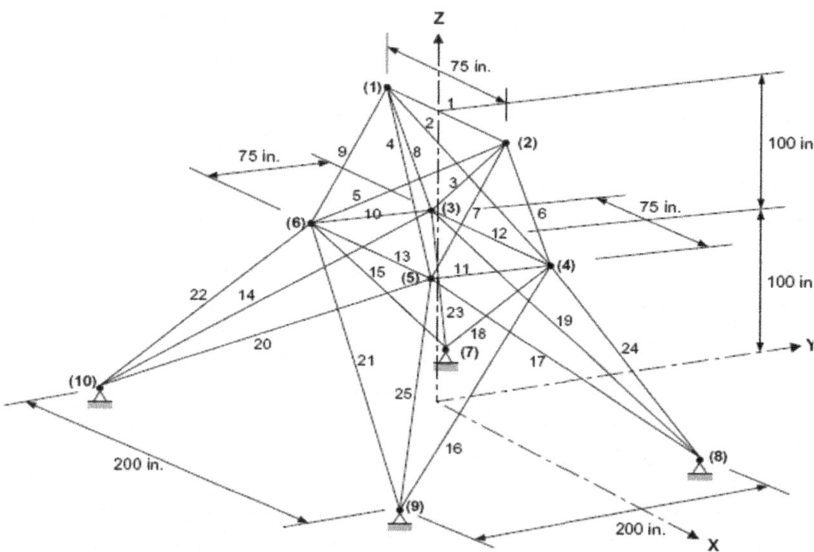


Fig. 6.20 A 25-bar spatial truss structure

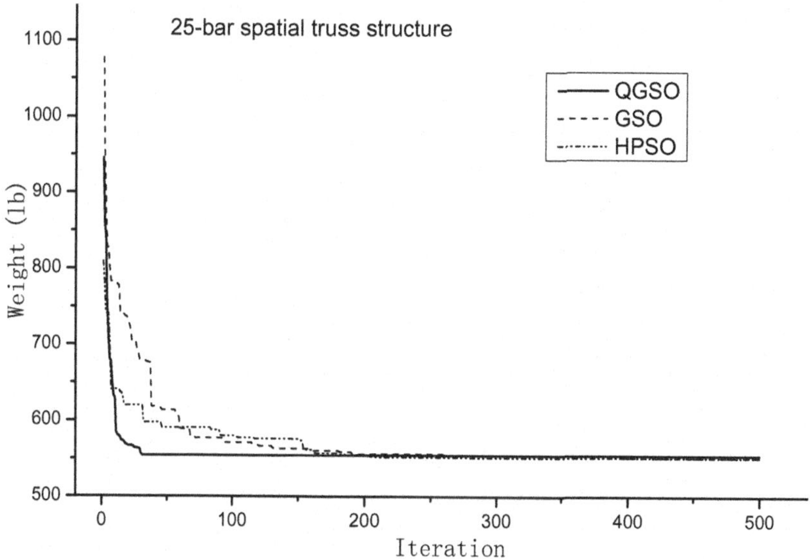


Fig. 6.21 Comparison of convergence rates for the 25-bar spatial truss structure

**Table 6.18** Comparison of optimal designs for the 25-bar spatial truss structure

| Variables ($in^2$) | Wu [24] | Li [14] HPSO | Li [9] GSO | QGSO |
|---|---|---|---|---|
| $A_1$ | 0.307 | 0.111 | 0.111 | 0.111 |
| $A_2$~$A_5$ | 1.990 | 2.130 | 2.130 | 2.380 |
| $A_6$~$A_9$ | 3.130 | 2.880 | 2.880 | 2.880 |
| $A_{10}$~$A_{11}$ | 0.111 | 0.111 | 0.111 | 0.111 |
| $A_{12}$~$A_{13}$ | 0.141 | 0.111 | 0.111 | 0.111 |
| $A_{14}$~$A_{17}$ | 0.766 | 0.766 | 0.766 | 0.602 |
| $A_{18}$~$A_{21}$ | 1.620 | 1.620 | 1.620 | 1.457 |
| $A_{22}$~$A_{25}$ | 2.620 | 2.620 | 2.620 | 2.880 |
| Weight (lb) | 556.43 | 551.14 | 551.14 | 554.38 |

***(4) A 52-Bar Planar Truss Structure***

A 52-bar planar truss structure, shown in Fig. 6.22, has been analysed by Wu [24] Lee [28] and Li [14]. The members of this structure are divided into 12 groups: (1) $A_1$~$A_4$, (2) $A_5$~$A_6$, (3) $A_7$~$A_8$, (4) $A_9$~$A_{10}$, (5) $A_{11}$~$A_{14}$, (6) $A_{15}$~$A_{18}$, and (7) $A_{19}$~$A_{22}$. The material density is 7860.0 kg/m$^3$ and the modulus of elasticity is $2.07\times10^5$ MPa. The members are subjected to stress limitations of ±180 MPa. Both of the loads, $P_x$ =100 kN, $P_y$ =200 kN are considered. The discrete variables are selected from the Table 6.16. A maximum number of 3,000 iterations is imposed.

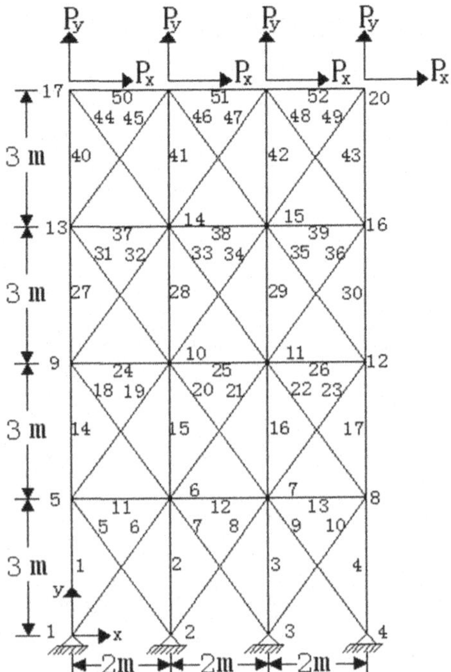


**Fig. 6.22** A 52-bar planar truss structure

**Table 6.19** Comparison of optimal designs for the 52-bar planar truss structure

| Variables (mm$^2$) | Wu [24] | Lee [28] | Li [14] HPSO | GSO | QGSO |
|---|---|---|---|---|---|
| $A_1$~$A_4$ | 4658.055 | 4658.055 | 4658.055 | 4658.060 | 4658.060 |
| $A_5$~$A_{10}$ | 1161.288 | 1161.288 | 1161.288 | 1161.290 | 1161.290 |
| $A_{11}$~$A_{13}$ | 645.160 | 506.451 | 363.225 | 363.230 | 494.190 |
| $A_{14}$~$A_{17}$ | 3303.219 | 3303.219 | 3303.219 | 3303.220 | 3303.220 |
| $A_{18}$~$A_{23}$ | 1045.159 | 940.000 | 940.000 | 940.000 | 940.000 |
| $A_{24}$~$A_{26}$ | 494.193 | 494.193 | 494.193 | 494.190 | 494.190 |
| $A_{27}$~$A_{30}$ | 2477.414 | 2290.318 | 2238.705 | 2238.710 | 2238.710 |
| $A_{31}$~$A_{36}$ | 1045.159 | 1008.385 | 1008.385 | 1008.380 | 1008.380 |
| $A_{37}$~$A_{39}$ | 285.161 | 2290.318 | 388.386 | 641.290 | 494.190 |
| $A_{40}$~$A_{43}$ | 1696.771 | 1535.481 | 1283.868 | 1283.870 | 1283.870 |
| $A_{44}$~$A_{49}$ | 1045.159 | 1045.159 | 1161.288 | 1161.290 | 1161.290 |
| $A_{50}$~$A_{52}$ | 641.289 | 506.451 | 792.256 | 494.190 | 494.190 |
| Weight (kg) | 1970.142 | 1906.76 | 1905.495 | 1903.365 | 1902.605 |

Table 6.19 and Fig. 6.23 give the comparison of optimal design results and convergence rates of 52-bar planar truss structure respectively. From Table 6.19 and Fig. 6.23, it can be observed that the QGSO algorithm not only achieves the best optimal result but also has the fastest convergent rate.

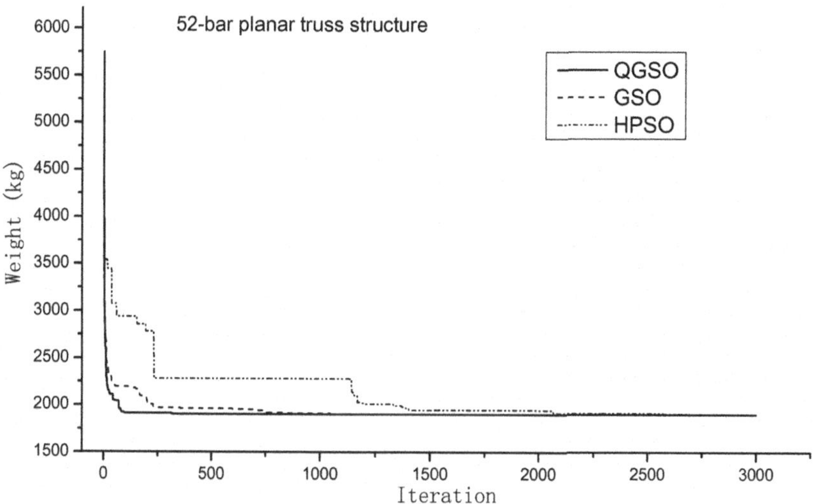


**Fig. 6.23** Comparison of convergence rates for the 52-bar planar truss structure

### *(5) A 72-Bar Spatial Truss Structure*

A 72-bar spatial truss structure, shown in Fig. 6.24, has been studied by Wu [24] Lee [28] and Li [14, 9]. The material density is 0.1 lb/in.$^3$ and the modulus of elasticity is 10,000 ksi. The members are subjected to stress limitations of ±25 ksi. The uppermost nodes are subjected to displacement limitations of ±0.25 in. both in *x* and *y* directions. Two load cases are listed in Table 6.20. There are 72 members, which are divided into 16 groups, as follows: (1) $A_1$~$A_4$, (2) $A_5$~$A_{12}$, (3) $A_{13}$~$A_{16}$, (4) $A_{17}$~$A_{18}$, (5) $A_{19}$~$A_{22}$, (6) $A_{23}$~$A_{30}$ (7) $A_{31}$~$A_{34}$, (8) $A_{35}$~$A_{36}$, (9) $A_{37}$~$A_{40}$, (10) $A_{41}$~$A_{48}$, (11) $A_{49}$~$A_{52}$, (12) $A_{53}$~$A_{54}$, (13) $A_{55}$~$A_{58}$, (14) $A_{59}$~$A_{66}$ (15) $A_{67}$~$A_{70}$, (16) $A_{71}$~$A_{72}$. The discrete variables are selected from the Table 6.16. A maximum number of 1,000 iterations is imposed. Table 6.21 gives the optimal results and Fig. 6.25 gives the convergence rates.

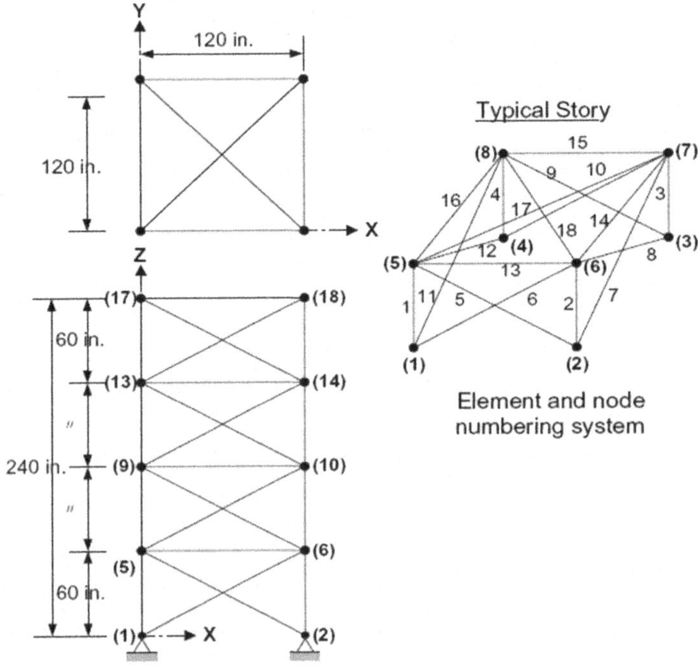


**Fig. 6.24** The 72-bar spatial truss structure

**Table 6.20** The load cases for the 72-bar spatial truss structure

| Nodes | Load Case 1 | | | Load Case 2 | | |
|---|---|---|---|---|---|---|
| | $P_X$ (kips) | $P_Y$ (kips) | $P_Z$ (kips) | $P_X$ (kips) | $P_Y$ (kips) | $P_Z$ (kips) |
| 17 | 5.0 | 5.0 | -5.0 | 0.0 | 0.0 | -5.0 |
| 18 | 0.0 | 0.0 | 0.0 | 0.0 | 0.0 | -5.0 |
| 19 | 0.0 | 0.0 | 0.0 | 0.0 | 0.0 | -5.0 |
| 20 | 0.0 | 0.0 | 0.0 | 0.0 | 0.0 | -5.0 |

**Table 6.21** Comparison of optimal designs for the 72-bar spatial truss structure (case 1)

| Variables ($in^2$) | Wu [24] | Lee [28] | Li [14] HPSO | Li [9] GSO | QGSO |
|---|---|---|---|---|---|
| $A_1$~$A_4$ | 1.5 | 1.9 | 2.1 | 3.0 | 2.0 |
| $A_5$~$A_{12}$ | 0.7 | 0.5 | 0.6 | 1.5 | 0.5 |
| $A_{13}$~$A_{16}$ | 0.1 | 0.1 | 0.1 | 0.1 | 0.1 |
| $A_{17}$~$A_{18}$ | 0.1 | 0.1 | 0.1 | 0.1 | 0.1 |
| $A_{19}$~$A_{22}$ | 1.3 | 1.4 | 1.4 | 2.6 | 1.3 |
| $A_{23}$~$A_{30}$ | 0.5 | 0.6 | 0.5 | 1.5 | 0.5 |
| $A_{31}$~$A_{34}$ | 0.2 | 0.1 | 0.1 | 0.1 | 0.1 |
| $A_{35}$~$A_{36}$ | 0.1 | 0.1 | 0.1 | 0.1 | 0.1 |
| $A_{37}$~$A_{40}$ | 0.5 | 0.6 | 0.5 | 1.6 | 0.5 |
| $A_{41}$~$A_{48}$ | 0.5 | 0.5 | 0.5 | 1.4 | 0.5 |
| $A_{49}$~$A_{52}$ | 0.1 | 0.1 | 0.1 | 0.1 | 0.1 |
| $A_{53}$~$A_{54}$ | 0.2 | 0.1 | 0.1 | 0.4 | 0.1 |
| $A_{55}$~$A_{58}$ | 0.2 | 0.2 | 0.2 | 0.4 | 0.2 |
| $A_{59}$~$A_{66}$ | 0.5 | 0.5 | 0.5 | 1.6 | 0.6 |
| $A_{67}$~$A_{70}$ | 0.5 | 0.4 | 0.3 | 1.3 | 0.4 |
| $A_{71}$~$A_{72}$ | 0.7 | 0.6 | 0.7 | 1.3 | 0.6 |
| Weight (lb) | 400.66 | 387.94 | 388.94 | 967.68 | 385.54 |

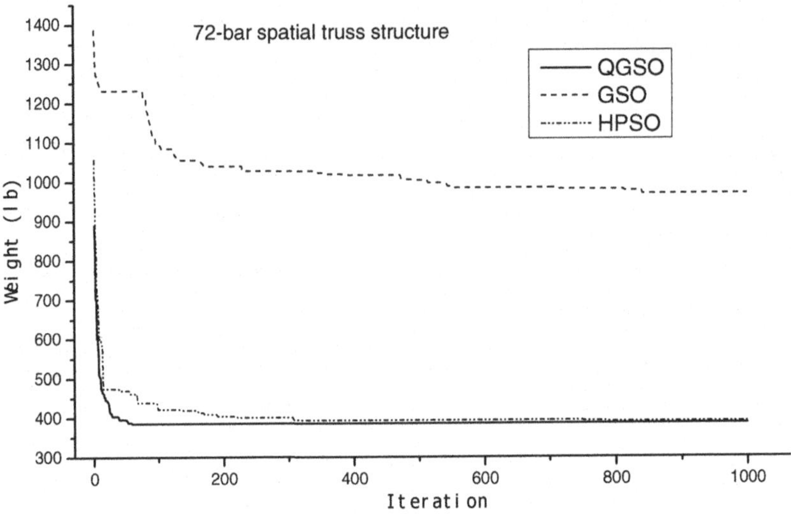


**Fig. 6.25** Comparison of convergence rates for the 72-bar spatial truss structure

It can be seen from Fig. 6.25 that the QGSO algorithm gets the optimal solution after 1000 iterations and shows a fast convergence rate and the best result, especially during the early iterations. For the GSO algorithm, it gets into the local minima.

### *(6) The Weight Optimization of Grid Spherical Shell Structure*

A double-layer grid steel shell structure with 83.6m span, 14.0m arc height and 1.5 shell thickness is shown in Fig. 6.26. The elastic module is 210 GPa and the density is 7850 kg/m3. There are 6761 nodes and 1834 bars in this shell. The 1834 bars were divided into three groups, which were upper chord bars, lower chord bars and belly chord bars. All chords were thin circular tubes and their sections were limited to Chinese Criterion GB/T8162-1999, which has 379 types of size to choose. The circumference nodes of lower chords are constrained. 50 kN vertical load is acted on each node of upper chords. The maximum permit displacement for all nodes is 1/400 of the length of span, that is ±0.209 m. The maximum permit stress for all chord bars is ±215 MPa. The stability of compressive chords is considered according to Chinese Standard GB50017-2003. The maximum slenderness ratio for compressive chords and tensile chords are 180 and 300 respectively.

For each algorithm, the population size is set to at 50 and the maximum number of iteration is limited to 200. For the GSO algorithm, 20% of the population is selected as rangers; the initial head angle $\varphi_0$ of each individual is set to be $\pi/4$. The constant a is given by round $(\sqrt{n+1})$. The maximum pursuit angle $\theta_{max}$ is $\pi/a^2$. The maximum turning angle $\alpha$ is set to be $\pi/2a^2$.For the HPSO algorithm, the inertia weight (w) is starting at 0.9 and ending at 0.4 by linearity descending. The acceleration constants $c_1$ and $c_2$ is set to be 0.8. The passive congregation coefficient $c_3$ is 0.6. For the QGSO algorithm, when target goes forward, the

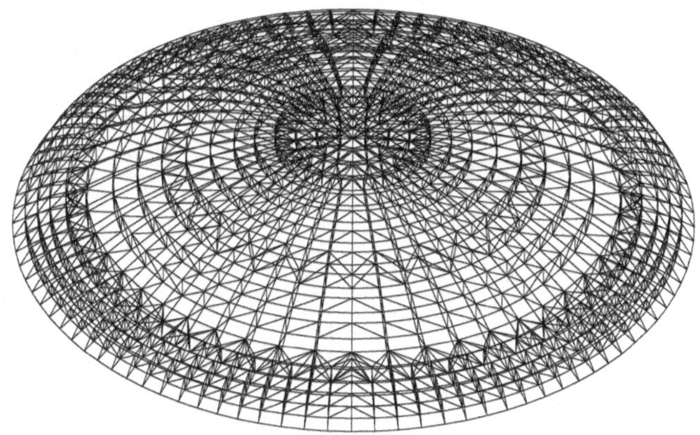

**Fig. 6.26** The double layer reticulated spherical shell structure

parameters are set in this: information transfer factor $w_1 = w_2 = 4$, selected probability $w_3 = 0.2$, component mutation probability $w_4 = 0.5$.Otherwise the parameters are set in this: information transfer factor $w_1 = 0.8$, $w_2 = 1.5$, selected probability $w_3 = 0.35$, component mutation probability $w_4 = 0.6$.

Considering that there are 379 types of size to choose in this structure which feasible region is much bigger than the one in the truss structures, the mutation factor in the ranger searching will not meet the requirement and the ranger performance will be changed into:

$$X_i^{k+1} = leftflag \times (P_i^k - 1) + changeflag \times P_g^k \tag{6.18}$$

Where $(P_i^k - 1)$ is called the mutation operation and the mutation direction is set into -1because -1 is the negative gradient direction of the integer space. While adding the gradient information into the hybrid process, the ranger can search more efficiently. Scrounger searches as equation (6.12).

When handling constraint, no more group classify, for the members out of the stress or the displacement boundary, they will be punished by given a biggish value.

With the above modification, QGSO algorithm is more suited for practical engineering optimum problems.

The optimization results are shown in Table 6.22. The convergence velocity is shown in Fig. 6.27. It can be seen from Fig. 6.27 that QGSO can be used effectively to optimize the complicated engineering structures and can obtained the best optimization solution.

**Table 6.22** The optimal solution for the double layer reticulated spherical shell structure

| Algorithms | Upper chord bars | Lower chord bars | Belly chord bars | Weight (kg) |
|---|---|---|---|---|
| HPSO | φ108×4 | φ83×3.5 | φ89×3.5 | 148811.71 |
| GSO | φ108×4 | φ95×3.5 | φ95×4 | 163954.70 |
| QGSO | φ83×4 | φ76×3.5 | φ102×3.5 | 139107.97 |

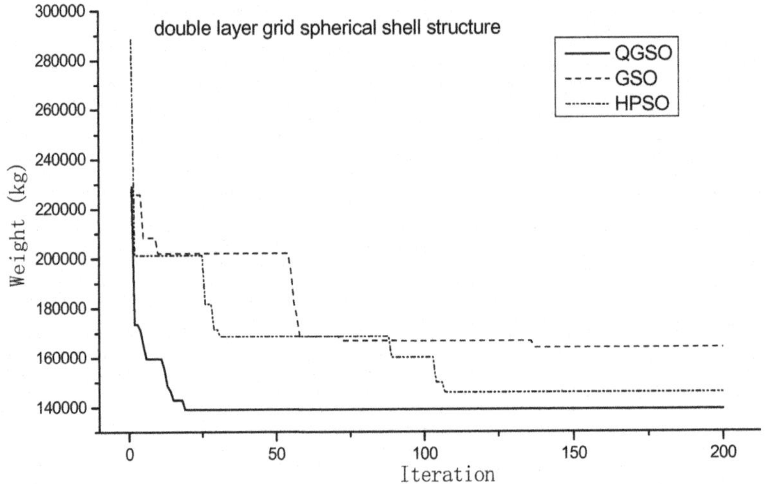


**Fig. 6.27** The convergence rate of the HPSO for the double layer grid spherical shell structure

The compared results show that the QGSO has a markedly superior preferable convergence rate and accuracy. It means that the QGSO algorithm can find the best solution in the shortest time. So the QGSO algorithm provides an effective and feasible approach for the structural optimization problems, especially for the practical large scale structures.

### *6.6.2 Conclusion*

The QGSO algorithm is based on the basic principle of GSO. It is also a group intelligent random optimization algorithm but has a markedly superior performance on convergence rate and accuracy than GSO.

Compared with HPSO, QGSO has a little better convergence accuracy, and in the aspect of convergence rate, the QGSO converges more quickly than the HPSO especially in the high-dimensional optimization problems.

When used for discrete variable problems, the QGSO expresses good properties than the GSO or the HPSO does, especially in term of convergence rate.

The quick convergence of QGSO makes it particularly attractive for sophisticate engineering structures, and is a banausic optimization algorithm.

## 6.7 Stability Studies

With the development of intelligent optimization algorithms, many traditional mathematical optimization algorithms have been replaced by this new algorithm in structural optimization. The merit of the new optimization algorithms is that they do not require conventional mathematical assumptions and posses better global search abilities than the traditional optimization algorithms. But the new algorithm also has its disadvantages. Without the conventional mathematical assumption, it is hard to testify whether the bionic optimization algorithms convergence to the optimal result at each optimization computation. So the stability of the bionic optimization algorithms is very important.

It is not difficult to find these bionic algorithms come with a random number, so there is a certain randomicity while using it and we use statistical methods in studying this randomicity. By applying different optimization algorithms for the same problem in continuous computing times, we can find out a relatively stable algorithm by comparing the average, standard deviation values and curves.

## 6.8 The Quick Group Search Optimizer with Passive Congregation (QGSOPC)

Based on the quick group search optimizer (QGSO), this section presents a quick group search optimizer with passive congregation (QGSOPC) and it will be verified by a planar truss, two spatial trusses and a double-layer grid steel shell structure with discrete variables. Compared with QGSO, GSO and HPSO algorithms, the results show that the QGSOPC algorithms have the best stability of convergence rate and accuracy.

As the PSOPC algorithm, QGSOPC is a hybrid QGSO with passive congregation as follow:

$$X_i^{k+1} = Floor(X_i^k + w_1 r(P_g^k - X_i^k) + w_2 r(P_i^k - X_i^k) + w_5 r(R_i^k - X_i^k)) \tag{6.19}$$

$$X_i^{k+1} = leftflag \times (P_i^k - 1) + changeflag \times P_g^k \tag{6.20}$$

Where Ri is a member selected randomly from the group, $W_5$ is the passive congregation coefficient. The other parameters and equations is the same as QGSO algorithm.

In this section, four pin connected structures commonly used in literature are selected as benchmark problems to test the QGSOPC. The examples given in the simulation studies include:

(1) A 25-bar spatial truss structure;
(2) A 52-bar planar truss structure;
(3) A 72-bar spatial truss structure;
(4) A double-layer grid steel shell structure;

The QGSOPC, QGSO, GSO, HPSO schemes are applied respectively to the examples and the results are compared in order to evaluate the performance of the modified algorithm. For all the algorithms, the population size is set to at 50 and compute in 50 times continuously. For the GSO algorithm, 20% of the population is selected as rangers; the initial head angle $\varphi_0$ of each individual is set to be $\pi/4$. The constant $a$ is given by round $(\sqrt{n+1})$. The maximum pursuit angle $\theta_{max}$ is $\pi/a^2$. The maximum turning angle $\alpha$ is set to be $\pi/2a^2$.For the HPSO algorithm, the inertia weight (w) is starting at 0.9 and ending at 0.4 by linearity descending. The acceleration constants $C_1$ and $C_2$ are set to be 0.8. The passive congregation coefficient $C_3$ is 0.6. For the QGSO algorithm, when target goes forward, the parameters are set in this: information transfer factors $w_1$= $w_2$=4, selected probability $w_3$=0.2, component mutation probability $w_4$=0.65. Otherwise the parameters are set in this: information transfer factors $w_1$=0.8, $w_2$=1.5, selected probability $w_3$=0.35, component mutation probability $w_4$=0.85. For the QGSOPC algorithm, $w_5$ is set to be 0.6 when target goes forward, otherwise is set to be 2.0. The other parameters are the same as the QGSO algorithm.

### *6.8.1 Numerical Examples*

#### *(1) A 25-Bar Spatial Truss Structure*

A 25-bar spatial truss structure, shown in Fig. 6.20, has been studied by Wu [24] and Li [14]. The material density is 0.1 lb/in.$^3$ and the modulus of elasticity is 10,000 ksi. The stress limitations of the members are ±40000 psi. All nodes in three directions are subjected to displacement limitations of ±0.35 in. The structure includes 25 members, which are divided into 8 groups, as follows: (1) $A_1$, (2) $A_2$~$A_5$, (3) $A_6$~$A_9$, (4) $A_{10}$~$A_{11}$, (5) $A_{12}$~$A_{13}$, (6) $A_{14}$~$A_{17}$, (7) $A_{18}$~$A_{21}$ and (8) $A_{22}$~$A_{25}$. The discrete variables are selected from the American Institute of Steel Construction (AISC) Code, which is shown in Table 6.16. The loads are shown in Table 6.17. A maximum number of 500 iterations is imposed for three cases.

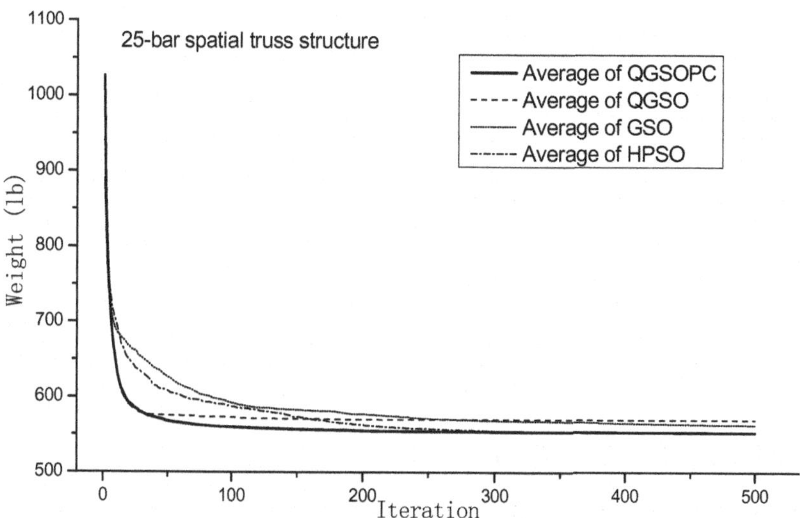


**Fig. 6.28** Convergence rates of the 25-bar spatial truss structure (Mean in 50 times)

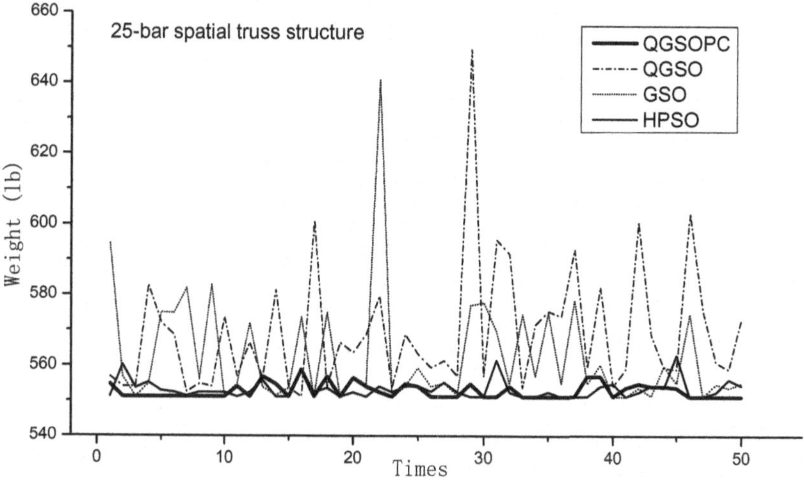


**Fig. 6.29** Convergence stability for the 25-bar spatial truss structure optimal design

Fig. 6.28 is the convergence rates and Fig. 6.29 is the stability of the four algorithms. Table 6.23 gives the optimal results.

**Table 6.23** Optimization results of the 25-bar spatial truss structure

| Optimization algorithm | Mean in 50 times | Standard deviation | The Best solution in 50 times | The Worse solution in 50 times |
|---|---|---|---|---|
| QGSOPC | 552.700 | 2.1411 | 551.137 | 558.844 |
| QGSO | 569.210 | 18.1144 | 551.137 | 649.482 |
| GSO | 562.764 | 15.7826 | 551.137 | 641.157 |
| HPSO [26] | 553.248 | 2.6717 | 551.137 | 562.944 |

It can be seen from Fig. 6.28 that the QGSOPC algorithm has the best convergence rate. Fig. 6.29 shows QGSOPC has the best convergence stability. Data in Table 6.23 prove the stability of the QGSOPC algorithm.

### *(2) A 52-Bar Planar Truss Structure*

A 52-bar planar truss structure, shown in Fig. 6.21, has been analysed by Wu [24] Lee [28] and Li [14]. The members of this structure are divided into 12 groups: (1) $A_1$~$A_4$, (2) $A_5$~$A_6$, (3) $A_7$~$A_8$, (4) $A_9$~$A_{10}$, (5) $A_{11}$~$A_{14}$, (6) $A_{15}$~$A_{18}$, and (7) $A_{19}$~$A_{22}$. The material density is 7860.0 kg/m$^3$ and the modulus of elasticity is 2.07×105 MPa. The members are subjected to stress limitations of ±180 MPa. Both of the loads, $P_x$ =100 kN, $P_y$ =200 kN are considered. The discrete variables are selected from the Table 6.16. A maximum number of 3,000 iterations is imposed.

Fig. 6.30 is the convergence rates and Fig. 6.31 is the stability of the four algorithms. Table 6.24 gives the optimal results.

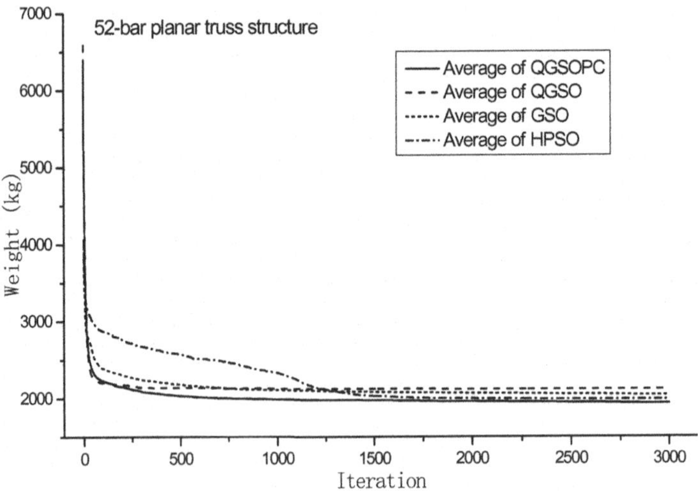


**Fig. 6.30** Comparison of convergence rates for the 52-bar planar truss structure optimal design (Mean in 50 times)

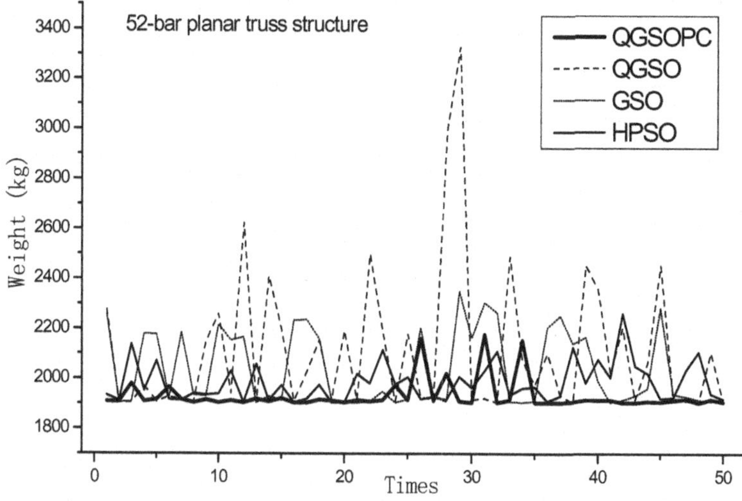


**Fig. 6.31** Comparison of convergence accuracy for the 52-bar planar truss structure optimal design

**Table 6.24** Compare of optimization results of the 52-bar spatial truss structure

| Optimization algorithm | Mean in 50 times | Standard deviation | The Best solution in 50 times | The Worse solution in 50 times |
|---|---|---|---|---|
| QGSOPC | 1929.95 | 62.7928 | 1902.605 | 2177.479 |
| QGSO | 2112.664 | 292.2927 | 1902.605 | 3326.917 |
| GSO | 2037.638 | 149.1894 | 1902.605 | 2349.302 |
| HPSO [26] | 1982.464 | 77.9442 | 1903.365 | 2261.448 |

From Fig. 6.30 and Fig. 6.31, it is showed that QGSOPC algorithm has the best convergence stability. Results in Table 6.24 proves this conclusion.

### *(3) A 72-Bar Spatial Truss Structure*

A 72-bar spatial truss structure, shown in Fig. 6.24, has been studied by Wu [24] Lee [28] and Li [14]. The material density is 0.1 lb/in.$^3$ and the modulus of elasticity is 10,000 ksi. The members are subjected to stress limitations of ±25 ksi. The uppermost nodes are subjected to displacement limitations of ±0.25 in. both in $x$ and $y$ directions. Two load cases are listed in Table 6.20. There are 72 members, which are divided into 16 groups, as follows: (1) $A_1$~$A_4$, (2) $A_5$~$A_{12}$, (3) $A_{13}$~$A_{16}$, (4) $A_{17}$~$A_{18}$, (5) $A_{19}$~$A_{22}$, (6) $A_{23}$~$A_{30}$ (7) $A_{31}$~$A_{34}$, (8) $A_{35}$~$A_{36}$, (9) $A_{37}$~$A_{40}$, (10) $A_{41}$~$A_{48}$, (11) $A_{49}$~$A_{52}$, (12) $A_{53}$~$A_{54}$, (13) $A_{55}$~$A_{58}$, (14) $A_{59}$~$A_{66}$ (15) $A_{67}$~$A_{70}$, (16) $A_{71}$~$A_{72}$. The discrete variables are selected from the Table 6.16. A maximum number of 1,000 iterations is imposed.

Fig. 6.32 is the convergence rates and Fig. 6.33 is the stability of the four algorithms. Table 6.25 gives the optimal results.

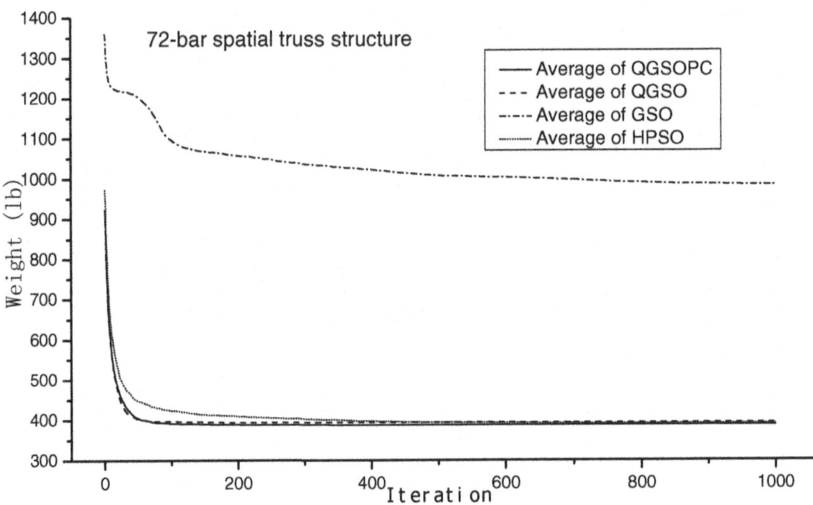


**Fig. 6.32** Convergence rates for the 72-bar truss structure optimal design (Mean in 50 times)

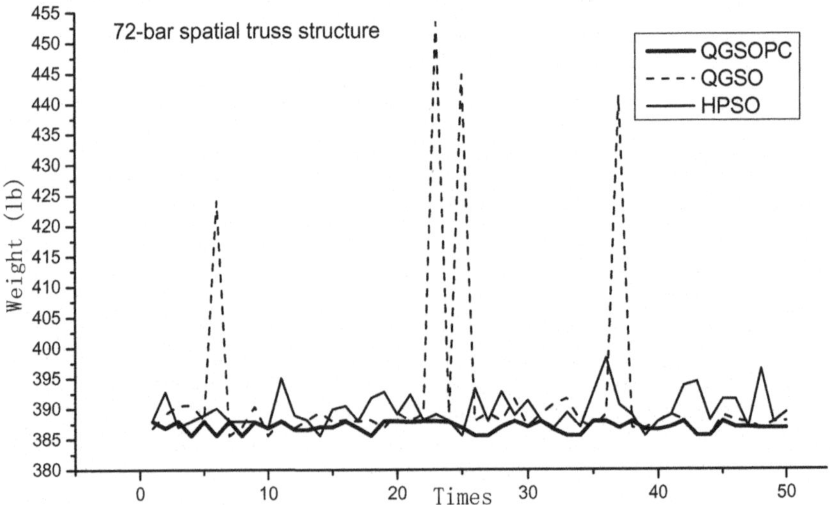


**Fig. 6.33** Convergence stability for the 72-bar truss structure optimal design

**Table 6.25** Optimization results of the 25-bar spatial truss structure

| Optimization algorithm | Mean in 50 times | Standard deviation | The Best solution in 50 times | The Worse solution in 50 times |
|---|---|---|---|---|
| QGSOPC | 386.9935 | 0.890642 | 385.543 | 387.943 |
| QGSO | 392.3947 | 14.67056 | 385.543 | 453.336 |
| GSO | 983.5541 | 9.228858 | 971.652 | 1004.429 |
| HPSO [14] | 389.7171 | 2.802837 | 385.543 | 398.264 |

Fig. 6.32 shows that QGSOPC and QGSO have the fastest convergence rate; GSO is trapped in local minimum. So the stability of GSO algorithm was not presented in the Fig. 6.33. It is shown in the Fig. 6.32, the most stable algorithm is QGSOPC. Data in the Table 6.25 prove the stability of the QGSOPC algorithm.

### *(4) A Double-Layer Grid Steel Shell Structure*

A double-layer grid steel shell structure with 83.6 m span, 14.0 m arc height and 1.5 shell thickness is shown in Fig. 6.26. The elastic module is 210 GPa and the density is 7850 kg/m$^3$. There are 6761 nodes and 1834 bars in this shell. The 1834 bars were divided into three groups, which were upper chord bars, lower chord bars and belly chord bars. All chords were thin circular tubes and their sections were limited to Chinese Criterion GB/T8162-1999, which has 379 types of size to choose. The circumference nodes of lower chords are constrained. 50 kN vertical load is acted on each node of upper chords. The maximum permit displacement for all nodes is 1/400 of the length of span, that is ±0.209 m. The maximum permit stress for all chord bars is ±215 MPa. The stability of compressive chords is considered according to Chinese Standard GB50017-2003. The maximum slenderness ratio for compressive chords and tensile chords are 180 and 300 respectively.

For the QGSOPC algorithm, $W_5$ is set to be 0.6 when target goes forward, otherwise is set to be 2.0. The other parameters are the same as the QGSO algorithm. And as to test the efficiency of the QGSOPC algorithm for this structure, the population size of QGSOPC is set to be 25 which is half of the other three algorithms.

Fig. 6.34 is the convergence rates of the four intelligent algorithm. Table 6.26 gives the optimal results.

From Fig. 6.33 it is showed that the QGSOPC algorithm has the best global search capacity and can find the best solution in the shortest time. Besides, the population size of QGSOPC is half of other three algorithms. Table 6.26 shows QGSOPC finds the lightest structure.

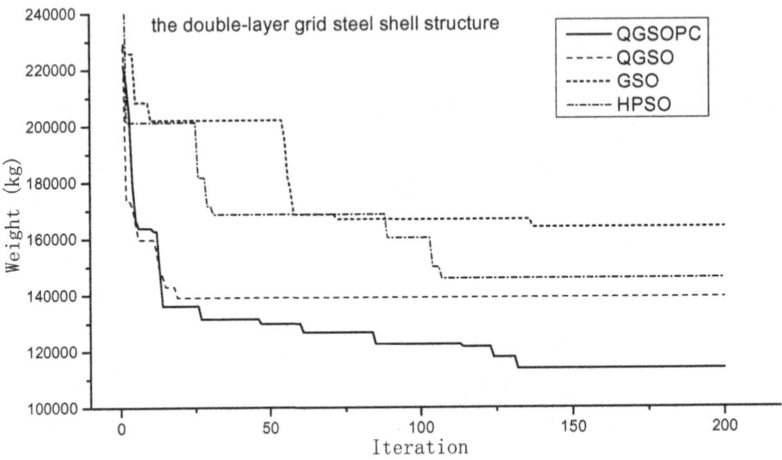


**Fig. 6.34** Convergence rates for the double-layer grid steel shell structure optimal design

**Table 6.26** Optimization results of the double-layer grid steel shell structure

| Algorithms | Upper chord bars | Lower chord bars | Belly chord bars | Weight (kg) |
|---|---|---|---|---|
| HPSO | φ108×4 | φ83×3.5 | φ89×3.5 | 148811.71 |
| GSO | φ108×4 | φ95×3.5 | φ95×4 | 163954.70 |
| QGSO | φ83×4 | φ76×3.5 | φ102×3.5 | 139107.97 |
| QGSOPC | φ76×3 | φ76×3 | φ89×3.5 | 113794.05 |

### *6.8.2 Conclusion*

From the numerical examples showed in this section, it is triumphant to add the passive congregation model to the QGSO. Because QGSOPC algorithm emphasises on information transmission and as to getting forward in a society, there is important to communicate with other members besides the best position it had been and the best member in the group. Therefore, information of the other members decides you will not be over-confidence with the best individual. And the key to prevent the algorithm from going into the local minimum is the passive congregation coefficient. Also using the target progress or not decision-making system is a good respond to the individual transmission of information between members. So with QGSOPC algorithm, no matter in convergent rate or accuracy of convergence, there is the best stability to ensure the optimization results.

## 6.9 Conclusion Remark

Based on the group search optimizer (GSO) and the particle swarm optimizer with passive congregation (PSOPC), a quick group search optimizer (QGSO) and a quick group search optimizer with passive congregation (QGSOPC) are presented in this chapter. The QGSO and QGSOPC algorithms can handle the constraint problems with discrete variables efficiently. The QGSOPC has more efficient search ability, faster convergent rate and less iterative times to find out the optimum solution.

It is successful to use QGSOPC in section optimization of truss structures. It is desired that QGSOPC can be used for topology optimization or other structural optimal design tasks in the future.

Currently, QGSOPC is only used in the single-objective optimization design, it is also desired that it can be further applied to multi-objective optimization design in the follow-up research work.

## References


1. Coello Coello, C.A.: Theoretical and numerical constraint-handling techniques used with evolutionary algorithms: a survey of the state of the art. Computer Methods in Applied Mechanics and Engineering 191(11-12), 1245–1287 (2002)
2. Dorigo, M., Di Caro, G., Gambardella, L.: Ant algorithms for discrete optimization. Artificial Life 5(3), 137–172 (1999)
3. Kennedy, J., Eberhart, R.C.: Particle swarm optimization. In: IEEE International Conference on Neural Networks, vol. 4, pp. 1942–1948. IEEE Press, Los Alamitos (1995)
4. Barnard, C.J., Sibly, R.M.: Producers and scroungers: a general model and its application to captive flocks of house sparrows. Animal Behavior 29(2), 543–550 (1981)
5. Kaveh, A., Hassani, B., Shojaee, S., Tavakkoli, S.M.: Structural topology optimization using ant colony methodology. Engineering Structures 30(9), 2559–2565 (2008)
6. Angeline, P.: Evolutionary optimization versus particle swarm optimization: philosophy and performance difference. In: Proceeding of the Evolutionary Programming Conference, San Diago, USA (1998)
7. Kennedy, J., Eberhart, R.C.: Swarm intelligence. Morgan Kaufmann, San Francisco (2001)
8. He, S., Wu, Q.H., Wen, J.Y., Saunders, J.R., Paton, R.C.: A particle swarm optimizer with passive congregation. BioSystem 78(1-3), 135–147 (2004)
9. Li, L.J., Liu, F., Xu, X.T., Liu, F.: The group search optimizer and its application to truss structure design. Advances in Structural Engineering 13(1), 43–51 (2010)
10. Qin, G., Liu, F., Li, L.J.: A quick group search optimizer with passive congregation and its convergence analysis. In: 2009 International Conference on Computational Intelligence and Security, Beijing, pp. 249–253 (2009)
11. Shen, H., Zhou, Y.H., Niu, B., Wu, Q.H.: An improved group search optimizer for mechanical design optimization problems. Progress in Natural Science 19(1), 91–97 (2009)
12. Snyman, J.A., Nielen, S., Roux, W.J.: A dynamic penalty function method for the solution of structural optimization problems. Applied Mathematical Modelling 18(8), 453–460 (1994)

13. Li, L., Liu, F.: Harmony particle swarm algorithm for structural design optimization. In: Geem, Z.W. (ed.) Harmony Search Algorithms for Structural Design Optimization. Studies in Computational Intelligence, vol. 239, pp. 121–157. Springer, Heidelberg (2009)
14. Li, L.J., Huang, Z.B., Liu, F.: A heuristic particle swarm optimization method for truss structures with discrete variables. Computers and Structures 87(7-8), 435–443 (2009)
15. Qin, G., Liu, F., Li, L.J.: A quick group search optimizer and its application to the optimal design of double layer grid shells. In: 2nd International Symposium on Computational Mechanics, 12th International Conference on Enhancement and Promotion of Computational Methods in Engineering and Science, HK-Macau (2009) Paper number 309
16. Lee, K.S., Geem, Z.W.: A new structural optimization method based on the harmony search algorithm. Computers and Structures 82(9-10), 781–798 (2004)
17. Schmit Jr., L.A., Farshi, B.: Some approximation concepts for structural synthesis. AIAA J. 12(5), 692–699 (1974)
18. Rizzi, P.: Optimization of multiconstrained structures based on optimality criteria. In: AIAA/ASME/SAE 17th Structures, Structural Dynamics and Materials Conference, King of Prussia, PA (1976)
19. Li, L.J., Huang, Z.B., Liu, F.: A heuristic particle swarm optimizer (HPSO) for optimization of pin connected structures. Computers and Structures 85(7-8), 340–349 (2007)
20. Khot, N.S., Berke, L.: Structural optimization using optimality criteria methods. In: Atrek, E., Gallagher, R.H., Ragsdell, K.M., Zienkiewicz, O.C. (eds.) New Directions in Optimum Structural Design, John Wiley, New York (1984)
21. Adeli, H., Kumar, S.: Distributed genetic algorithm for structural optimization. Journal of Aerospace Engineering ASCE 8(3), 156–163 (1995)
22. Sarma, K.C., Adeli, H.: Fuzzy genetic algorithm for optimization of steel structures. Journal of Structural Engineering ASCE 126(5), 596–604 (2000)
23. He, D.K., Wang, F.L., Mao, Z.Z.: Study on application of genetic algorithm in discrete variables optimization. Journal of System Simulation 18(5), 1154–1156 (2006)
24. Wu, S.J., Chow, P.T.: Steady-state genetic algorithms for discrete optimization of trusses. Computers and Structures 56(6), 979–991 (1995)
25. Rajeev, S., Krishnamoorthy, C.S.: Discrete optimization of structures using genetic algorithm. Journal of Structural Engineering, ASCE 118(5), 1123–1250 (1992)
26. Ringertz, U.T.: On methods for discrete structural constraints. Engineering Optimization 13(1), 47–64 (1988)
27. Zhang, Y.N., Liu, J.P., Liu, B., Zhu, C.Y., Li, Y.: Application of improved hybrid genetic algorithm to optimize. Journal of South China University of Technology 33(3), 69–72 (2003) (in Chinese)
28. Lee, K.S., Geem, Z.W., Lee, S.H., Bae, K.W.: The harmony search heuristic algorithm for discrete structural optimization. Engineering Optimization 37(7), 663–684 (2005)

# Chapter 7
# Group Search Optimizer and Its Applications on Multi-objective Structural Optimal Design

**Abstract.** There are some problems in multi-objective optimization of engineering structures, such as, the difficulties in dealing with the constraints, the complexity of program and the low computational efficiency. To solve these problems, an improved group search optimizer, named Multi-objective Group Search Optimizer (MGSO), and combined with Pareto solutions theory is presented in this chapter. Different types of examples are employed to evaluate the performance of MGSO, including truss structures and frame structures with continuous variables or discrete variables. The calculation results show the feasibility, practicality and superiority of MGSO in structure optimal design. As a stochastic algorithm, MGSO has excellent performance in terms of convergence rate. Only the best individual is needed to be selected and partial constraints are needed to be checked to find the producer, thus a great deal of computational time is saved. The MGSO is of obvious advantages for complex engineering problems especially for high-dimensional ones.

## 7.1 Introduction

Multi-objective optimal problems are the kind of problem that has multiple objective functions. In the practical engineering, there are many optimal problems that have relations with multiple objectives. The results of multi-objective optimization problems are all of those potential solutions that the components of the corresponding objective vectors cannot be all simultaneously improved. This is known as the concept of Pareto optimality. The designer chooses the final solution from these Pareto-optimal solutions according to the practical situations [1]. Classical optimization methods suggest converting the multi-objective optimization problem to a single-objective optimization problem by emphasizing one particular Pareto-optimal solution at a time. When such a method is used to find multiple objective optimal solutions, it has to be applied many times, hopefully finding a different solution at each simulation run. With the development of computer technology, more and more stochastic optimization algorithms are applied to multi-objective optimization. There is a rapid development of genetic algorithm [2-5] and Particle Swarm Optimization (PSO) algorithm [6-9] for multi-objective

L. Li & F. Liu: Group Search Optimization for Applications in Structural Design, ALO 9, pp. 207–246.
springerlink.com 

structural optimal design. The disadvantages of genetic algorithm are that it has slow convergence rate in some cases and is complex in programming. It is also easy to converge to local optimal solution in the process of evolution. For PSO it is not only needed to find the overall best individual (*gbest*) in the group, but also need to find the best individual of the history (*pbest*). For the practical complex structures, it is a very time-consuming process. This paper introduces a novel intelligence optimization algorithm named Group Search Optimizer (GSO) [10], and it is improved to a multi-objective optimization algorithm. This improved algorithm converges fast, only needs to find the best individual of a group at each iteration, and does not need to check the constraints for all individuals. Therefore, it can save a lot of computing time and get better results. For complex engineering problems, It has obvious advantages. There will be possible application of the method to stochastic optimization for robustness, involving mean value and variance of the objective function just as Doltsinis reported in reference [11].

## 7.2 Multi-objective Optimization Concepts

### 7.2.1 *Multi-objective Optimization Problem*

A general multi-objective optimization problem (MOP) includes a set of $n$ parameters (decision variables), a set of $k$ objective functions, and a set of $m$ constraints. Objective functions and constraints are functions of the decision variables. The optimization goal is as follows:

$$\begin{aligned} &\textit{maximize} \quad \mathbf{y} = f(\mathbf{x}) = (f_1(\mathbf{x}), f_2(\mathbf{x}), \ldots, f_k(\mathbf{x})) \\ &\textit{subject to} \quad \mathbf{e}(\mathbf{x}) = (e_1(\mathbf{x}), e_2(\mathbf{x}), \ldots, e_m(\mathbf{x})) \le 0 \\ &\textit{where} \quad \mathbf{x} = (x_1, x_2, \ldots, x_n) \in \mathbf{X} \\ &\qquad \mathbf{y} = (y_1, y_2, \ldots, y_k) \in \mathbf{Y} \end{aligned} \tag{7.1}$$

where $\mathbf{x}$ is the decision vector, $\mathbf{y}$ is the objective vector, $\mathbf{X}$ is denoted as the decision space, and $\mathbf{Y}$ is called the objective space. The constraints $\mathbf{e}(\mathbf{x}) \le 0$ determine the objective set of feasible solutions.

### 7.2.2 *Feasible Set*

The feasible set $X_f$ is defined as the set of decision vectors $\mathbf{x}$ that satisfy the constraint $\mathbf{e}(\mathbf{x})$:

$$X_f = \{\mathbf{x} \in \mathbf{X} \mid \mathbf{e}(\mathbf{x}) \le 0\} \tag{7.2}$$

The image of $X_f$, i.e., the feasible region in the objective space, is denoted as:

$$Y_f = f(X_f) = \cup_{\mathbf{x} \in X_f} \{f(\mathbf{x})\}. \tag{7.3}$$

Considering the generality, a maximization problem is assumed here. For minimization or mixed maximization/minimization problems the definitions presented in this section are similar.

### 7.2.3 For Any Two Objective Vectors u and v:

$$\begin{aligned} \mathbf{u} = \mathbf{v} \quad & \textit{iff} \quad \forall i \in \{1,2,\ldots,k\} : u_i = v_i \\ \mathbf{u} \geq \mathbf{v} \quad & \textit{iff} \quad \forall i \in \{1,2,\ldots,k\} : u_i \geq v_i \\ \mathbf{u} > \mathbf{v} \quad & \textit{iff} \quad \mathbf{u} \geq \mathbf{v} \wedge \mathbf{u} \neq \mathbf{v} \end{aligned} \tag{7.4}$$

The relations '$\leq$' and '$<$' are defined similarly.

### 7.2.4 Pareto Optimality

The concept of Pareto optimality was formulated by Vilfredo Pareto in 19[th] century. A decision vector $\mathbf{x} \in X_f$ is said to be non-dominated regarding a set $A \subseteq X_f$ if for any vector $\mathbf{a}$ :

$$\mathbf{a} \in A : \mathbf{a} \succ \mathbf{x} \tag{7.5}$$

Moreover, $\mathbf{x}$ is said to be Pareto optimal if $\mathbf{x}$ is non-dominated regarding $X_f$ .

### 7.2.5 Pareto Dominance

For any two decision vectors $\mathbf{a}$ and $\mathbf{b}$, that $\mathbf{a}$ dominates $\mathbf{b}$ is denoted as:

$$\mathbf{a} \succ \mathbf{b} \quad \textit{iff} \quad f(\mathbf{a}) > f(\mathbf{b})$$

That $\mathbf{a}$ weakly dominates $\mathbf{b}$ is denoted as:

$$\mathbf{a} \succeq \mathbf{b} \quad \textit{iff} \quad f(\mathbf{a}) \geq f(\mathbf{b})$$

That $\mathbf{a}$ is indifferent from $\mathbf{b}$ is denoted as:

$$\mathbf{a} \sim \mathbf{b} \quad \textit{iff} \quad f(\mathbf{a}) \ngeq f(\mathbf{b}) \wedge f(\mathbf{b}) \ngeq f(\mathbf{a}) \tag{7.6}$$

The definitions for a minimization problem ( $\prec, \preceq, \sim$ ) are analogical.

### 7.2.6 Pareto Optimal Set and Pareto Front

Let $A \subseteq X_f$ . The function $p(A)$ gives the set of non-dominated decision vectors in $A$ :

$$p(A)=\{\mathbf{a}\in A|\mathbf{a}\} \tag{7.7}$$

The set $p(A)$ is the non-dominated set regarding $A$. The corresponding set of objective vectors $f(p(A))$ is the non-dominated front set regarding $A$ . Furthermore, the set $X_p = p(X_f)$ is called the Pareto optimal set and the set $Y_p = f(X_p)$ is denoted as the Pareto optimal front.

### 7.2.7 *Pareto Constraint-Dominance*

For multi-objective optimization problems with constrains, a solution $\mathbf{u}$ is said to Pareto constraint-dominate a solution $\mathbf{v}$, if only one of the following three conditions is true. Firstly, the solution $\mathbf{u}$ is feasible and solution $\mathbf{v}$ is not. Secondly, solution $\mathbf{u}$ and $\mathbf{v}$ are both infeasible, but solution $\mathbf{u}$ has a smaller constraint violation degree. Thirdly, solutions $\mathbf{u}$ and $\mathbf{v}$ are both feasible and solution $\mathbf{u}$ dominates solution $\mathbf{v}$.

### 7.2.8 *The Measure of Multi-objective Optimization Results*

Unlike single-objective optimization problem, a multi-objective optimization problem has two goals [1]. Firstly, optimization can converge closer to the true Pareto front. Secondly, try to get the widest and most uniform distribution of Pareto front of non-dominated solutions.

## 7.3 Group Search Optimizer

### 7.3.1 *Group Search Optimizer for Continuous Variables*

Group search optimizer is based on the Producer-Scrounger (PS) model of biological models [10, 12]. PS model assumes that the group members search either for 'finding' (producer) or for 'joining' (scrounger) opportunities. GSO also employs 'rangers' who perform random walks to avoid entrapment in local minima. The GSO algorithm just needs to find the best member as producer who is followed by the other members except the rangers. The members are sorted by the fitness values without constraints, and then the constraints are checked by sequence. For this method, not all the members' constraints are needed to be checked to find the producer, so GSO can save a great deal of computational time.

At each iteration, a group member, located in the most promising area, conferring the best fitness value, is chosen as the producer. It then stops and scans the environment to seek resources (optima). Scanning is an important component of search orientation; it is a set of mechanisms by which animals move sensory receptors and some times their bodies or appendages so as to capture information from the environment. Scanning can be accomplished through physical contact or by visual, chemical, or auditory mechanisms. In the GSO, vision, as the main scanning mechanism used by many animal species, is employed by the producer. To perform visual searches, many animals encode a large field of view with retinas having variable spatial resolution, and then use high-speed eye movements to direct the highest resolution region towards potential target locations. In our GSO algorithm, basic scanning strategies is employed. The scanning field of vision is generalized to a $n$-dimensional space, which is characterized by maximum pursuit angle $\alpha_{max} \in \mathbf{R}^{n-1}$ and maximum pursuit distance $l_{max} \in \mathbf{R}^1$.

The population of the GSO algorithm is called a group and each individual in the group is called a member. In an n-dimensional search space, the $i_{th}$ member at the $k_{th}$ searching bout (iteration) has a current position $X_i^k \in \mathbf{R}^n$, a head angle $\varphi_i^k = (\varphi_{i1}^k, ..., \varphi_{i(n-1)}^k) \in \mathbf{R}^{n-1}$ and a head direction $D_i^k(\varphi_i^k) = (d_{i1}^k, ..., d_{in}^k) \in \mathbf{R}^n$, which can be calculated from $\varphi_i^k$ via a Polar to Cartesian coordinate transformation:

$$d_{i1}^k = \prod_{p=1}^{n-1} \cos(\varphi_{ip}^k) \tag{7.8}$$

$$d_{ij}^k = \sin(\varphi_{i(j-1)}^k) \cdot \prod_{p=i}^{n-1} \cos(\varphi_{ip}^k) \tag{7.9}$$

$$d_{in}^k = \sin(\varphi_{i(n-1)}^k) \tag{7.10}$$

At the $k_{th}$ iteration the producer $X_p$ behaves as follows:

(1) The producer will scan at zero degree and then scan laterally by randomly sampling three points in the scanning field[13, 14]: one point at zero degree:

$$X_z = X_p^k + r_1 l_{max} D_p^k(\varphi^k) \tag{7.11}$$

one point in the left hand side hypercube:

$$X_l = X_p^k + r_1 l_{max} D_p^k(\varphi^k - r_2 \theta_{max} / 2) \tag{7.12}$$

and one point in the right hand side hypercube:

$$X_r = X_p^k + r_1 l_{max} D_p^k(\varphi^k + r_2 \theta_{max} / 2) \tag{7.13}$$

where $r_1 \in \mathbf{R}^1$ is a normally distributed random number with mean 0 and standard deviation 1 and $r_2 \in \mathbf{R}^{n-1}$ is a random sequence in the range (0, 1).

(2) The producer will then find the best point with the best resource (fitness value). If the best point has a better resource than its current position, then it will fly to this point. Otherwise, it will stay in its current position and turn its head to a new angle:

$$\varphi^{k+1} = \varphi^k + r_2 \alpha_{max} \tag{7.14}$$

where $\alpha_{max}$ is the maximum turning angle.

(3) If the producer cannot find a better area after a iterations, it will turn its head back to zero degree:

$$\varphi^{k+a} = \varphi^{k} \tag{7.15}$$

where $a$ is a constant. At the $k_{th}$ iteration the area copying behavior of the $i_{th}$ scrounger can be modeled as a random walk towards the producer:

$$X_i^{k+1} = X_i^k + r_3(X_p^k - X_i^k) \tag{7.16}$$

where $r_3 \in \mathbf{R}^n$ is a uniform random sequence in the range (0, 1).

Besides the producer and the scroungers, a small number of rangers have also been introduced into our GSO algorithm. In nature, group members often have different searching and competitive abilities. Subordinates, who are less efficient foragers than the dominant, will be dispersed from the group. This may result in ranging behavior. The ranging animals-rangers, perform search strategies, which include random walks and systematic search strategies to locate resources efficiently. In the GSO algorithm, random walks, which are thought to be the most efficient searching method for randomly distributed resources, are employed by rangers. If the $i_{th}$ group member is selected as a ranger, at the $k_{th}$ iteration, it generates a random head angle $\varphi_i$ :

$$\varphi^{k+1} = \varphi^{k} + r_2\alpha_{max} \tag{7.17}$$

where $\alpha_{max}$ is the maximum turning angle; and it chooses a random distance:

$$l_i = a \cdot r_1 l_{max} \tag{7.18}$$

and move to the new point:

$$X_i^{k+1} = X_i^k + l_i D_i^k(\varphi^{k+1}) \tag{7.19}$$

## 7.3.2 Group Search Optimizer for Discrete Variables

A structural optimization design problem with discrete variables can be formulated as a nonlinear programming problem. In the size optimization for a truss structure, the cross-section areas of the truss members are selected as the design variables. Each of the design variables is chosen from a list of discrete cross-sections based on production standard. The objective function is the structure weight. The design cross-sections must also satisfy some inequality constraints equations, which restrict the discrete variables. The optimization design problem for discrete variables can be expressed as follows:

$$\min f\left(x^1, x^2, \ldots, x^d\right), \quad d = 1, 2, \cdots, D$$

subjected to:

$$g_q\left(x^1,x^2,\ldots,x^d\right)\le 0,\quad d=1,2,\cdots,D,\quad q=1,2,\cdots,M$$

$$x^d\in S_d=\left\{X_1,X_2,\cdots,X_p\right\}$$

where $f\left(x^1,x^2,\ldots,x^d\right)$ is the truss's weight function, which is a scalar function. And ( $x^1,x^2,\ldots,x^d$ ) represent a set of design variables. The design variable $x^d$ belongs to a scalar $S_d$ , which includes all permissive discrete variables $\left\{X_1,X_2,\ldots X_p\right\}$ . The inequality $g_q\left(x^1,x^2,\ldots,x^d\right)\le 0$ represents the constraint functions. The letter $D$ and $M$ are the number of the design variables and inequality functions respectively. The letter $p$ is the number of available variables.

Considering the areas of cross-sections aren't continuum, when the GSO algorithm is used to optimize problems with discrete variables, a mapping function is usually created to make the discrete section areas correspond to the continuum integers from small to large. Suppose a discrete set $S_d$ with $p$ discrete variables, by arranging from small to large:

$$S_d=\{X_1,X_2,\cdots,X_j,\cdots X_p\},\quad 1\le j\le p$$

Employ a mapping function $h(j)$ to replace the discrete values of $S_d$ with its serial numbers like this:

$$h(j)=X_j$$

The discrete values were replaced by the serial numbers to keep the searching with continuum values and avoid declining of search efficiency. Suppose that there are $n$ members in the search space with $D$ dimension. And the position of the $i_{th}$ member is denoted with vector $\mathbf{x}_i$ as:

$$\mathbf{x}_i=(x_i^1,x_i^2,\cdots,x_i^d,\cdots,x_i^D)\;,\quad 1\le d\le D,\quad i=1,\cdots,n$$

in which, $x_i^d\in\{1,2,\cdots,j,\cdots,n\}$ corresponds to the discrete variables $\{X_1,X_2,\cdots,X_j,\cdots X_p\}$ by mapping function $h(j)$. After that, all of the members will search in the continuum space which is the integer space. Each component of vector $\mathbf{x}_i$ is integer.

## 7.4 Multi-objective Group Search Optimizer (MGSO) and Calculation Procedure

The GSO algorithm can not be directly applied to multi-objective optimization problems although it has been applied successfully to single-objective optimization problems [15-17]. Since the multi-objective optimization problems are essentially different from the single objective optimization problems. The former is usually a set of a group or several groups of solutions, while the latter is only a single solution or a group solution. However, the successful application of genetic algorithms and particle swarm algorithm to multi-objective optimization problems, and the similarity of GSO algorithm to PSO and to GA show that GSO may deal with multi-objective optimization problem efficiently.

### *7.4.1 Key Issues of Multi-objective GSO Algorithm*

The difference between MGSO and GSO is in the comparison rules of the fitness of individuals. In MGSO, the fitness of individuals is compared by domination relations. Then, an optimal Pareto set is archived. A producer is selected from any individuals in the optimal Pareto set archived. An external archive is needed to be setup. The main objective of the external archive is to store a history of non-dominated individuals found along the search process, and try to keep those produces a well-distributed Pareto front. The key issues of MGSO are as follows:

(1) How to maintain and update the external archives to get a well-distributed Pareto front.

The key issue in the external archive management is to decide whether a new solution should be added to or not. When a new solution is found during MGSO evolutionary process, it is compared with each of the solution in the archive. If this new solution is dominated by an individual in the archive (i.e., the solution is dominated by the archive), then such solution is discarded, otherwise, the solution is stored in the archive. If there are solutions in the archive that are dominated by the new element, then such solutions are removed. If the archive reaches its maximum allowable capacity after adding the new solution, a decision has to be made regarding removal of one of its individuals. Several density estimation methods are proposed for multi-objective evolutionary algorithms to maintain the archive size, whenever the archive reaches maximum allowed capacity. In this article, the crowding distance calculation proposed by Deb [4] is used. To get an estimation of the density of solutions surrounding a particular solution in the population, the average distance of two points on either side of this point along each of the objectives are calculated. This quantity serves as an estimation of the perimeter of the cuboids formed by using the nearest neighbors as the vertices (call this the crowding distance). The crowding-distance computation requires sorting the population according to each objective function value in ascending order of magnitude. Thereafter, for each objective function, the boundary solutions (solutions with smallest and largest function values) are assigned an infinite distance value. All other intermediate solutions are assigned a distance value equal to the absolute normalized difference in the function values of two adjacent solutions. This calculation is continued for other objective functions. The overall crowding-distance value is calculated as the sum of individual distance values corresponding to each objective. The crowding distance is calculated as:

$l = |I|$

$for \quad each \quad i, \quad set\, I[i]_{\text{distance}} = 0$

$for \quad each \quad objective \quad m\,,\; I = sort(I, m)$

$I[1]_{\text{distance}} = I[l]_{\text{distance}} = \infty$

$for \quad i = 2\ to(l-1)$

$$I[i]_{\text{distance}} = I[i]_{\text{distance}} + {(I[i+1].m - I[i-1].m)}\Big/{\left(f_m^{\max} - f_m^{\min}\right)}$$

where, $l$ is the series number sorted according to some objective function. $I$ is the number of population. *sort (I, m)* is the series of an individual respect to the *m*th objective function value. $I[i\text{-}1]_{.m}$ and $I[i\text{+}1]_{.m}$ represent two individuals adjacent to the ith individual when sorted by *m*th objective function value. $f_m^{\max}$ and $f_m^{\min}$ is the maximum and minimum value of the *m*th objective function respectively. $I[i]_{distance}$ represents the crowding distance value of the ith individual.

(2) How to select an individual from the external archive as the producer.

If the crowding distances of the individuals in the external archive are all infinite, then randomly select an individual as a producer. If there are individuals whose crowding distances are not infinite in the external archive, then randomly select an individual whose crowding distance is not infinite as the producer.

(3) The mechanism to handle constraints

The mechanism proposed in this paper to handle constraints is to ensure a feasible solution better than the infeasible solutions. The fitness of individuals who violate the constraint is given to infinity (*inf*) or 0, in order to ensure the individuals that do not violate constraints are better than the individuals that violate constrains. This approach is not only effective but also very simple.

## 7.4.2 The Structure of Multi-objective Search Group Optimizer

Step 1: Randomly initialize positions and head angles of all members.

Step 2: Check if each initial individual is in the feasible region, if it is not in the feasible region, then reinitialize the individual.

Step 3: Setup an external archive for collection of non-dominated solution, and setup the maximum capacity of the external archive $M$ . When the number of non-dominated solutions $N$ does not exceed $M$ , the non-dominated solutions are directly copied to the external archive. If the non-dominated solutions are more than $M$ , then the crowding distances of non-dominated solutions are sorted by decreasing sequence, and the ( $N$ - $M$ ) particles at the back will be deleted. An individual in the external selection is randomly chosen as a producer.

Step 4: Perform producing. Firstly, the producer scans at zero degree and then scans laterally by randomly sampling three points in the scanning field as equation (7.11), (7.12) and (7.13). Secondly, if any point violates the constraints, it will be replaced by the producer's previous position. Thirdly, compare the new points with the original point. If any new point dominates the original point, the producer will jump to this point. If the new points and the original point are non-dominant relationship, then randomly select a new point as a producer. If all the new points are dominated by the original point, the producer will remain in its original point, and switch the direction to the next iteration according to equation (7.14). If the producer can not find a better point after a iterations, it will turn its head back to zero degree according to equation (7.15).

Step 5: Perform scrounging. Randomly select 80% from the rest members to perform scrounging, and follow the producer according to equation (7.16).

Step 6: Perform ranging. For the rest members, they will perform ranging according to equation (7.17), (7.18) and (7.19).

Step 7: Check feasibility. Check whether any member of the group violates the constraints. If it does, use the mechanisms to handle it.

Step 8: Calculate the fitness value of current members. Find all non-dominated solutions, update the external archive, and update the producer and search angle.

Step 9: Check whether the maximum number of iterations is reached, if it is, then stop the calculation, otherwise, go to step 4 to continue the computation.

## 7.5 Truss Optimization

Structural optimization is consistent with the structure from the use of functional requirements and meets the structural strength, stability and rigidity of all feasible design, the relative standard by designers to find the optimal solution. It enables designers to shift from passive to active analysis of the design. Optimization can be divided into: a given type of structure, materials, layout and geometry of the case, the optimal size of the component sections, making the structure of the smallest quality or the most economical, this optimization is called size optimization; if the structure's shape is uncertain in advance, compared to structural shape optimization; further optimize the layout of the structure, this is the topology optimization.

There are generally three ways for the design of complex structures: First, to simplify the overall structure into separate components to optimize, the method can not accurately take into account the overall effect of the structure; Second, build a calculation model, repeated a spreadsheet, but there is the larger workload and a design review enumeration algorithm is feasible by design, not the optimal design in theory; The third way is applied finite element optimization module to optimize the complex structures. However, for the complex steel structure, its stability (the overall stability of structural systems, the overall stability of components and local stability of plates) is also an important factor in the success of optimized design and it will lead to the fail optimization of finite element software.

In the past few decades, structural optimization has become a hot research project Truss structure are widely used in engineering, because of its low cost, light weight and simple construction, etc. Its applications related to bridge design, building structure and the grid frame and other aspects. It is playing an increasingly important role in the actual construction. For the truss structure, in the case of a given, material, topological layout and shape, the optimal size of each bar of the cross section called size optimization to the lightest and most economical structure. In the size optimization variables are cross-sectional areas of the bars. In the field of structural optimization, the objective function of single-objective optimization is generally the total weight of the structure, natural frequency and deflection and so on. For the optimization of space truss design, designers are often most concerned about the weight of the structure optimization and base frequency. There are many single-objective optimization of truss design results, but multi-objective optimization results is less. Since Zadeh introduced into concept of multi-scale optimization into engineering optimization, Stadler sum of the multi-scale optimization in mechanics, more and more multi-objective optimization results emerge. In reference [18], a detailed description of traditional methods about multi-objective optimization of space truss was introduced.

Recently, genetic algorithm (GA), simulated annealing (SA) and particle swarm optimization (PSO) algorithm and other heuristic optimization algorithms began to rise in the field of engineering optimization. There are some key parameters in GA (such as population size, crossover and mutation rate) which are not easy to determine, and GA has big calculate volume, large time-consuming and other shortcomings. SA simulates annealing process of solids, using Metropolis acceptance criteria, and with a group parameter called the cooling schedule to control algorithm process, and obtained the approximate optimal solution. This algorithm also does not require the continuity and differentiability of the function. It can handle the programming problems which have continuous - discrete -integer nonlinear variables. When used in optimization it can provide a better solution. When SA is used for structural optimization, there is much re-analysis, large computational capacity, low efficiency, and parameters are controlled with difficulty. For particle swarm optimization, its computation is required to verify whether each particle violates constraint during each iteration. For the complex structure of actual projects, this is a very time-consuming process of calculation and analysis.

In this paper, MGSO algorithm is applied to the size and shape optimization of truss structure, and it is explained with examples respectively.

### 7.5.1 The Size Optimization of Truss

The size optimization of truss is in the case of a given type of structure, materials, layout and geometry, optimizing the size of the component sections, making the structure of the smallest quality or the most economic. In this chapter, it will design the optimal structure that satisfies all constraints in the premise of the lightest weight of the structure, while the maximum displacement of the node is minimum. To n-bars truss structure system, the basic parameters (including the elastic modulus, material density, the maximum allowable stress, etc.) are known, and the optimization's target is that identify the optimal n-bar truss section area so that the structure is of the smallest total weight, the smallest maximum nodes' displacement, in a given load conditions. Therefore, mathematical models can be expressed as:

$$A = [A_1, A_2, \ldots, A_n]^T$$

$$f_1 = \min W = \sum_{i=1}^{n} \rho_i A_i L_i$$

$$f_2 = \min\left\{\max\left\{u_x, u_y, u_z\right\}\right\}$$

Subject to: $g_i^\sigma(A) = [\sigma_i] - \sigma_i \geq 0 \quad (i = 1, 2, \ldots, K)$

$$A_{\min} \leq A \leq A_{\max} \tag{7.20}$$

Where $A = [A_1, A_2, \ldots, A_n]^T$ are design variables for the cross section, $n$ is the group number of design variables for the cross section bar, $f_1$ and $f_2$ are two objective functions of multi-objective optimization. $W$ is the total weight of the structure, $L_i$, $A_i$ and $\rho_i$ were the bar length, section area, and density of the $i^{th}$ group bars, respectively. For a specific node, $u_x$, $u_y$ and $u_z$ is the displacement of the $x$, $y$ and $z$ direction, respectively. $g_i^\sigma(A)$ are the stress constraints. $[\sigma_i]$ and $\sigma_i$ are allowable stress and the worst stress of the $i^{th}$ group bars under various conditions, and $K$ is the total number of bars. $A_{\min}$ and $A_{\max}$ are the minimum and maximum section size, respectively.

### 10-Bar Truss

The schematic representation of a 10-bar truss structure is shown in Fig. 7.1. A tip load of 444.5 kN is applied to the truss. The stress limit is 172.25 MPa in both tension and compression for all members. Young's modulus is specified as $6.89 \times 10^4$ MPa, and the material density is $2.768 \times 10^3$ kg/m$^3$. Node 2, 3, 5 and 6 are allowed to move only in the $y$ direction. $f_1$ and $f_2$ are design objectives, $f_1$ is the total mass of the structure $W$, and $f_2$ is the maximum displacements $\delta$ which is the maximum displacement of the node 2, 3, 5 and 6 along the load direction. Rod stress is less than permitted stress, and the bar area is between 6.452 mm$^2$ and 258064 mm$^2$.

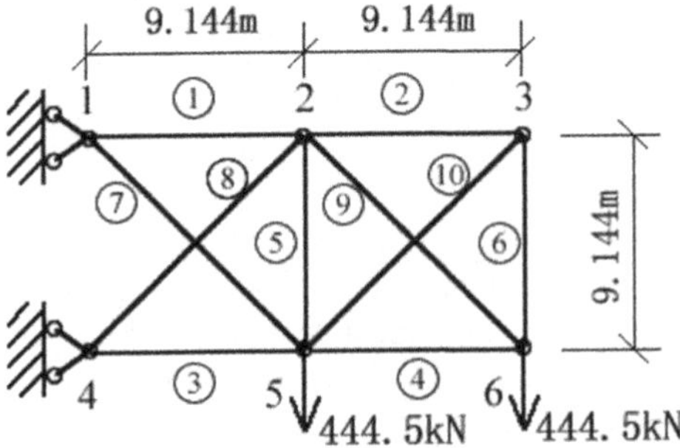


**Fig. 7.1** The 10-bar planar truss structure

This example is the plane truss structure optimization problem with continuous variable. The variable is the cross-sectional area, and the variable dimension is 10. The population of the GSO is 300, and the capacity of the external archive is 50.

The minimum cross-sectional area is 6.452 mm$^2$, and the maximum cross-sectional area is 25806.4 mm$^2$. The constraint is that the stress of bars must meet the allowable stress. The objective functions are the total mass of the structure $W$ and the maximum displacement δ of the node 2, the node 3, the node 5 and the node 6, and the displacement is along the load direction. The ideal optimization result is the smallest of total weight and the maximum of the minimum displacement. Using Matlab for finite element analysis, respectively, through 50, 150 and 250 independent iterations, the results are shown in Fig. 7.2, Fig. 7.3 and Fig. 7.4. The Pareto optimal front after 250 iterations is shown in Fig. 7.5.

The Pareto optimal front of the elite set after 50 iterations is shown in Fig. 7.2 , and the boundary values are as follows: the (max $W$, min δ) is (6741.1 kg, 0.0266 m),and the (min $W$, max $\delta$) is (1568.6 kg, 0.1111 m). So, after 50 iterations the ideal solution (min $W$, minδ) is (1568.6 kg, 0.0266 m). Seeing from Fig. 7.2, MGSO algorithm has not been uniform Pareto optimal front after 50 iterations.

The Pareto optimal front of the elite set after 150 iterations is shown in Fig. 7.3. The Pareto front boundary values are as follows: the (max $W$, min $\delta$) is (6604.6 kg, 0.0263 m), and the(min $W$, max$\delta$) is (1050.3 kg, 0.1352 m). Compared Fig. 7.3 with Fig. 7.2, after 50 iterations the ideal solution (min $W$, min $\delta$) is (1568.6 kg, 0.0266 m), and after 150 iterations the ideal solution (min $W$, min $\delta$) is (1050.3 kg, 0.0263 m). Whether for the target min $W$ or the target min $\delta$, the Pareto optimal front after 150 iterations is better than the Pareto optimal front after 50 iterations. Seeing from Fig. 7.3, after iteration 150 the Pareto optimal front of the elite set is smooth, but uneven, and there is fault zone in the Pareto optimal front. According to the computer system of crowding distance, in that region there are no feasible Pareto optimal solutions or the Pareto optimal solutions are too stacked.

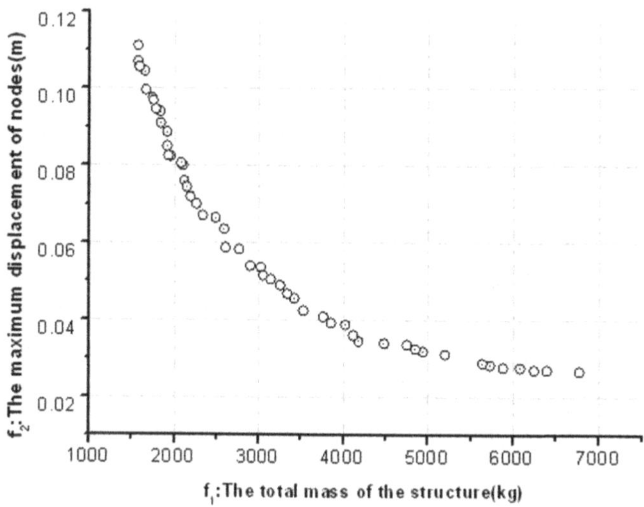


**Fig. 7.2** The Pareto optimal front of the elite set after 50 iterations

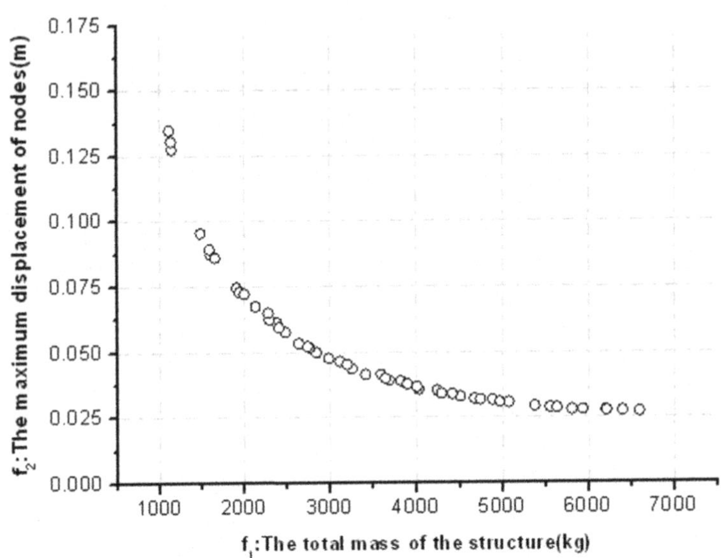


**Fig. 7.3** The Pareto optimal front of the elite set after 150 iterations

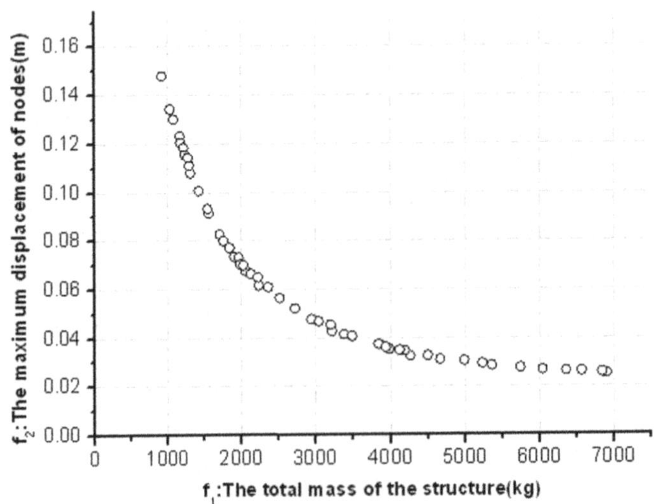


**Fig. 7.4** The Pareto optimal front of the elite set after 250 iterations

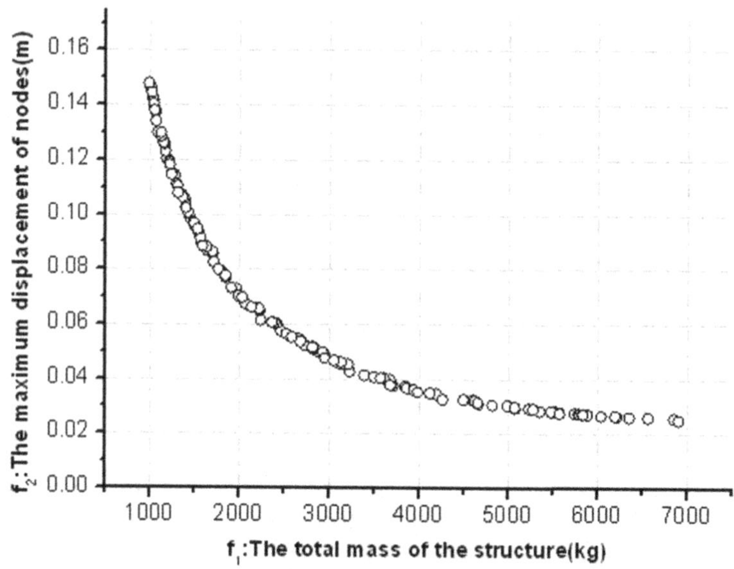


**Fig. 7.5** The Pareto optimal front after 250 iterations

The Pareto optimal front of the elite set after 250 iterations is shown in Fig. 7.4. The Pareto front boundary values are as follows: the (max $W$, min $\delta$) is (6906.9 kg, 0.0251 m), and the(min $W$, max$\delta$) is (920.2167 kg, 0.1479 m). Therefore, after 250 iterations the ideal solution (920.2167 kg, 0.0251 m) has significantly improved, compared with the ideal solution (1050.3 kg, 0.0263 m) after 150 iterations. From Fig. 7.4, the Pareto optimal front after 250 iterations is of a wider distribution than 150 iterations, and the distribution does not appear fault. Thus, the Pareto optimal front obtained by MGSO algorithm distributes widely, evenly and smoothly.

The Pareto optimal front corresponding to all the Pareto optimal solutions after 250 iterations is shown in Fig. 7.5. The results of this paper are basically the same with the SM-MOPSO's results in the reference [18], and MGSO can get more uniform and wider Pareto optimal front than MOPSO reported in reference [19], which proves MGSO algorithm is feasibility and superiority. Table 7.1 lists the comparison of three algorithms SM-MOPSO, MOPSO and MGSO after 250 iterations for computing 5 times randomly. It can be seen form Table 7.1, the MGSO is better than MOPSO, and the optimal result of SM-MOPSO is a little bit better than MGSO, but the calculation of MGSO is much simpler and it has much higher efficiency than SM-MOPSO. At the same time, it is known that the computing performance of MGSO is every stable.

**Table 7.1** Comparison of three algorithms' results

| Algorithm | The weight target's minimum /kg | | The displacement target's minimum /m | |
|---|---|---|---|---|
| | The best value | The worst value | The best value | The worst value |
| Chuang [19] MOPSO | 883 | 1346.5 | 0.025 | 0.025 |
| An [18] SM-MOPSO | 720.34 | 743.32 | 0.025 | 0.025 |
| MGSO | 920.22 | 940.45 | 0.0251 | 0.0257 |

### *7.5.2 The Shape Optimization of Truss*

After years of exploration and research, the optimization of only considering the truss section has matured. The shape optimization of truss refers to optimize the location of space truss's nodes. In the design of truss shape optimization, discrete variables, such as the section size of bars, and continuous variable, such as the nodes' coordinates, are mixed. It not only increases the difficulty of solving the problem, but also it is often difficult to process optimization convergence. Currently, traditional methods usually use hierarchical optimization method to deal with such issues. However, most of hierarchical optimization methods usually find the optimal solution in the conventional mathematical programming, which is easy to fall into local optimal solution.

In this chapter, the total weight of the truss and the special node's displacement are the objective functions. The cross-section dimensions and coordinate variables are changed to a unified design variables dealing with the difficulties of different types design variables.

The shape optimization of truss is, in the case of a given type of structure, materials, layout and geometry, optimizing the size of the component sections, making the structure of the smallest quality or the most economic. In this chapter, the aim is to design the optimal structure that satisfies all constraints in the premise of the lightest weight of the structure, while the maximum displacement of the node is minimum. For a $n$-bars truss structure system, the basic parameters, including the elastic modulus, material density, the maximum allowable stress, etc., are known, and the optimization target is to identify the optimal cross-sectional area of bars and the optimal nodes' location so that the structure is of the smallest total weight, the smallest maximum nodes' displacement, in a given load conditions. Therefore, mathematical models can be expressed as:

$$
\begin{aligned}
& f_1 = \min W = \sum_{i=1}^{M} \rho_i A_i l_i (X) \\
& f_2 = \min U = \min\left\{\max\left\{u_x, u_y, u_z\right\}\right\} \\
& s.t. \quad g_n(A, X) \le 0 \quad n = 1, 2, \ldots, n_G
\end{aligned}
$$

$$A_i \in S = \{S_1, S_2, ..., S_{n_s}\} \quad i = 1, 2, ..., M; \quad k = 1, 2, ..., n_c$$
$$X_k^{\min} \le X_k \le X_k^{\max}$$

$$A = [A_1, A_2, ..., A_M]^T \quad X = [X_1, X_2, ..., X_{nc}]^T \tag{7.21}$$

Where $A = [A_1, A_2, ..., A_M]^T$ are design variables for the cross section and they are discrete variables. $f_1$ and $f_2$ are two objective functions of multi-objective optimization, $M$ is the total number of bars, and $n_c$ is the number of nodes' coordinates. $g_n\ (n = 1, 2, ..., n_G)$ are the constraints of stress and displacement, $n_G$ is the total number of constraints, $S$ is discrete set of bars' cross-section, and $n_s$ is the number of discrete variables in $S$. $L_i$、 $A_i$ and $\rho_i$ are the bar length, the section area, and the density of the $i^{th}$ group bars, respectively, and $X_k^{\max}$ and $X_k^{\min}$ are the maximum and minimum nodes' coordinate.

## 25-Bar Truss

In this section, the performance of the MOGSO is studied on a 25-bar truss shown in Fig. 7.6. The materials are aluminum alloy, density is 2.768×103 kg/m$^3$ , and elastic modulus is 6.888×1010 N/m$^2$. $f_1$and $f_2$ are design objectives, $f_1$ is the total mass of the structure $W$, and $f_2$ is the maximum displacements $\delta$ which is the maximum displacement of all nodes .Rod stress limit is 172.25 MPa in both tension and compression for all members. Discrete values considered for this example are taken from the set S={0.072, 0.091, 0.112, 0.142, 0.174, 0.185, 0.224, 0.284, 0.348, 0.615, 0.697, 0.757, 0.860, 0.960, 1.138, 1.382, 1.740, 1.806, 2.020, 2.300, 2.460, 3.100, 3.840, 4.240, 4.640, 5.500, 6.000,

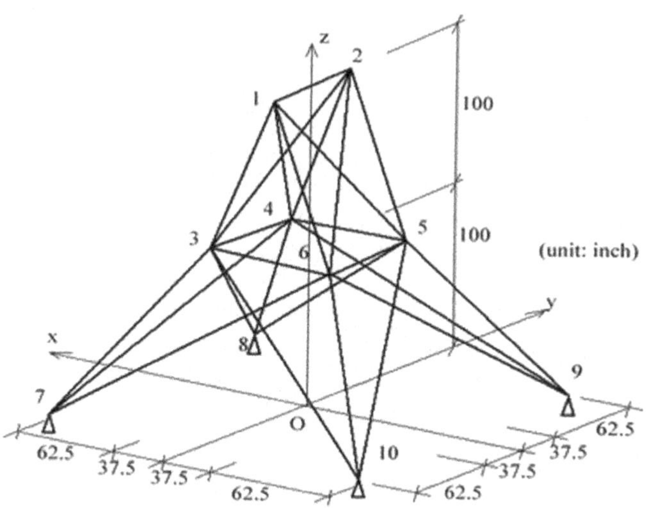


**Fig. 7.6** The 25-bar spatial truss structure

7.000, 8.600, 9.219, 11.077, 12.374} ($\times 10^{-3}$ m$^2$). The coordinates of node 3, 4, 5, and 6 are spatial variables, and they are $x_1$, $x_2$ and $x_3$; the coordinates of node 7, 8, 9, and10 are planar variables, and they are $x_4$ and $x_5$; the coordinates of node 1 and 2 are unchanged. Boundaries of sizing variables are: $0.508 \le x_1 \le 1.524$, $1.016 \le x_2 \le 2.032$, $2.286 \le x_3 \le 3.302$, $1.016 \le x_4 \le 2.032$, $2.540 \le x_5 \le 3.356$ (m). Structural loading conditions are in Table 7.2.

**Table 7.2** The load cases of a 25-bar spatial truss

| Node | $F_x$ (*N*) | $F_y$ (*N*) | $F_z$ (*N*) |
|---|---|---|---|
| 1 | 4448 | -44482 | -44482 |
| 2 | 0.0 | -44482 | -44482 |
| 3 | 2224 | 0.0 | 0.0 |
| 6 | 2669 | 0.0 | 0.0 |

This example belongs to mixed variables optimal problem of space truss structure, and the variables are the cross-sectional areas of the bars and the coordinates of the specified nodes. The cross-sectional areas of the bars are divided into 8 groups, so the cross-sectional areas' variables are 8-dimensional variable, and the coordinates' variables are 5-dimensional. In this case, variables dimension are 13, including eight discrete variables and five continuous variables. The number of population is 300, and the capacity of elite set *M* is set to 50. After 100, 200 and 500 iterations, the results are shown in Fig. 7.7, Fig. 7.8 and Fig. 7.9 respectively. Fig. 7.10 is the Pareto optimal front results.

Fig. 7.7 indicates the Pareto optimal front of the elite set after 100 iterations. The boundary values are as follows: the (*max W, min* $\delta$) is (969.47086 lb, 0.06177 in.), and the (*min W, max* $\delta$) is (89.2046 lb, 1.05158 in.*)*. After 100 iterations the ideal solution (min *W, min* $\delta$) is (89.2046 lb, 0.06177 in.). From the Fig. 7.7, it is known that the Pareto optimal front of the elite set after 100 iterations is widely distributed, but the curve is not smooth, and the result has not converged.

The Pareto optimal front of the elite set after 200 iterations is shown in Fig. 7.8. The boundary values are as follows: the (*max W, min* $\delta$) is (1032.5620 lb, 0.05637 in.), and the (*min W, max* $\delta$) is (83.53463 lb, 1.5616 in.). Compared Fig. 7.7 with Fig. 7.8, after 100 iterations the min *W* is 89.2046 lb, and after 200 iterations the min *W* is 83.53463 lb. For 100 iterations and 200 iterations the total mass *W* is not a big change, but after 200 iterations the *max* $\delta$ is 1.5616 in., much greater than the *max* $\delta$ after 100 iterations which is 1.05158 in.. In the multi-objective optimization design, the region is called as the sensitive areas where an objective function along with other subtle changes has dramatic changes. The solutions should avoid being selected from the sensitive areas, and generally the final optimal solution is selected in the gentle curve segment. After 200 iterations the min δ is 0.05637 in., far less than the min δ after 100 iterations which is 0.06177 in.. Although the Pareto optimal

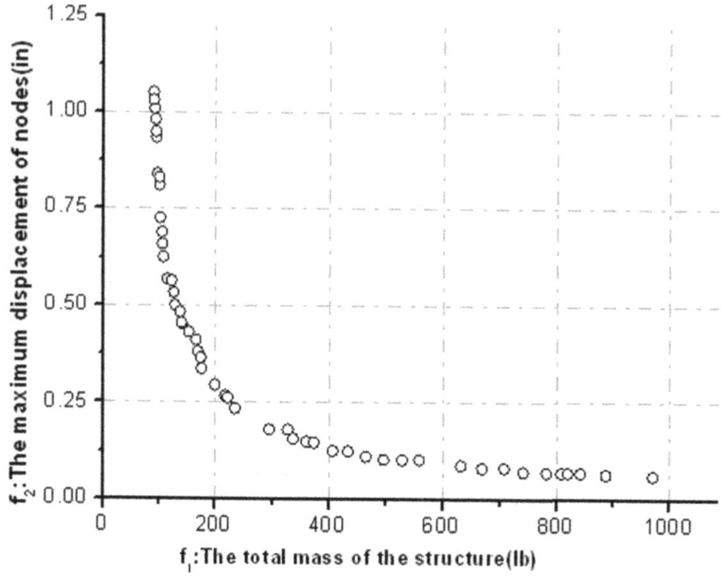


**Fig. 7.7** The Pareto optimal front of the elite set after 100 iterations

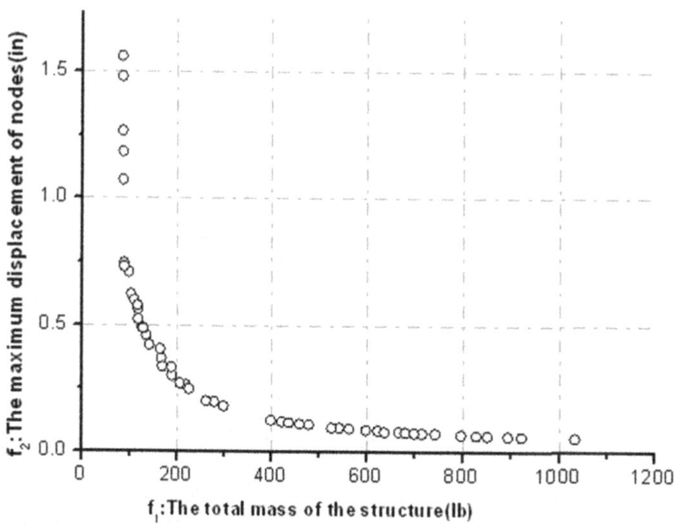


**Fig. 7.8** The Pareto optimal front of the elite set after 200 iterations

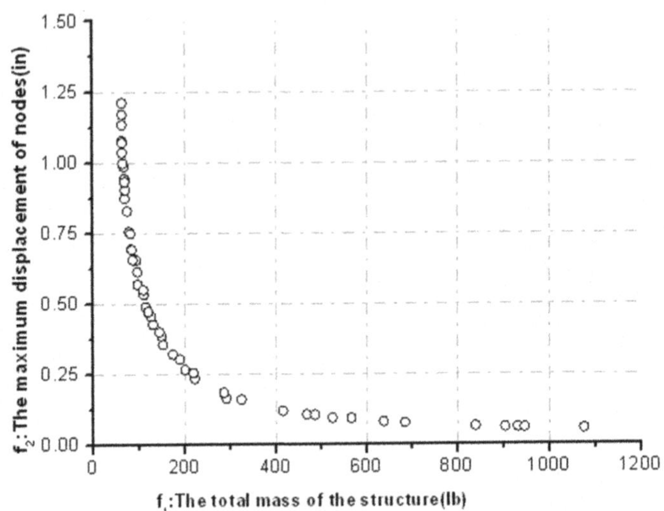


**Fig. 7. 9** The Pareto optimal front of the elite set after 500 iterations

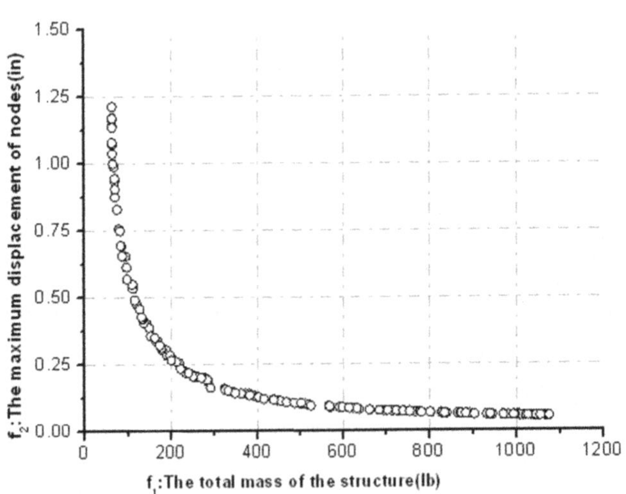


**Fig. 7.10** The Pareto optimal front after 500 iterations

front appears sensitive areas, the distribution of Pareto optimal front is much wider, indicating the good local search capability of MGSO.

Fig. 7.9 is the Pareto optimal front of the elite set after 500 iterations, and the boundary values are as follows: the (*max W, min* $\delta$) is (1074.93181 lb, 0.0549 in.), and the (*min W, max* $\delta$) is (63.6239 lb, 1.2099 in.) . Compared Fig. 7.8 with Fig. 7.9, the *min W* and the *min* $\delta$ are reduced significantly, and the distribution's uniformity of the Pareto optimal front is of a marked increase, indicating that MGSO is an excellent multi-objective algorithm.

Fig. 7.10 indicates the Pareto optimal front corresponding to all Pareto optimal solutions after 500 iterations. From Fig. 7.10, it is known that MGSO can get a wide and uniform Pareto optimal front, and for structural shape optimization MGSO is a good algorithm. Following the comparison of the results of SM-MOPSO and MOPSO cited in the literature [18, 19], it is obtained that MGSO has apparent correctness and superiority.

### 40-Bar Truss

Taking the planar 40 truss bridge for an example, the structure is shown in Fig. 7.11. All bars are the same materials, density is 7800 kg/m$^3$, and elastic modulus is $1.9613\times10^{11}$ N/m$^2$. The allowable stress is $\pm1.5691\times10^8$ N/m$^2$ for all members. At node 2, node 3, node 4, node 5, node 6 and node 7, the load P = $9.80665\times10^4$ N. This example is designed for multi-objective optimization, and the objective functions are the total weight of the structure $W$ and the maximum vertical displacement δ of node 9, node 10, node 11, node 12, node 13, node 14, node 15 and node 16. In other words, $\delta = max\ (D_9, D_{10}, D_{11}, D_{12}, D_{13}, D_{14}, D_{15}, D_{16})$, and $D_9$, $D_{10}$, $D_{11}$, $D_{12}$, $D_{13}$, $D_{14}$, $D_{15}$ and $D_{16}$ are the vertical displacements of node 9, node 10, node 11, node 12, node 13 , node 14, node 15 and node 16. Bars' stress is less than allowable stress, and the bar cross-sectional area is between the upper and lower limits. The minimum area is 0.001 m$^2$, and the maximum area is 0.05 m$^2$. At the same time the cross-sectional areas of all bars must be multiple of 0.0005 m$^2$. Node 9, node 10, node 11, node 12, node 13, node 14, node 15 and node 16 can be moved along the vertical direction and the boundary of coordinates' variables is $1\ m \le x_i \le 5\ m$.

This example is of continuous variables and discrete variables optimization problems. According to the symmetry of the structure, the variables of cross-sectional areas are 19-dimensional which are discrete variables, and the variables of nodes' coordinates are four-dimensional position which are continuous variables. Therefore, in this case the variables are 23-dimensional, including 19-dimensional discrete variables and four-dimensional continuous variables. The number of population is 300, and the capacity of the elite set $M$ is 50. The results are shown in Fig. 7.12, Fig. 7.13 and Fig. 7.14 after 100, 200 and 500 iterations respectively.

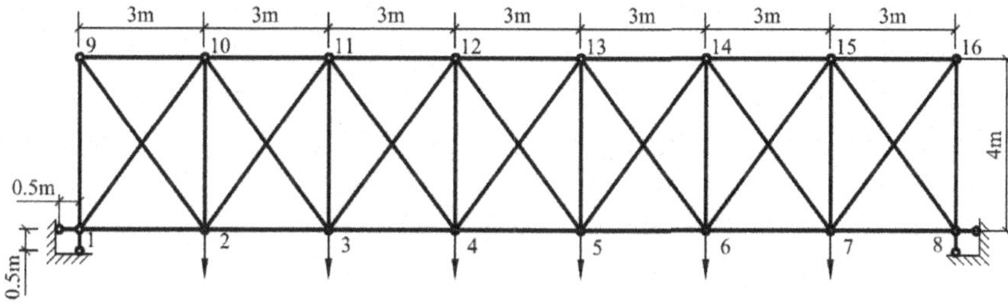


**Fig. 7.11** The 40-bar planar truss structure

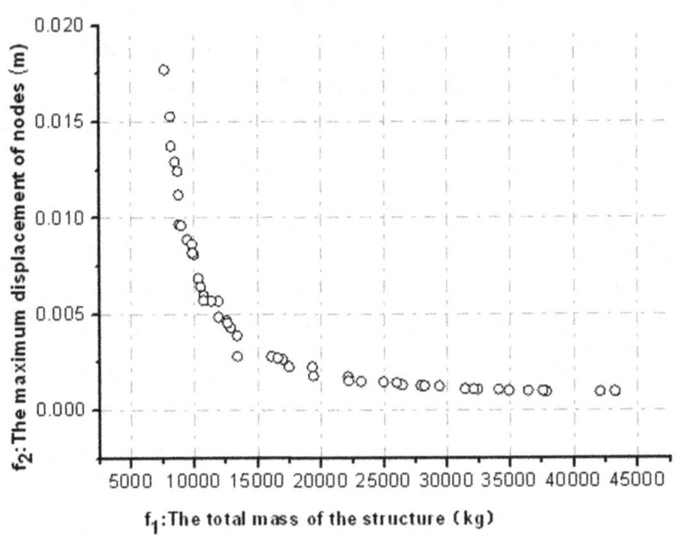


**Fig. 7.12** The Pareto optimal front of the elite set after 100 iterations

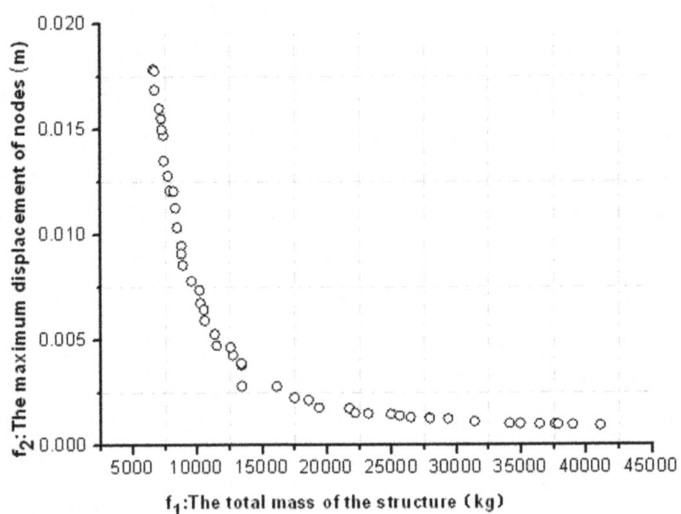


**Fig. 7.13** The Pareto optimal front of the elite set after 200 iterations

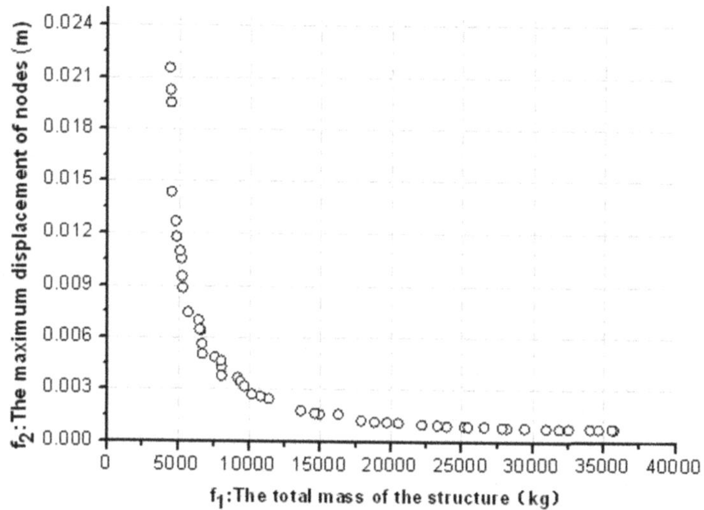


**Fig. 7.14** The Pareto optimal front of the elite set after 500 iterations

The Pareto front of the elite set after 100 iterations is shown in Fig. 7.12. The boundary values are as follows: the (*max W*, *min* $\delta$) is (43256.26285 kg, $9.62636\times10^{-4}$ m), and the (*min W*, *max* $\delta$) is ( 7677.27588 kg, 0.01771 m). It is known that the Pareto front after 100 iterations has large width but it is unevenly distributed.

Fig. 7.13 shows the Pareto front of the elite set after 200 iterations. The boundary values are as follows: the (*max W*, *min* $\delta$) is (41071.45054 kg, $9.37465\times 10^{-4}$ m), and the (*min W*, *max* $\delta$) is (6609.92518 kg, 0.0178 m). After 200 iterations the *min W* is 6609.92518 kg, and the *min W* is 7677.27588 kg for 100 iterations. So, it is known that the Pareto optimal front after 200 iterations is wider and more uniform than the one after 100 iterations.

The Pareto front of the elite set after 500 iterations is shown in Fig. 7.14. The boundary values are as follows: the (*max W*, *min* $\delta$) is (35704.71038 kg, $7.68596\times10^{-4}$ m), and the (*min W*, *max* $\delta$) is (4326.07903 kg, 0.0215 m). Compared Fig. 7.14 with Fig. 7.13, the Pareto optimal front after 500 iterations has wider distribution, but the improvement of uniformity is not obvious.

The Pareto optimal front of all the Pareto optimal solutions after 500 iterations is displayed in Fig. 7.15. The width and the uniformity of the Pareto optimal front show the feasibility of MGSO for shape optimization.

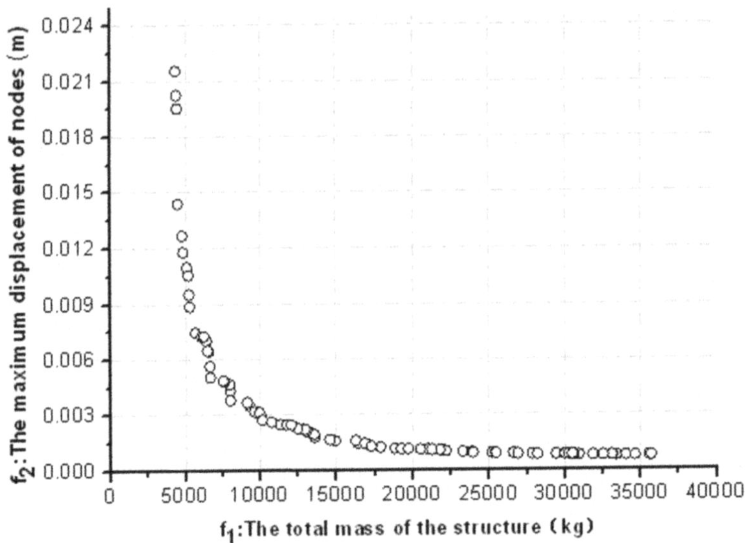


**Fig. 7.15** The Pareto optimal front after 500 iterations

### *7.5.3 The Dynamic Optimization of Truss*

In this item, the dynamic optimization of truss will focus on designing the optimal structure that satisfies all constraints in the premise of the lightest weight of the structure, while fundamental frequency of the structure is maximum. For a $n$-bar truss structure system, the basic parameters, including the elastic modulus, material density, the maximum allowable stress, etc., are known, and the optimization's target is to identify the optimal cross-sectional area of bars and the optimal nodes' location so that the structure is of the smallest total weight and the maximum fundamental frequency in a given load conditions. The mathematical models can be expressed as:

$$\text{min. } f_1 = W = \sum_{i=1}^{M} \rho_i A_i l_i \tag{7.22}$$

$$\begin{aligned} &\text{max. } f_2 = \omega \\ &s.t. \quad g_n(A, X) \le 0 \quad n = 1, 2, \ldots, n_G \\ &A_i \in S = \{S_1, S_2, \ldots, S_{n_S}\} \quad i = 1, 2, \ldots, M \\ &X_k^{\min} \le X_k \le X_k^{\max} \qquad k = 1, 2, \ldots, n_c \end{aligned}$$

Where $A = [A_1, A_2, \ldots, A_M]^T$ are design variables for the cross section and they are discrete variables. $f_1$ and $f_2$ are two objective functions of multi-objective optimization, $M$ is the total number of bars, and $n_c$ is the number of nodes' coordinates variables. $W$ is the total weight of the structure, and $\omega$ is the fundamental frequency of the structure. $g_n(A,X)$ are the constraints of stress and displacement, $n_G$ is the total number of constraints, $S$ is discrete set of bars' cross-section, and $n_s$ is the number of discrete variables in $S$. $L_i$, $A_i$ and $\rho_i$ are the bar length, section area, and density of the $i^{th}$ group bars, respectively, and $X_k^{\max}$ and $X_k^{\min}$ are the maximum and minimum nodes' coordinate.

As the computational complexity of the frequency, the dynamic optimization has much double counting, and it is more time-consuming than the static optimization. To save computing time, in this article the dynamic optimization takes the less number of iterations, setups the less capacity of elite set $M$, compared to the static optimization.

### 10-Bar Truss

Taking the planar 10-bar truss structure for an example, the structure is shown in Fig. 7.16. All bars are the same materials, density is $7.68\times10^3$ kg/m$^3$, and elastic modulus is $2.1\times10^{11}$ N/m$^2$. The allowable stress is $\pm1\times10^8$ N/m$^2$ for all members. The position of node 1 can be moved along the vertical direction, the locations of node 2 and node 3 can be moved along the horizontal and vertical directions. The vertical coordinates variables of node 1 is $x_1$; the horizontal and vertical coordinates variables of node 2 are $x_2$ and $x_3$; the horizontal and vertical coordinates variables of node 3 are $x_4$ and $x_5$, where -2.5 m $\leq x_1 \leq$ 2.4 m, 0.1 m $\leq x_2 \leq$ 2.5 m, -2.5 m $\leq x_3 \leq$ 2.4 m, 2.5 m $\leq x_4 \leq$ 5 m, $0 \leq x_5 \leq$ 2.4 m. There is vertical load 100 kN, at node 5 and node 6 respectively. Bars' stress is less than allowable stress, and the bar cross-sectional area is between the upper and lower limits. The minimum area is 0.001 m$^2$, and the maximum area is 0.01 m$^2$. At the same time the cross-sectional areas of all bars must be multiple of 0.0005 m$^2$.

This example is a continuous and discrete variable mixed optimization problems of the plane truss structure. The bars' cross-sectional areas are discrete variables, whose number are 10; the nodes' coordinate variables are continuous variables, whose number are 5. The population of the GSO is 300, and the capacity of the external archive is 30. The multi-objective optimization results are listed in Fig. 7.17, Fig. 7.18 and Fig. 7.19, after 50 and 100 iterations respectively.

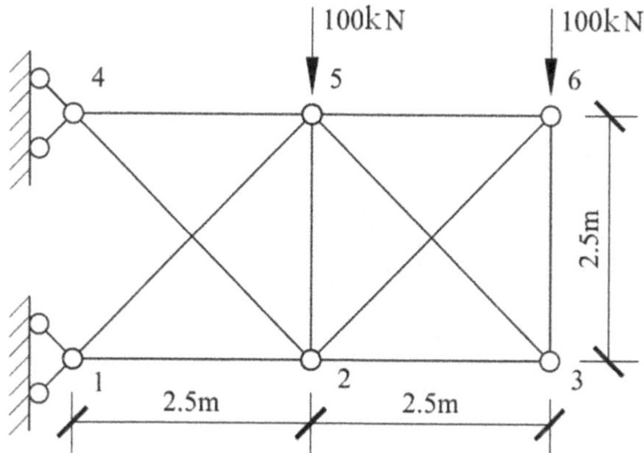


**Fig. 7.16** The 10-bar planar truss structure

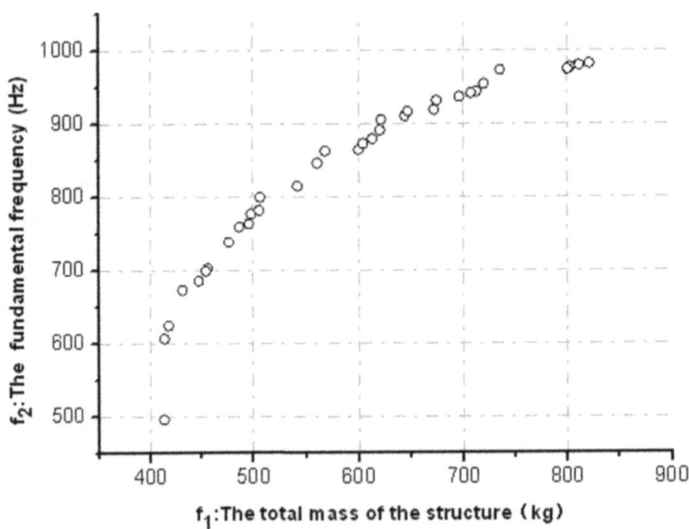


**Fig. 7.17** The Pareto optimal front of the elite set after 50 iterations

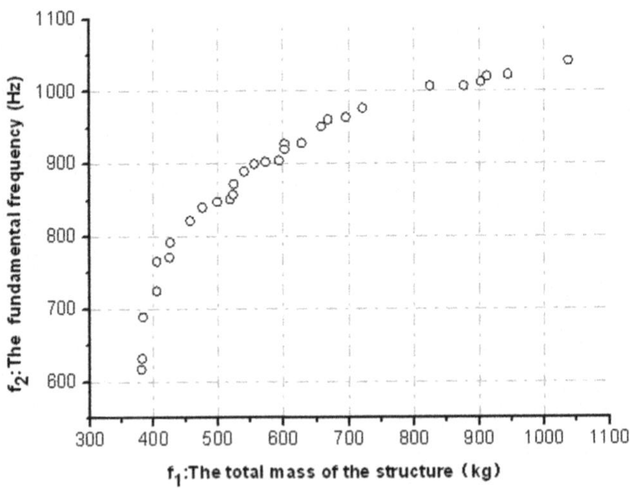


**Fig. 7.18** The Pareto optimal front of the elite set after 100 iterations

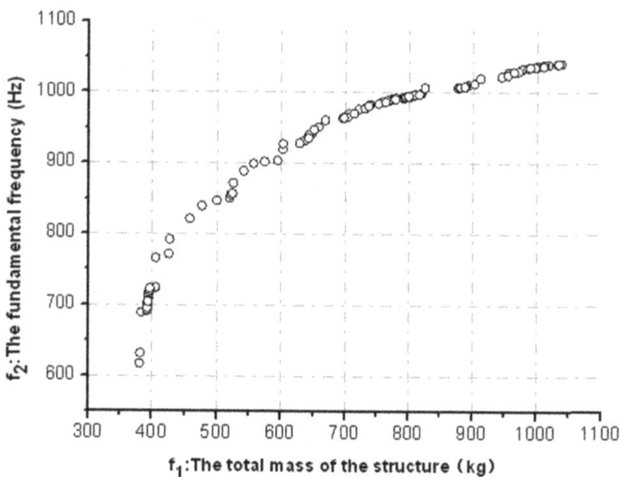


**Fig. 7.19** The Pareto optimal front after 100 iterations

After 50 iterations, the Pareto optimal front of the elite set for the maximum fundamental frequency and the lightest weight of the 10-bar truss is shown in Fig. 7.17. The boundary values are as follows: the (*max W, maxω*) is (821.1588 kg, 980.8905 Hz), and the (*min W, minω*) is (412.9979 kg, 496.1717 Hz). It is known from Fig. 7.17 that the Pareto optimal solutions are widely distributed after 50 iterations, and the ideal optimal result (*min W, maxω*) is (412.9979 kg, 980.8905 Hz). Although only after 50 iterations, the Pareto optimal solutions distributes widely and quite uniformly. It is said that MGSO is a good algorithm for searching. By comparison with single-objective optimization results, the result of dynamic optimization of MGSO is verified and is correct.

The Pareto optimal front of the elite set after 100 iterations is shown in Fig. 7.18. The boundary values are as follows: the (*max W, maxω*) is (1037.345 kg, 1041.1 Hz), and the (*min W, minω*) is (380.5196 kg, 616.7485 Hz). After 100 iterations the ideal solution (*min W, maxω*) is (380.5196 kg, 1041.1 Hz). Compared with the ideal solution (412.9979 kg, 980.8905 Hz) after 50 iterations, both optimization goals *min W* and *maxω* have been significantly improved. It indicates that the MGSO algorithm is still a good performance for the complex dynamic optimization problems.

The Pareto optimal front of the all Pareto optimal solutions after 100 iterations is listed in Fig. 7.19. It is known that MGSO algorithm can get uniform Pareto optimal front, converge speedily, and be an ideal multi-objective optimization algorithm for highly nonlinear dynamic optimization problem.

### 40-Bar Truss

Taking the planar 40 truss bridge for an example, the structure is shown in Fig. 7.20. All bars are the same materials, density is 7800 kg/m$^3$, and elastic modulus is $1.9613 \times 10^{11}$ N/m$^2$. The maximum allowable displacement of all nodes [δ] is

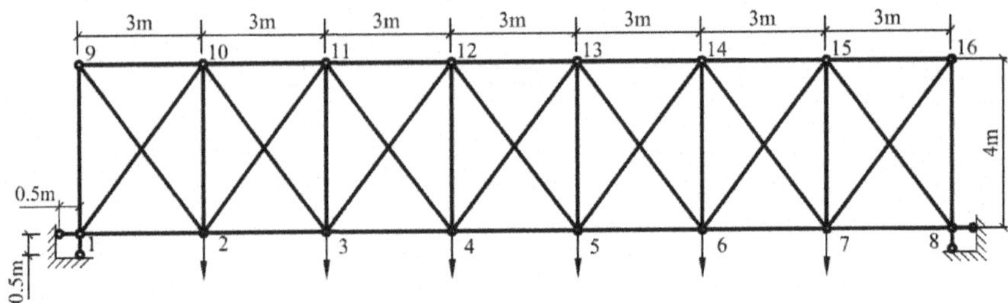


**Fig. 7.20** The 40-bar planar truss structure

0.035m. The maximum allowable stress is $1.5691\times10^8$ N/m$^2$ for all members. At node 2, node 3, node 4, node 5, node 6 and node 7, the load P is $9.80665 \times 10^4$ N. This example is designed for multi-objective optimization, and the objective functions are the total weight of the structure $W$ and the fundamental frequency of the structure $\omega$. The ideal structure is that the total mass is minimum and the fundamental frequency is maximum. Bars' stress is less than allowable stress, and the bar cross-sectional area is between the upper and lower limits. The minimum area is 0.001 m$^2$, and the maximum area is 0.05 m$^2$. At the same time the cross-sectional areas of all bars must be multiple of 0.0005 m$^2$. Node 9, node 10, node 11, node 12, node 13, node 14, node 15 and node 16 can be moved along the vertical direction and the boundary of coordinates' variables is $1\ \text{m} \le x_i \le 5\text{m}$.

This example is also a continuous variables and discrete variables mixed optimization problems. According to the symmetry of the structure, the variables of cross-sectional areas are 19-dimensional which are discrete variables, and the variables of nodes' coordinate positions are 4-dimensional which are continuous variables. Therefore, in this case the variables are 23-dimensional, including 19-dimensional discrete variables and 4-dimensional continuous variables. The number of GSO population is 300, and the capacity of the elite set $M$ is 30. After 50 iterations, the multi-objective optimization results are listed in Fig. 7.21.

The Pareto optimal front after 50 iterations is shown in Fig. 7.21 The boundary values are as follows: the ($max\ W$, $max\ \omega$) is (10255.1303 kg, 197.3186 Hz), and the ($min\ W$, $min\ \omega$) is (5501.8 kg, 137.5998 Hz). The span of structural quality is from 5501.8 kg to 10255.1303 kg, and it shows that the Pareto front after 50 iterations is wide spread. MGSO is a feasible algorithm for dynamic multi-objective optimization algorithm. The example indicates that the dynamic optimization is much more complex than static optimization, its computing time is growing exponentially, and the procedure is much more difficult than static optimization. For the practical significance of dynamic optimization, but obviously, it is very worthy of study.

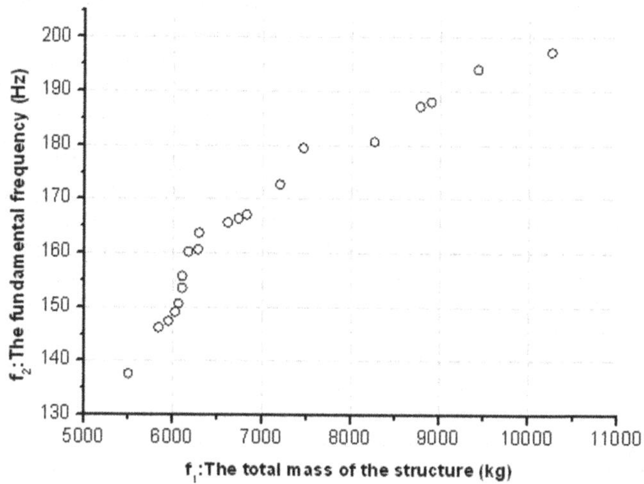


**Fig. 7.21** The Pareto optimal front after 100 iterations

## 7.6 The Optimization of Frame

In addition to all the advantages of a traditional steel structures, modern light steel structures have more prominent features, which mainly include: light envelope, constant load, high-strength material, new structural system, standardized design, factory production, easy lifting by small machines, labor intensity light, fast construction speed, low construction cost and so on. Common steel structures used in modern engineering structures were steel truss structures, as well as frame structures.

In the field of structural optimization, genetic algorithms and particle swarm optimization algorithm, etc., are also widely used, including the weight of truss structure optimization, shape optimization and topology optimization.

For solving highly nonlinear problems, such algorithms demonstrated strong vitality and applicability than the traditional full stress criteria and mathematical programming method, and they were more suitable for complex structures. Research in this item mainly focused on the frame structures. Many research progresses have been achieved such as frame section optimization with genetic algorithm and the section optimal design of frame based on semi-rigid assumptions and seismic by Saka and Kameshki, etc.. Camp optimized the cross-section of framework with ant colony algorithm, etc.

Unit as rigid frame structure, there are two mathematical models for the optimization. The first mathematical model is as follows:

$$\text{min. } f_1 = W = \sum_{i=1}^{M} \rho_i A_i l_i$$

$$\text{min. } f_2 = E = \sqrt{\sum_{j=1}^{m} \frac{1}{2} \left(\sum_{k=1}^{s} P_{kj} \delta_{kj}\right)^2}$$

$$\text{s.t. } g_i^{\sigma} = [\sigma_i] - \sigma_i \geq 0, \quad (i = 1, 2, \cdots n)$$
$$A_i \in D \qquad (i = 1, 2, \cdots, n)$$
$$D = [A_1, A_2, \cdots, A_n]^T \tag{7.23}$$

Where $A_i$ are design variables for the cross sections, $f_1$ and $f_2$ are two objective functions of multi-objective optimization. $g_i^{\sigma}$ are the stress constraints. $n$ is the number of design variables for the cross section bar, $W$ is the total weight of the structure, $E$ is the total dynamic strain energy, and $L_i$, $A_i$ and $\rho_i$ are the bar length, section area, and density of the $i^{th}$ group bars, respectively. Section $D$ is the discrete set.

The second mathematical model is as follows:

$$f_1 = \min W = \sum_{i=1}^{n} \rho_i A_i L_i$$
$$f_2 = \min\{\max U\}$$
$$st. \quad g_i^{\sigma} = [\sigma_i] - \sigma_i \geq 0 \quad (i = 1, 2, ..., n)$$
$$A_i \in D(i = 1, 2, ..., n) \tag{7.24}$$
$$D = [A_1, A_2, ..., A_n]^T$$

Where $A_i$ are design variables for the cross section, $f_1$ and $f_2$ are two objective functions of multi-objective optimization. $g_i^{\sigma}$ are the stress constraints. $n$ is the number of design variables for the cross section bar, $W$ is the total weight of the structure, and $L_i$, $A_i$ and $\rho_i$ are the bar length, section area, and density of the $i^{th}$ group bars, respectively. $U$ is the displacement value of given node $j$ on the direction under the condition to allow the most negative displacement, and Section $D$ is the discrete set.

Considering the stress constraints, the actual stress of the components of the framework is mainly on account of the role of bending and axial force, neglecting shear effects, so the maximum stress by the combination of bending and axial force, combined stress is expressed as follows:

$$\sigma_k = \left|\frac{N_k}{A_k}\right| + \left|\frac{(M_k)_x}{(W_k)_x}\right| + \left|\frac{(M_k)_y}{(W_k)_y}\right| \tag{7.25}$$

Where $N_k$ is the axial force of the component, $A_k$ is cross-sectional area of the component, $(M_k)_x$ and $(M_k)_y$ are the bending moment of strong axis and weak axis around component section, $(W_k)_x$ and $(W_k)_y$ are flexural modulus corresponds to the strong axis and weak axis.

### 7.6.1 The One Bay Eight Storey Planar Frame

The geometry of structure is shown in Fig. 7.22. The elastic modulus $E$ is $2.9 \times 10^4$ ksi, and the density $\rho$ is 0.283 lb/in$^3$. In this example the bars have not the allowable

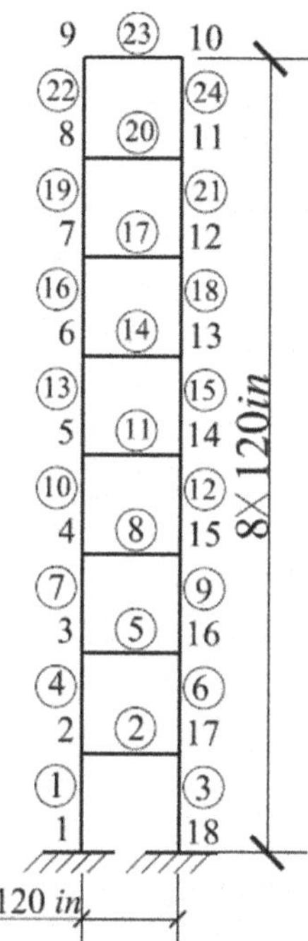


**Fig. 7.22** One bay eight storey planar frame

stress limit, and only the node 9 and the node 10 have the maximum displacement limit on the $x$ direction, which is $[\delta] = \pm 2$ in.. The cross-section variables are grouped into: $A_1$(1, 3, 4, 6), $A_2$(7, 9, 10, 12), $A_3$(13, 15, 16, 18), $A_4$(19, 21, 22, 24), $A_5$(2, 5), $A_6$(8, 11), $A_7$(14, 17), $A_8$(20, 23). The selection of discrete variables use the AISC specification of the U.S. (AISC 2001), reported in reference [19]. The model is only one load case: there is a vertical force -100 kips at node 2 to 17, and there are horizontal forces 0.272 kips, 0.544 kips, 0.816 kips, 1.088 kips, 1.361 kips, 1.633 kips, 1.905 kips and 2.831 kips at node 2 to 9, respectively. The objective functions are the total weight $W$ of the structure and the total dynamic strain energy $E$ of the structure.

This example is a discrete variable optimization problem of the plane frame structure. The variables are the cross-sectional areas of the frame' bars, and the variables' dimension is 8. The objective functions are the total weight $W$ and the total dynamic strain energy $E$, and the ideal situation of the optimized structure is the lightest total weight *min W* and the minimum total dynamic strain energy *min E*. The number of GSO population is 300, and the capacity of the elite set $M$ is set to 50. The results are shown in Fig. 7.23, Fig. 7.24, Fig. 7.25 and Fig. 7.26, after 100, 200 and 500 iterations respectively.

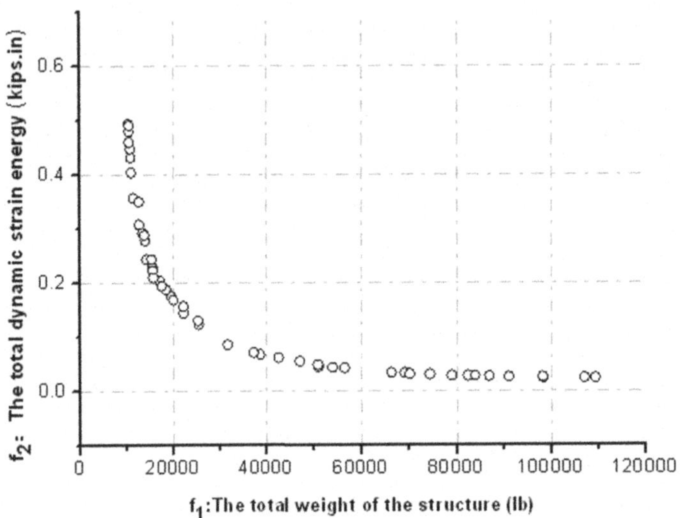


**Fig. 7.23** The Pareto optimal front of the elite set after 100 iterations

After 100 iterations the Pareto optimal front of the elite set is placed in Fig. 7.23 , and the boundary values are as follows: the (*max W*, *min E*) is (109351.2 lb, 0.0235 kips. in), and the (*min W*, *max E*) is (10327.9152 lb, 0.4925 kips. in).

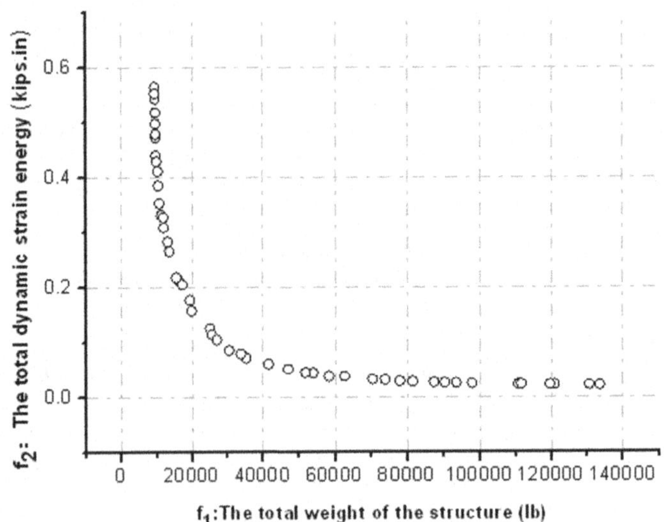


**Fig. 7.24** The Pareto optimal front of the elite set after 200 iterations

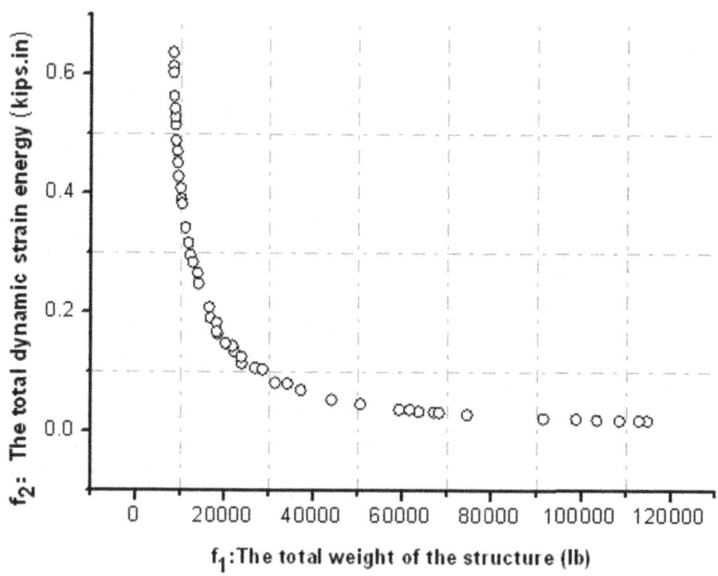


**Fig. 7.25** The Pareto optimal front of the elite set after 500 iterations

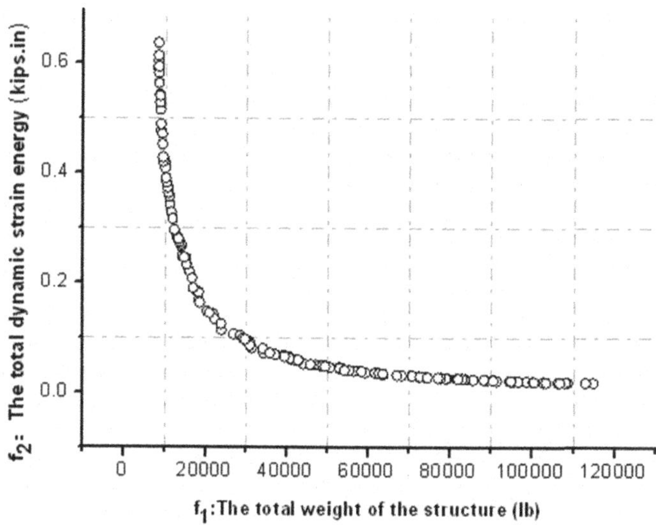


**Fig. 7.26** The Pareto optimal front after 500 iterations

The Pareto optimal front of the elite set after 200 iterations is displayed in Fig. 7.24. The boundary values are as follows: the (*max W*, *min E*) is (133517.136 lb, 0.0219 kips. in); the (*min W*, *max E*) is (9362.772 lb, 0.5633 kips. in). Compared Fig. 7.24 and Fig. 7.23, the *min W* and the *min E* after 200 iterations are much less, and the Pareto optimal front is much wider than the results after 100 iterations. It shows that MGSO has good global search capability, and not easily falls into local optimum.

The Pareto optimal front of the elite set after 500 iterations is listed in Fig. 7.25. The boundary values are as follows: the (*max W*, *min E*) is (114723.672 lb, 0.01829 kips. in); the (*min W*, *max E*) is (8204.0568 lb, 0.63551 kips. in). From Fig. 7.25, it is known that after 500 iterations the *min W* and the *min E* have significantly been improved, and the Pareto optimal front is more uniform. MGSO converges rapidly, not only has good global search capability, but also has good local search capability.

The Pareto front corresponding to all the Pareto optimal solution after 500 iterations is shown in Fig. 7.26. After 500 iterations the ideal optimal solution (*min W*, *min E*) is (8204.0568 lb, 0.01829 kips. in), and the *min W* is similar to the single-objective optimization result reported in the literature [19]. It illustrates that MGSO is feasible and correct.

### 7.6.2 The Two Bay Five Story Planar Frame

Fig. 7.27 is a two-bay 5-story plane frame, load cases is shown in Table 7.3. Elastic modulus $E=2.058\times10^{11}$ N/m$^2$ ,density $\rho=7.8\times10^4$ N/m$^3$, allowable stress $[\sigma]=\pm 1.666\times10^8$ N/m$^2$, the allowable displacement on x-direction of 1, 2, 3 node $[\delta]=\pm$ 4.58 cm. Variable groups: $A_1$(1, 2, 3), $A_2$(4, 5, 6), $A_3$(7, 8, 9), $A_4$(10, 11, 12), $A_5$(13, 14, 15), $A_6$(16), $A_7$(17), $A_8$(18), A9(19), A10(20), A11(21), A12(22), A13(23), A14(24), $A_{15}$(25). Cross sections are selected from Table 7.4.

This example is a discrete variable optimization problem of the plane frame structure. The variables are the cross-sectional areas of the frame' bars, and the variables' dimension is 15. The variables are grouped into: $A_1$(1, 2, 3), $A_2$(4, 5, 6 ), $A_3$(7, 8, 9), $A_4$(10, 11, 12), $A_5$ (13, 14, 15), $A_6$ (16), $A_7$ (17), $A_8$ (18), $A_9$ (19), $A_{10}$ (20), $A_{11}$ (21), $A_{12}$ (22), $A_{13}$ (23), $A_{14}$ (24), $A_{15}$ (25). The objective functions are the total weight $W$ and the maximum level drift $\delta$, and the ideal situation of the optimized structure is the lightest total weight *min W* and the minimum maximum level drift *min* $\delta$. The number of GSO population is 300, and the capacity of the elite set $M$ is set to 50. The results are shown in Fig. 7.28, Fig. 7.29, Fig. 7.30 and Fig. 7.31, after 100, 200 and 500 iterations respectively.

After 100 iterations the Pareto optimal front of the elite set is placed in Fig. 7.28 , and the boundary values are as follows: the (*max W*, *min* $\delta$) is (165158.8848 N, 1.00924 cm), and the (*min W*, *max* $\delta$ ) is (77943.9492 N, 4.51741 cm). From Fig. 7.28, it is known that MGSO can get a uniform and smooth Pareto optimal front after 100 iterations.

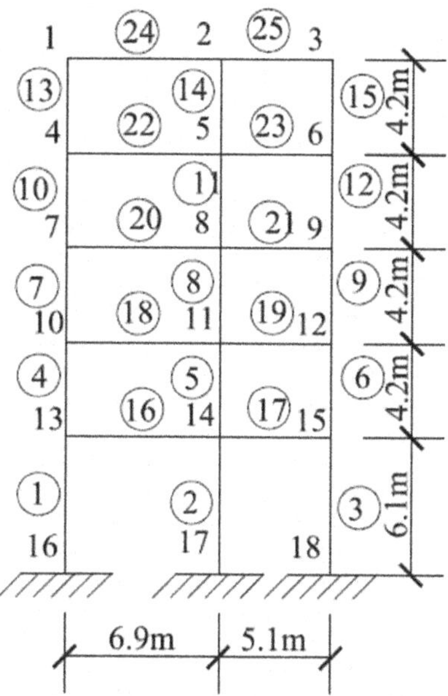


**Fig. 7.27** Two bay five story planar frame

**Table 7.3** Load case of two bay five storey planar frame

| Load combination | | | | |
|---|---|---|---|---|
| Load type | Add site | Load type | Load direction | Load size |
| Constant load G | Component 16-25 | Uniform load | -y | 11.76*kN*/*m* |
| | Node 1, 3 | Concentrated load | -y | 19.6*kN* |
| | Node 4, 6, 7, 8, 9, 12, 13, 15 | Concentrated load | -y | 40.2*kN* |
| Live load Q | Component 16-25 | Uniform load | -y | 10.78*kN*/*m* |
| Wind load W | Node 1 | Concentrated load | +x | 5.684*kN* |
| | Node 4 | Concentrated load | +x | 7.252*kN* |
| | Node 7 | Concentrated load | +x | 6.664*kN* |
| | Node 10 | Concentrated load | +x | 5.978*kN* |
| | Node 13 | Concentrated load | +x | 6.272*kN* |
| Load combination | | G+0.9(Q+W)<br>G+W<br>G+Q | | |

**Table 7.4** Cross-section group for two bay five storey frame

| Section number | Sectional area ($cm^2$) | Flexural modulus($cm^3$) | Sectional moment of inertia ($cm^4$) |
|---|---|---|---|
| 1 | 51.38 | 282.83 | 2545.50 |
| 2 | 57.66 | 356.08 | 3560.80 |
| 3 | 63.67 | 435.25 | 4787.70 |
| 4 | 69.81 | 537.46 | 6710.20 |
| 5 | 79.81 | 579.13 | 7239.10 |
| 6 | 80.04 | 678.13 | 9505.10 |
| 7 | 91.24 | 731.20 | 10236.80 |
| 8 | 97.00 | 938.83 | 15021.30 |
| 9 | 109.80 | 1007.10 | 16113.50 |
| 10 | 121.78 | 1319.35 | 23748.20 |
| 11 | 136.18 | 1405.75 | 25303.40 |
| 12 | 150.09 | 1757.77 | 35155.40 |
| 13 | 166.09 | 1864.44 | 37288.70 |
| 14 | 182.09 | 1971.10 | 39422.10 |

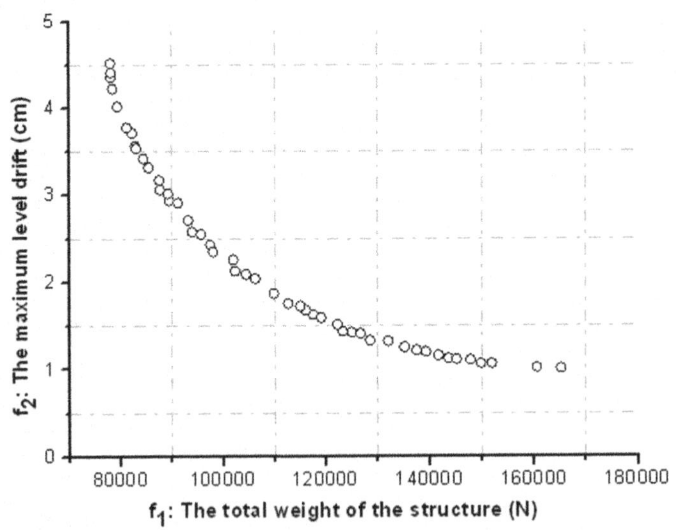


**Fig. 28** The Pareto optimal front of the elite set after 100 iterations

The Pareto optimal front of the elite set after 200 iterations is showed in Fig. 7.29. The boundary values are as follows: the (*max W*, *min* $\delta$) is (172434.249 N, 0.9660 cm), and the (*min W*, *max* $\delta$) is (71284.4964 N, 5.8630 cm).

Compared with Fig. 7.28, it is known from Fig. 7.29 that the Pareto optimal front after 200 iterations has wider distribution than the Pareto optimal front after 100 iterations. The *min W* after 200 iterations is 71284.4964 N, which is smaller compared with the *min W* =77943.9492 N of 100 iterations. Although the uniformity of the Pareto optimal front does not increase, the distributed span increases significantly. So MGSO is an effective multi-objective approach.

The Pareto optimal front of the elite set after 500 iterations is showed in Fig. 7.30. The boundary values are as follows: the (*max W*, *min δ*) is (174772.962 N, 0.9580 cm), and the (*min W*, *max δ*) is (68989.2606 N, 5.7551 cm).

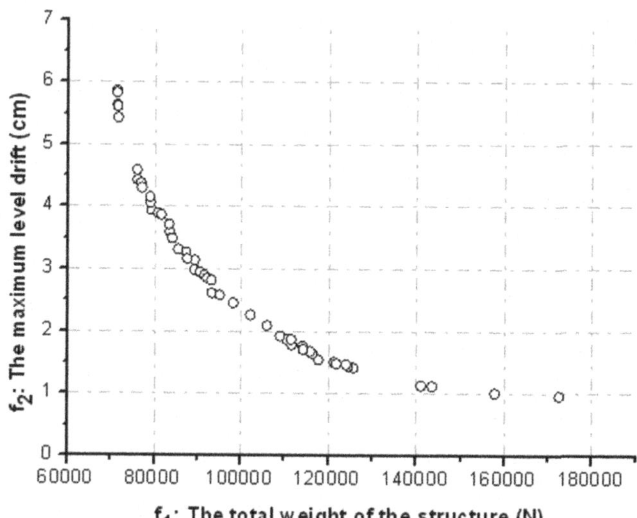

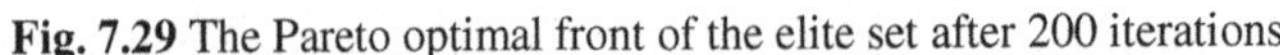
**Fig. 7.29** The Pareto optimal front of the elite set after 200 iterations

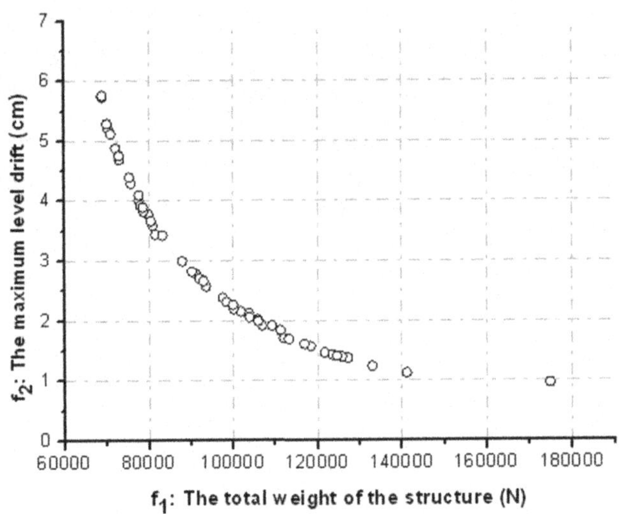


**Fig. 7.30** The Pareto optimal front of the elite set after 500 iterations

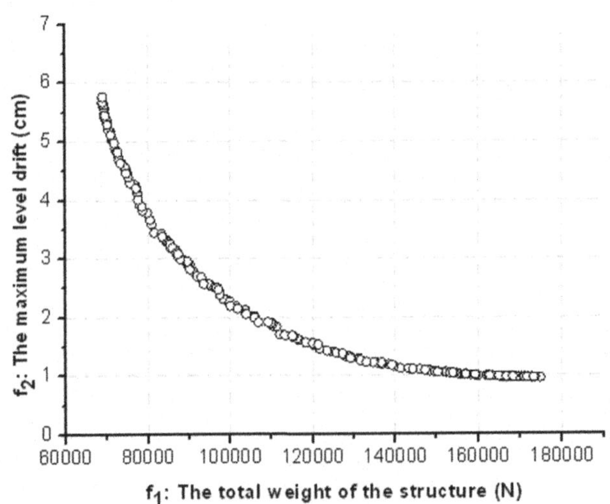


**Fig. 7.31** The Pareto optimal front after 500 iterations

Compared Fig. 7.29 with Fig. 7.30, it is noticed that the distribution of the Pareto front does not increase significantly, but the uniformity is good.

The Pareto front corresponding to all the Pareto optimal solution after 500 iterations is shown in Fig. 31. The Pareto optimal front after 500 iterations is uniform, and MGSO is an effective multi-objective algorithm.

## 7.7 Conclusions

A new multi-objective group search optimizer which uses Pareto dominance and a crowding-distance based selection mechanism is presented in this paper. It is used for size optimization, shape optimization and dynamic optimization of truss structures and frame structures with continuous and discrete mixed variables. Seven numerical examples show the robust performance of the MGSO technique. Calculation results show that MGSO uses smaller populations, less generations and achieves better convergence rate, MGSO Pareto optimal solution distributes well, MGSO has simple constraint handling property, and has an obvious advantage for high-dimensional optimization problem. Meanwhile, its program is easy to implement, It is desired that MGSO is an efficient, practical multi-objective optimizer for the structural optimization of complex projects.

## References


1. Cui, X.X.: Multiobjective evolutionary algorithms and their applications. National Defense Industry Press (2006)
2. Knowles, J.D., Come, D.W.: Approximating the nondominated front using the Pareto archived evolution strategy. Evolutionary Computation 8(2), 149–172 (2000)
3. Shin, S.Y., Lee, I.H., Kim, D., et al.: Multi-objective evolutionary optimization of DNA sequences for reliable DNA computing. IEEE Transactions on Evolutionary Computation 9(2), 143–158 (2005)
4. Kalyanmoy, D., Samir, A., Amrit, P., et al.: A fast elitist non-dominated sorting genetic algorithm for multi-objective optimization: NSGA-II. IEEE Transactions on Evolutionary Computation 6(2), 182–197 (2002)
5. Osman, M.S., ABO-Sinna, M.A., Mouse, A.A.: An iterative co-evolutionary algorithm for multiobjective optimization problem with nonlinear constraints. Applied Mathematics and Computation 183, 373–389 (2006)
6. Zhou, Z., Dai, G.M., Fang, P., Chen, F.J., Tan, Y.: An improved hybrid multi-objective particle swarm optimization algorithm. Springer, Heidelberg (2008)
7. Ho, S.L., Yang, S.Y., Ni, G.Z., Lo Edward, W.C., Wong, H.C.: A particle swarm optimization based method for multi-objective design optimizations. IEEE Transactions on Magnetics 41(5), 1756–1759 (2005)
8. Baumgartner, U., Magele, C., Renhart, W.: Pareto optimality and particle swarm optimization. IEEE Transactions on Magnetics 40(2), 1172–1175 (2004)
9. Zhang, W.B., Lai, L.F.: Optimum design for the diaphragm spring with particle swarm optimization algorithm. Journal of Guangxi University (Natural Science Edition) 33(1), 40–44 (2008) (in Chinese)
10. He, S., Wu, Q.H., Saunder, J.R.: A novel group search optimizer inspired by animal behaviour ecology. In: 2006 IEEE Congress on Evolutionary Computation, Vancouver, BC, Canada, pp. 4415–4421 (2006)
11. Doltsinis, I., Kang, Z.: Robust design of structures using optimization methods. Computer Methods in Applied Mechanics and Engineering 193(23-26), 2221–2237 (2004)

12. Barnard, C.J., Sibly, R.M.: Producers and scroungers: a general model and its application to captive flocks of house sparrows. Animal Behavior 29(2), 543–550 (1981)
13. Carpenter, R.H.S., Williams, M.L.L.: Neural computation of log likelihood in control of saccadic eye movements. Nature 377(7), 59–62 (1995)
14. Liversedge, S.P., Findlay, J.M.: Saccadic eye movements and cognition. Trends in Cognitive Sciences 4(1), 6–14 (2004)
15. Li, L.J., Xu, X.T., Liu, F.: The group search optimizer and its application to truss structure design. Advances in Structural Engineering 13(1), 1–9 (2010)
16. He, S., Wu, Q.H., Saunders, J.R.: Group search optimizer: an optimization algorithm inspired by animal searching behaviour. IEEE Transactions on Evolutionary Computation 13(5), 973–990 (2009)
17. Wan, C.: The multi-objective group search optimizer and its application to the structural optimal design. Master's Thesis of Guangdong University of Technology, 25–66 (2010) (in Chinese)
18. An, W.G., Li, W.J.: SM-MOPSO—A hybrid algorithm. Journal of Northwestern Polytechnical University 22(5), 563–566 (2004)
19. Chuang, W.S.: A PSO-SA hybrid searching algorithm for optimization of structures. Master's Thesis of National Taiwan University, 116–126 (2007)

# Chapter 8
# Prospecting Swarm Intelligent Algorithms

As reviewed in previous chapters, there are only a few optimization algorithms inspired by animal behavior, including ACO, PSO and GSO. Although, PSO and GSO both are swarm intelligence (SI) optimization algorithms and draw inspiration from animal social forging behavior, both of them were initially proposed for continuous function optimization problems, then they were developed for discrete optimization problems, they have some obvious differences. It is not difficult to note from previous discussion that there are major difference between PSO and GSO. The first and the most fundamental one is that the PSO was originally developed from the models of coordinated animal motion. Animal swarm behavior, mainly bird flocking and fish schooling, serves as the metaphor for the design of PSO. The GSO was inspired by general animal searching behavior. A genetic social foraging model, e.g., PS model was employed as the framework to derive GSO. Secondly the producer of GSO is quite similar to the global best particle of PSO, the major difference is the producer performs producing, which is a search strategy that differs from the strategies performed by the scroungers and the dispersed members. While, in PSO each individual performs the same searching strategy. Thirdly, in GSO the producer remembers its head angle when it starts producing. In PSO each individual maintains memory to remember the best place it visited. Finally, unlike GSO, there is no dispersed group members that perform ranging strategy in PSO.

Although the GSO and the evolutionary algorithm (EA) were inspired by completely different disciplines, as a population-based algorithm GSO shares some similarity with some EAs. For example, they all use the concept of fitness to guide search toward better solutions. The scrounging behavior of scroungers in GSO is similar to the crossover operator, e.g., extended intermediate crossover of real-coded genetic algorithm. It can be seen from previous chapters that the GSO can get better results for some problems than other heuristic algorithms. The main reason is that GSO is more like a hybrid heuristic algorithm, and combines the searching strategy of direct search, such as EAs and PSO. Therefore it is possible for GSO to better handle problems that are difficult for a single heuristic optimization algorithm.

L. Li & F. Liu: Group Search Optimization for Applications in Structural Design, ALO 9, pp. 247–249.
springerlink.com

Based on their initial differences, PSO and GSO have inherent advantages and disadvantages and need to be continuously developed and improved. SI gradually shows its extensive use and strong vitality with over ten years' development. Its inherent parallelism features make it an effective solving method for dealing with the distributed processing problems based on the massive information. SI constantly makes progress and developments because of the advantages above. The natural system provides a useful elicitation for the artificial intelligence processing systems and the design of algorithms. SI as an important component of nature system has shown potential advantages such as flexibility, stability, distributed control and self-organizing capacity. An increasing complex of information processing requirements in engineering problems, especially the dynamic characteristics of outstanding issues, provide a broad space for the application research of SI. SI depends on probabilistic search algorithm, which is different from most other application algorithms that based on gradient optimization algorithms. Although, lots of evaluation functions are usually used in the probabilistic search algorithm, compared with the gradient method and traditional evolution algorithms, SI still has the obvious advantages.

However, the SI as an optimization algorithm, which inspires from the phenomenon of groups of organisms in nature, has no mature and strict mathematical theory as a guide. Therefore, it leaves much to be developed and improved for SI. For example, the algorithm model itself need to be improved in the search efficiency of solution space. The basic principle of the algorithm need rigorous mathematical demonstration. The parameters of algorithm have influence on the algorithm performance. The choosing and setting of parameters has no generally applicable method, and often relies on experience. All of these make the SI have strong dependence on the practical issues. There are cases in practical engineering which make the algorithms powerless on some problems. The SI algorithm need be improved greatly.

Nevertheless, The idea of SI algorithm comes from nature. It is established by simulating biological behavior. From this point, the synthetic predator search (SPS) algorithm inspired from the area-restricted searching behavior [1] can be added to the category of SI. The social and civilization (SC) algorithm inspired from the intra- and intersociety of animal societies, e.g., from human and social insect societies [2] is also a member of SI. SI algorithm is conceptually simple and easy to implement. It can be founded that SI is not sensitive to all of the parameter for different type of structures and shows the robustness of the algorithm. SI can handle a variety of large scale optimization problems which makes it particularly attractive for practical engineering optimization.

Besides, there is studies of simple single-celled organism for decades. Two different stochastic optimization algorithms were developed from the study of bacteria recently, one of them is the bacterial chemotaxis (BC) algorithm based on bacterial chemotaxis model [3], and the other is the bacterial foraging algorithm (BF) based on the chemotatic (foraging) behavior of *e. coli* bacteria [4]. Although only small-scale optimization problems were tackled by these two algorithms, the potential of bacterial-inspired algorithm remains to be explored.

Civil engineering structures such as buildings, bridges, stadiums, and offshore structures play an import role in our daily life. Constructing these structures requires lots of budget. Thus, how to cost-efficiently design structures satisfying all required design constraints is an important factor to structural engineers. Although traditional mathematical gradient-based optimal techniques have been applied to the design of optimal structures. While, many practical engineering optimal problems are very complex and hard to solve by traditional method. The intelligent based algorithms are very suitable for continuous and discrete design variable problems such as ready-made structural members and have been vigorously applied to various structural design problems and obtained good results. So the SI algorithms leaves much to be studied and desired. The readers can study the full spectrum of the algorithms form different points and apply the algorithms to their own research problems.

## References


1. Linhares, A.: Synthesizing a predatory search strategy for VLSI layouts. IEEE Transactions on Evolutionary Computation 3(2), 147–152 (1999)
2. Ray, T., Liew, K.M.: Society and civilization: an optimization algorithm based on the simulation of social behaviour. IEEE Transactions on Evolutionary Computation 7(4), 386–396 (2003)
3. Muller, S.D., Marchetto, J., Airaghi, S., Koumoustsakos, P.: Optimization based on bacterial chemotaxis. IEEE Transactions on Evolutionary Computation 6(1), 19–29 (2002)
4. Passino, K.M.: Biomimicry of bacterial foraging for distributed optimization and control. IEEE Control Systems Magazine 22(3), 52–67 (2002)